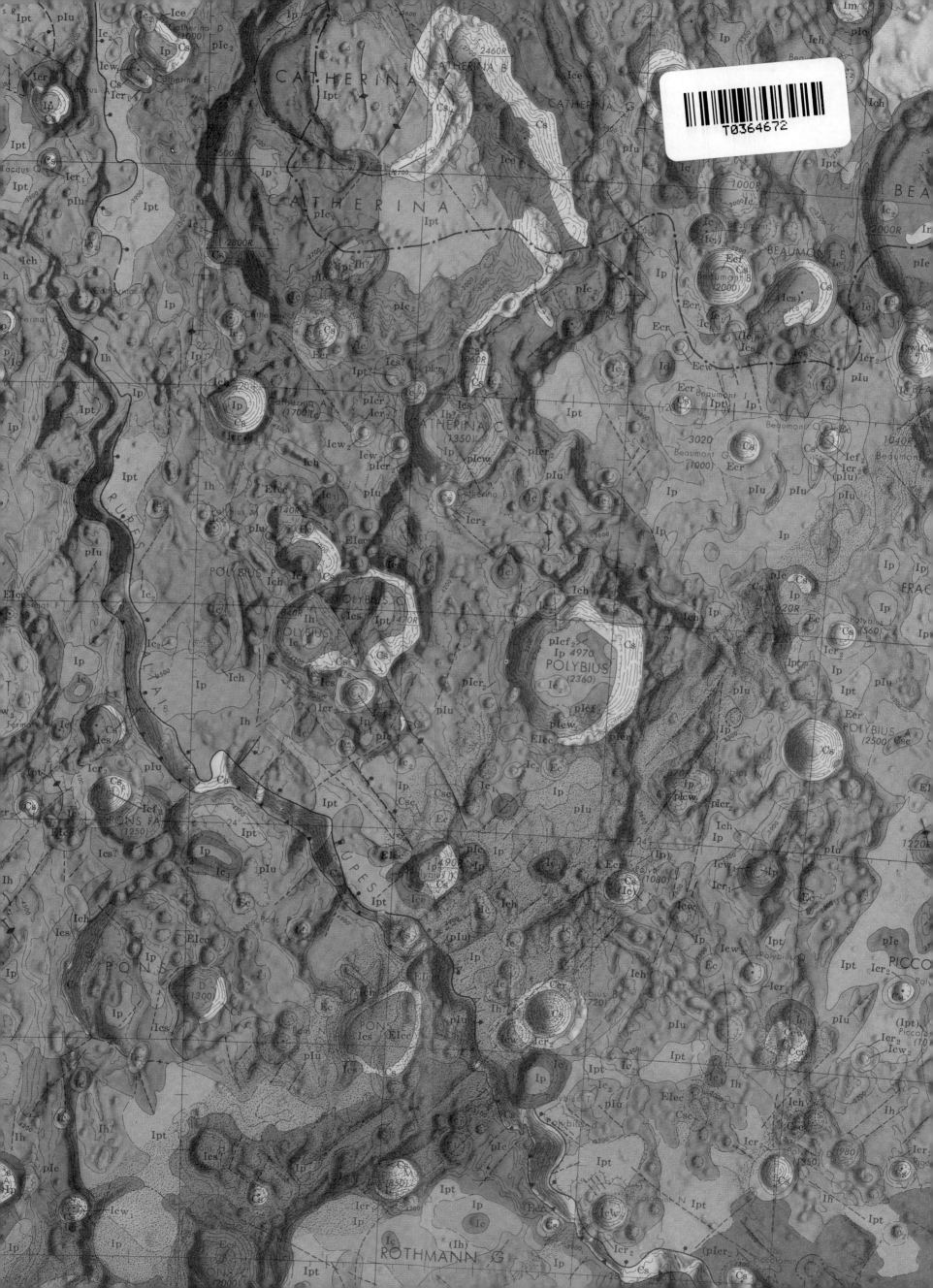

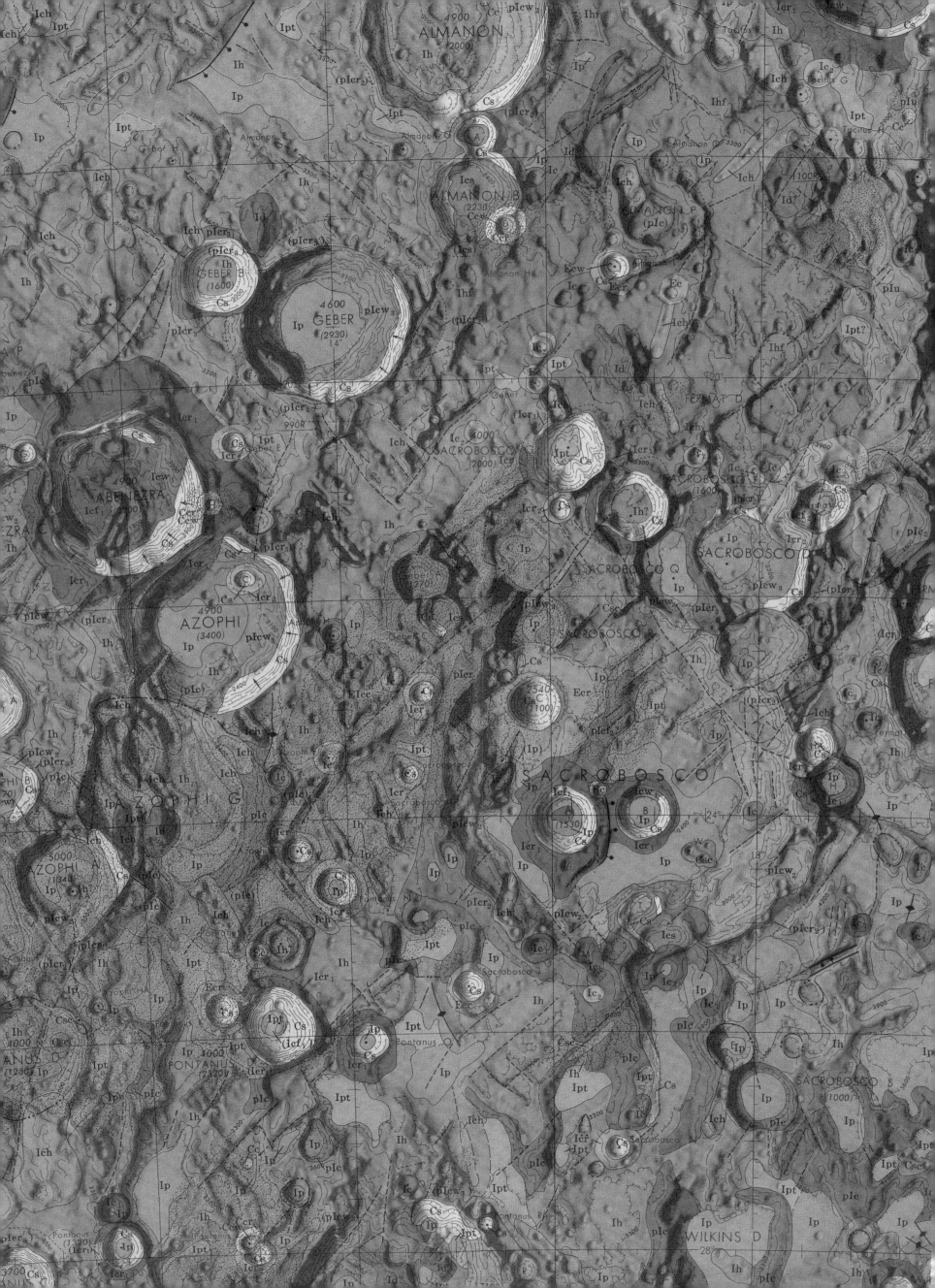

LUNAR

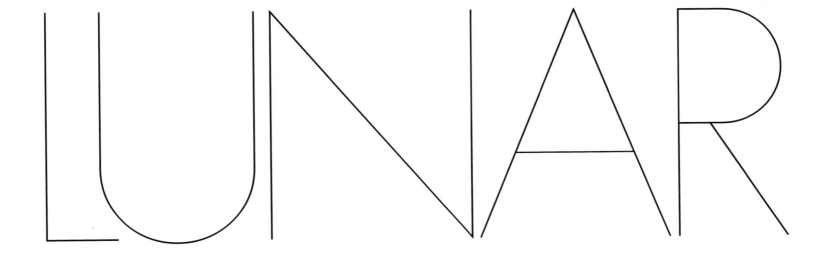

LUNAR

A History
of the Moon
in Myths, Maps
+ Matter

Incorporating
the USGS/NASA
Geologic Atlas,
1962–74

Matthew Shindell
Consultant Editor

Dava Sobel
Foreword

CONTENTS

FOREWORD—
DAVA SOBEL

I've been to the Moon, of course. Everyone has, at least vicariously, visited its stark landscapes, driven over its unmarked roads. Even so, I've never seen the Moon quite the way it appears here – a black-and-white world rendered in a riot of gorgeous colours.

The lush palette of this atlas charts a route through time, from the Moon's ancient, violent formation to its present quiescence. In each facet, or quadrangle, of the lunar surface, neighbouring features flaunt their different ages by virtue of colour contrasts rarely conjured in Nature. On one page, crimson rilles cut across azure ejecta blankets. On another, citron-chartreuse craters dot the mauve and lavender remains of catastrophes that shaped the Moon's layered terrain.

The twenty-two individuals who configured these forty-four maps started out as cartographers, illustrators, geographers or geologists, but their collaboration made selenographers of them all. Together, they explored the Moon via photographs taken through telescopes at Earthbound observatories and also by spacecraft. As they prepared the way for the first lunar astronauts, they marked the spots where early rockets to the Moon had either crashed or come to rest.

The vast lunar 'seas' depicted here, from Sinus Iridum (the Bay of Rainbows) to Oceanus Procellarum (the Ocean of Storms), assumed their Latin names long before the Moon admitted it could not hold water. Such misnomers, though, cannot be erased as mere mistakes. For a map of the Moon is a mix of many things – real faraway places tainted by longing and imagination, mountains and shadows shape-shifting in the changing light of scientific insight and technological advance.

A wealth of other imagery surrounds the maps in this volume. Photographs, paintings and stills from popular films serve to illustrate some thirty essays that look at lunar influences on myth, literature, religion, alchemy, astrology, fantasy, farming, philosophy, philately, fine art, cinema, geopolitics and more. No area of human concern, it seems, has gone untouched by the Moon.

Two of the essays specifically address the Moon's feminine essence. 'She' was worshipped as the goddess Selene in ancient Greece, as Luna in Rome, and by the Lakota people as Hanwi. The thirty-day cycle of her waxing and waning, which mirrored the flux of female physiology, lent the Moon's name – 'mene' in Greek – to both 'menstruation' and the calendar 'month'. Now a new space project, called Artemis after Apollo's twin sister, promises to send the first woman to the Moon.

Millennia in the future, when Earth's features have been altered by wind and water erosion, melting ice caps and continental drift, these maps of the Moon will still paint a fair picture of our closest planetary companion. On the arid, airless Moon, the largest tectonic force now in play is a slow, steady fall of micrometeorites that thicken the dust cover by about a millionth of a millimetre per year. This means the bootprints of the Apollo and Artemis teams may retain their crisp outlines indefinitely.

On the final Apollo mission, in 1972, a rear fender fell off the lunar rover. This mishap allowed tall 'rooster tails' of jagged Moon dust to fly up behind the vehicle and descend on its pilot and passenger. With some help from Houston, the men on the Moon fashioned a new fender from a few maps they no longer needed, fastened the replacement in place with duct tape, and went about their assigned tasks.

Fortunately, the full set of original Moon maps remains right here at hand – a lunatic vision come vibrantly real.

←
A crescent Earth rises above the lunar surface in one of the first photographs to feature both the Earth and the Moon. Captured by Lunar Orbiter I in August 1966, NASA's survey programme took over 400 photographs.

↳
The Moon card from the *Astronomia* constellation and planetary playing deck, which sits within the spring series of cards. Illustrated by Henry Corbould, the cards were created by Sir Francis Graham Moon in 1829.

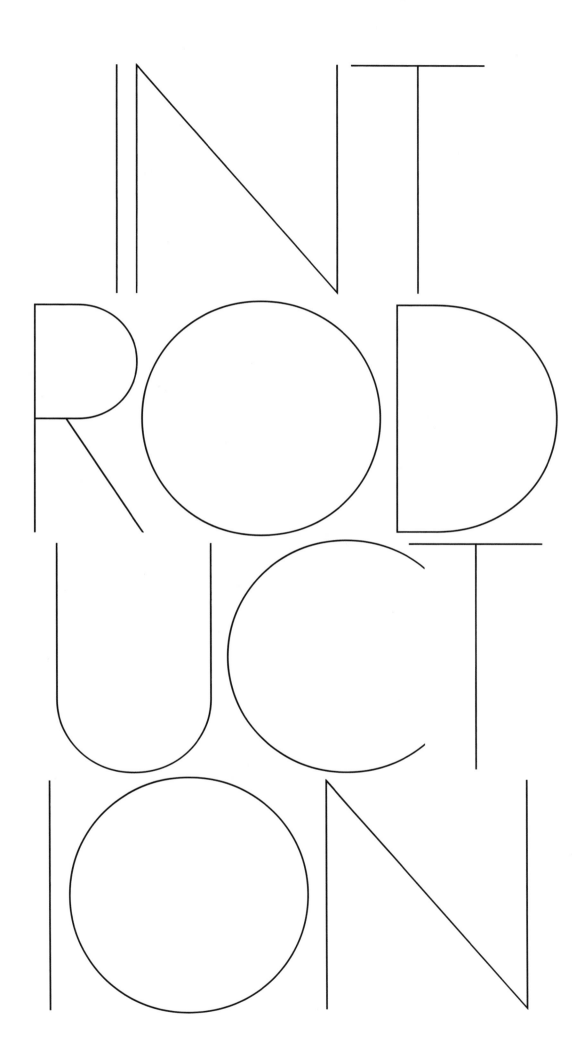

INTRODUCTION

INTRODUCTION— MYTHS, MAPS, MATTER

The Moon pre-dates the human world, and, for that matter, even the most distant human evolutionary ancestor. Its origin 4.5 billion years ago is intertwined with that of the Earth. Some scientists even see evidence that both the Moon and the Earth were the products of a dramatic and explosive impact between a stray Mars-sized planet they call Theia (named after the Titan mother of Selene, the ancient Greek Moon-goddess) and a planet that only properly became the Earth after the collision added parts of Theia to the planet and threw off material that congealed into the Moon.

The Moon is large, especially relative to the Earth. It is about half the diameter of Mars, and one-quarter the diameter of Earth, and this has led some to compare the Earth–Moon relationship to a dual-planet system. In short, the Moon is a proper other world. And yet, as distant and as distinct as it is from Earth, the Moon is connected to all of human history – to the very definition of humanity.

The Moon's surface has changed little during the 300,000 years that *Homo sapiens* has walked the Earth, and even less during recorded history. The Moon has been the same object, physically, but the way in which humans have seen it has changed dramatically over time and across cultures. The maps reproduced in this volume represent only one of the most recent understandings of the Moon, from a time when the Moon became the territory of human exploration and an extraterrestrial extension of geopolitical competition. While some of the knowledge presented in these maps has been superseded and the Cold War ended more than three decades ago, these maps stand as a testament to a historical moment in which the Moon's meaning and significance snapped into focus. As much as these maps depict the Moon, they also reflect the culture that made them, the knowledge and tools its scientists brought to bear, and the constellation of hopes and anxieties that drove their efforts.

While the US and Soviet lunar programmes mark the 1950s, 1960s and 1970s as an impressive moment for lunar fascination, this was far from the only time or place in which the Moon's glow was strongly felt. As the articles in this volume attest, the Moon has been important to people around the world since prehistoric times in ways that intersected with or were expressed through religious beliefs, divination and prognostication, the governance of kingdoms large and small, philosophical and scientific pursuit of the structure of the universe, medical practices and social and cultural identities, as well as art, poetry and music. The Moon itself may not have changed, but humans in different times and places certainly saw it in the light of their own ideas and preoccupations.

Prehistoric and ancient cultures may not have sent humans to the Moon, but they paid greater heed to, and found more meaning in, its movements than 20th- or 21st-century people. Some of the earliest archaeological evidence indicates that early humans paid close attention to the Moon and its phases, even as this same evidence gives us little idea of what sense they made of what they saw. The ancient Babylonians developed complicated astrological practices that formed the roots of Western and Indian astrology, adapted by people in different times and places to conform to their specific cultural contexts and understanding of the universe. Babylonian astrologers practised their art exclusively for kings. In later cultures astrologers became highly sought-after professionals, advising paying customers on a broad range of subjects.

Ancient Egyptians believed that a pantheon of gods lived in the heavens, several of which had lunar associations, and used the Moon and planets to track time and seasonal change. The Egyptian Moon-god Thoth measured time, regulated the seasons, and was also responsible for inventing writing. In ancient Greece and Rome, the Moon came to symbolize rhythms of female life, and was associated with various goddesses, including Selene, who was understood to help women during childbirth. In art and poetry, Selene sometimes represented female desire. Meanwhile, the Moon's association with Persephone also tied it to ideas about death. In ancient Mesoamerica the Moon caught the attention of artists and calendar-makers, and held a place at the centre of ceremonial and mathematical work.

Around the world, people marvelled at the sight of lunar and solar eclipses, and developed varying explanations based on their understanding of their universe. For most cultures, these were auspicious or frightening events. But eclipses also changed understandings, allowing astronomers rare glimpses of the Earth's shadow or the relative sizes of the Sun and Moon. Generations of observations, combined with mathematical tools, allowed astronomers to predict these events and demonstrate their regularity.

In the natural philosophy of ancient Greece and the medieval world, the Moon was part of an Earth-centred universe. It held a special place in this arrangement, located at the boundary between the imperfect earthly world and the perfect heavens. It was also understood to have a connection with the human body, influencing the humours (body fluids thought to affect health), as well as menstruation, just as it influenced the waters of the Earth.

The practice of using the Moon to track menstruation persisted into the modern era, practised via farmers' almanacs and, later, specialized women's almanacs.

The link between the Moon and madness – surviving in such terms as 'lunacy' and 'moonstruck' – dates at least as far back as ancient Greek ideas about the Moon's ability to moisten the brain or affect a person's dreamlife. The association of the Moon with madness even affected how and when patients were treated in asylums into the early 19th century. Even today werewolf stories remain a staple of popular culture, and the idea that the full Moon brings on insanity or criminal tendencies refuses to disappear.

The Moon remained important in natural philosophy, even after the Earth was (so to speak) taken out of the centre of the universe and placed in orbit around the Sun. This made the Moon a satellite of Earth, rather than a planet in its own right, and allowed astronomers to speak of other moons around other worlds. But this did little to diminish the Moon. Telescopic views of the Moon's surface produced by Galileo Galilei, and those who followed him, transformed the Moon into a world that could be mapped and compared to the Earth. Philosophers and scientists could speculate about life on the Moon and imagine journeys to touch its surface and look back at the Earth to witness its movements. These tales were only the most recent in a history of literary lunar imaginings that reaches back to ancient times. But the pace of lunar storytelling increased after Galileo's revelations.

These stories transformed, as did ideas about the influence of the Moon on the Earth, as the ancient and medieval understanding of nested heavenly spheres gave way to materialist understandings presented by Descartes and Newton – as the universe was converted from a plenum to vast and mostly empty 'space'. Such stories are found in literature and have been present in film since the advent of cinema, when Georges Méliès visually interpreted (albeit loosely) Jules Verne's 1865 novel, *From the Earth to the Moon*. Although absent from Verne's novel, Selenites – Moon people – appeared in Méliès' 1902 *A Trip to the Moon*. The idea that intelligent lifeforms inhabited the Moon was much older, but the concept was modernized in science fiction works such as H. G. Wells' 1901 *The First Men in the Moon*, or the earlier 'Great Moon Hoax' of 1835, a series of fictional articles that led readers of New York's *Sun* newspaper to believe that civilization had been discovered on the Moon.

The Moon's depiction in popular culture has continued to change, as it has gone from a place of mystery to a world of rock that humans have visited, if not conquered.

Still today, via space programmes, space visionaries and space science fiction, humans continue to tell stories about trips to the Moon. These forces aligned in early space age collaborations to sell lunar exploration to the public as a culturally valuable and technologically feasible pursuit. Through popular media, space was framed not just as a theatre of Cold War competition, but the place where the human future would be defined.

The Moon has also continued to fascinate fine artists. Not only have artists depicted the Moon itself, as well as mythical scenes of lunar goddesses, but the Moon has provided ethereal illumination for nocturnal landscapes. In the early 20th century, moonlit scenes allowed artists to resist the lure of the electric lighting that now competed with the Moon to rule the night. The Moon likewise found a place in surrealist art, and artists and filmmakers revelled in its purported effects on the human psyche, and its influence upon the unconscious mind. The Moon is a world of light and shadow, of serenity and malice, of stark contrasts

and subtle shades, an object simultaneously of nature and beyond its limits. It is also a landscape upon which multiple real and imagined voyages have been charted. Today, more than half a century past the Apollo landings, artists continue to explore the Moon, its influence on life and culture, and human designs for the lunar future.

For the artist, advancing technology may have threatened the Moon's supremacy. Meanwhile, for the astronomer, new photographic and spectrographic tools promised to further demystify the lunar surface. Eventually, photography did become a valuable tool in lunar astronomy, but in its early years the human eye remained the best tool for detailed viewing and mapping. In the popular realm, photography did allow telescopic images of the Moon to circulate broadly, fed a burgeoning market for popular science books that emerged in the 19th century, and also brought a more whimsical version of the Moon closer to people's lives. In *cartes de visite*, the Moon became a fanciful element of studio portraits.

Large Earth-based telescopes brought the eye into closer contact with the features of the Moon, but these behemoths were eventually surpassed by small camera-carrying spacecraft from the USA and USSR that raced to accomplish robotic firsts at the Moon and, in the process, helped to map the Moon's surface and understand its geology. Some of these spacecraft captured images of the Moon's far side for the first time, their images defying expectations; rather than the vast maria ('seas', actually dry lowlands) that defined the near side, the Moon's far side held mostly highland terrae. For a short few years, human astronauts carried hand-held cameras to document their exploits on the Moon, deposited scientific experiments and brought back rock and soil samples for study on Earth.

Humans have continued to explore the Moon since the heady period of lunar missions that produced the maps of this Lunar Atlas. They have done so with increasingly capable robotic orbiters, landers and rovers. While the first Space Race saw only two nations vie for lunar accomplishments, today lunar exploration is undertaken by European states, as well as China, India and Japan. Private companies have their own lunar aspirations too, and the US company Intuitive Machines landed the first commercial lander on the Moon in early 2024.

And, to paraphrase President John F. Kennedy, humans may set boots on the Moon again by the time this decade is out. Not because it is easy, but because national space programmes hope to build a new lunar economy and a sustainable and permanent programme of lunar and planetary exploration on the Moon. This time around, plans include international alliances, commercial partnerships and the construction of robust lunar infrastructure including new rockets, capsules, suits and roving vehicles, as well as bases of operation, robotic helpers and a Moon-orbiting space station. When this happens the humans sent to the Moon will be more diverse than the white men of the 1960s Apollo programme – a more representative sample of all humankind.

As the articles in this book demonstrate, the Moon has been and remains connected to almost every facet of human life. While it is impossible to predict precisely what is to come in the long relationship between humans and the Moon, what is clear is that the Moon will forever live in the human heart. Even if humans one day find themselves living on Mars, and calling themselves Martians, the Moon will likely have played an important role in getting them there. Phobos and Deimos will dance in their sky, but our Martian descendants will still tell stories about the Moon.

Matthew Shindell

INTRODUCTION— THE LUNAR ATLAS

In September 1961, the US Geological Survey (USGS) elevated a small, internal, NASA-funded Astrogeology Studies Group into a full new branch of the Survey.[1] Founded in 1879, the USGS had been shaped by the nation's expansion and shifting national priorities – from mapping newly acquired territories and their resources to, more recently, assessing uranium deposits on the Colorado Plateau in support of nuclear energy development. In 1961, with the advent of rocket-powered space exploration, NASA charting a course of robotic exploration of the Moon, and President John F. Kennedy advocating crewed missions to Earth's only natural satellite, the new Astrogeology Branch positioned the Survey for an era of extraterrestrial exploration.

This was an exciting time for planetary science inside and outside the USGS. The years after the Second World War had seen increased interest in the Moon and planets. A new cohort of scientists from various fields, not limited to astronomy, had begun to form around questions about the physical and chemical composition and evolution of the solar system.[2] They applied new atomic-age methods to the study of meteorites as well as terrestrial rocks and deposits.[3] They published new books about the Moon and planets that would influence a generation of planetary scientists.[4] And they eagerly anticipated the opportunity to study lunar and planetary surfaces up close.

Modern planetary science emerged in these years as a markedly interdisciplinary field. Out of all the disparate disciplines that made their claims on the Moon, geology would emerge as the lynchpin. That geology would play a central role in planetary exploration was not immediately obvious at the dawn of the space age. There were sceptics within the USGS who feared that space exploration might be a fad and not a permanent source of support for the Survey. There were critics outside the USGS who argued that geologists – whose science was defined by the study of the Earth and its processes – were not well-suited to extraterrestrial topics. Time has proven the sceptics wrong. As for the critics, they underestimated the rigour and adaptability of field geology and the range of questions it had evolved to answer.[5] In the early 1960s, however, the Astrogeology Branch geologists had yet to prove their utility to America's space programme.

One of the first tasks the Branch was given – and one that would occupy its staff for more than a decade – was

↑ Engraving of the near side of the Moon from *Selenographia* (1647) by Johannes Hevelius.

mapping the Moon for robotic and human exploration. Astronomers had been producing increasingly accurate maps of the Earth-facing side of the Moon since the 17th century, their progress roughly reflecting advances in the size and magnification power of telescopes, and evolving ideas about the solar system. These maps had captured geologically significant features, but they had not necessarily advanced geologic understanding of the Moon. In 1961, USGS geologists Eugene M. Shoemaker (1928–1997) and Robert J. Hackman (1923–1980) published a study of the Copernicus region of the Moon. Today, it is considered the first demonstration that geological methods – namely stratigraphy – could be applied to the lunar surface and that a geologic timescale could be established.[6] This paper and its proposed timescale, along with a prototype map of Copernicus produced by Shoemaker, became the basis for the mapping of the forty-four 1:1,000,000 scale rectangular maps, or 'quadrangles', of the Lunar Atlas presented in this book.

Produced between 1962 and 1974, these maps not only comprise the first complete and systematic geologic map of the Moon's near side, but also capture the evolution of knowledge of the Moon as well as the progress of lunar exploration during that period. The earliest maps were drawn based on what could be seen through the telescope; the USGS utilized photographs from US observatories such as Lick, McDonald, Mount Wilson and Yerkes, and USGS geologists spent nights observing their assigned

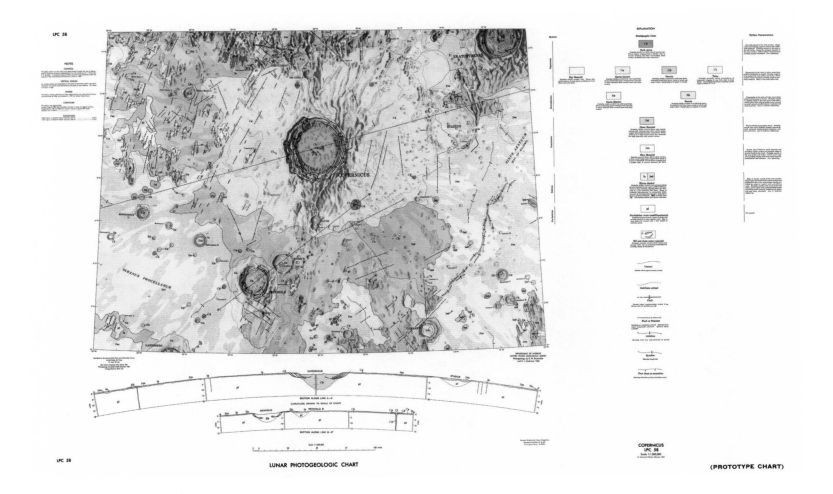

↑   Eugene Shoemaker and Robert J. Hackman's prototype map of the Copernicus region of the Moon, 1960.

quadrangles through telescopes, including the 30-inch (76-centimetre) reflector at their own observatory near Flagstaff, Arizona.[7] Later maps included details seen in images captured by robotic landers and orbiters. After the 20 July 1969 landing of the crewed Apollo 11 mission, and up to and including the final Apollo 17 mission in 1972, the geologists had access to even more orbital data as well as rock and soil samples collected and returned from the lunar surface. Although Apollo 17 astronaut Harrison 'Jack' Schmitt was the only trained geologist among the astronauts sent to the Moon, the USGS astrogeologists ran all the Apollo crew members through crash courses in geology, including field trips in which the astronauts practised the types of observations and sampling they would perform on the Moon.

Lunar Geology – An Evolving Understanding

When the geologist Thomas Mutch (1931–1980) published what can be considered the first textbook on lunar stratigraphy in 1970, he pointed out that the rapid progress of lunar exploration and mapping had been a valuable test of Shoemaker and Hackman's proposed lunar geologic timescale.[8] The timescale had held up well, at least in its basic principles, although there had been some need to reinterpret and redefine its stratigraphic units. This was no surprise, considering that Shoemaker and Hackman had developed their timescale based on the study of just one area of the Moon. Still, what they had been able to extract from their observations of the Copernicus region was impressive in its detail and accuracy.

They began with the assumption that the law of superposition would be as valid on the Moon as it is on Earth – that is, when one sees multiple layers of rock or sediment deposited on the surface, the oldest layer will be at the bottom and the youngest at the top (unless the surface has been deformed since deposition). What they saw around Copernicus were multiple overlapping layers, including the lava plains of the maria (so-called 'seas', actually dry lava-filled basins), craters displaying varying degrees of

degradation, and materials and features related to the explosive impacts that had created the craters. They noted that the ejecta thrown out of the crater Copernicus overlay the nearby craters Eratosthenes and Reinhold, meaning that their formation pre-dated that of Copernicus. In turn, these craters interrupted what must have been a pre-existing mare. In the Copernicus region, they noted what they considered to be four distinct layers and from them built a chronology. They added a fifth period to the beginning of the chronology, based on the assumption that the material of the Moon's primordial surface – formed when the Moon first congealed and cooled – might be preserved in some of the more complex terrain of the lunar hills and ridges. From oldest to youngest, they named the periods of their timescale Pre-Imbrian, Imbrian, Procellarian, Eratosthenian and Copernican.

The chronology of the Moon's geologic history is one of violent alteration. Shoemaker and Hackman named the Imbrian period after the materials exposed north of the

↑   Composite photograph of the Mare Imbrium composed of multiple photographs taken by NASA's Lunar Reconnaisance Orbiter.

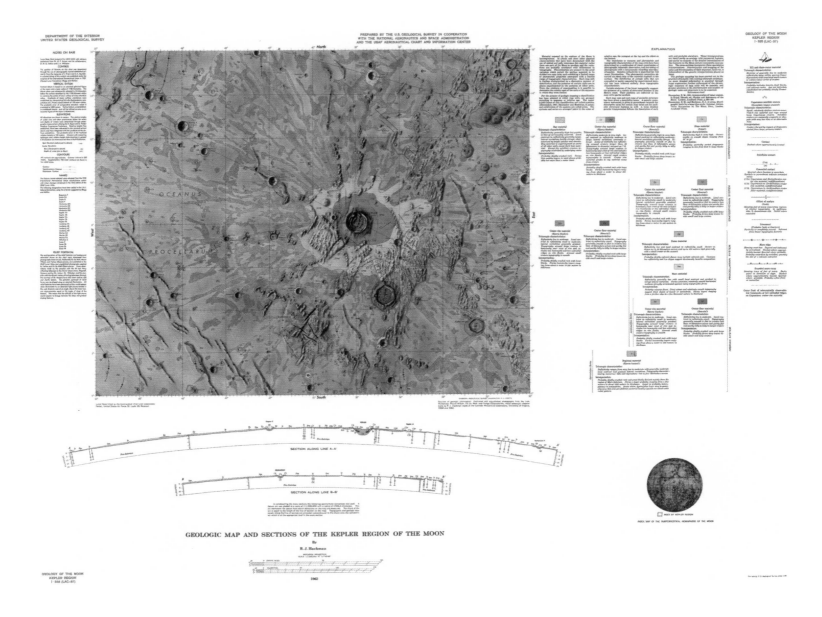

↑ The Kepler map was drawn by Robert J. Hackman and published
by the USGS in 1962. It was the first in a series of forty-four geologic
maps of the quadrangles of the near side of the Moon.

crater Copernicus, around the Montes Carpatus mountain
range (named after the Earth's Carpathian Mountains). The
hummocky terrain here lies south of Mare Imbrium (Sea of
Showers), and Shoemaker and Hackman proposed that
this was part of a large ejecta blanket produced by the
massive impact that excavated the Imbrium basin from the
Moon's original crust. The dark lava plains of the maria
defined the Procellarian period, named for the Oceanus
Procellarum (Ocean of Storms) mare. These plains had
formed when lava flooded portions of the Imbrian terrain,
and therefore were determined to be younger. Craters
such as Erastothenes that interrupt the smooth terrain of
the mare must have formed next, giving their name to
the Eratosthenian period. Finally, the excavation of fresh
craters such as Copernicus blanketed these older craters in
bright rays of ejecta material, and scarred the terrain with
secondary impact craters, defining the Copernican – the
Moon's most recent geologic period.[9]

Over the course of the mapping project, as more regions
and features were described and interpreted, these
periods were revised. The mapping project began with the
regions adjacent to Copernicus, attempting to apply
the now-defined stratigraphy of Copernicus to the
rest of the Moon in the hopes of refining it. The first six
quadrangles published were Kepler, Letronne, Riphaeus
Mountains, Timocharis, Montes Apenninus and Aristarchus,
all of which bordered Copernicus. Complications arose in
these nearby regions. In Letronne, it became clear that the
excavation of the Humorum basin, not that of Imbrium, was
responsible for producing the mountains in that region. This
raised the question of the relative ages of the Imbrian and
Humorum ejecta, a sequence which was not easily resolved

as evidence seen in different regions seemed contradictory.
These difficulties were compounded when the geologists
moved west and turned their attention to the Orientale
basin, which they soon concluded was one of the youngest
basins on the Moon. Ultimately the geologists determined
that these three basins, from oldest to youngest, could
be ordered Humorum, Imbrium, Orientale.[10] As evidence
from successive quadrangles was brought to bear on the
problem, the geologists did eventually move from the local
stratigraphy of Copernicus to a more global view of the
Moon's geologic history.

Today the lunar geologic timescale looks slightly different
from that first proposed in 1962. The original surface of
the Moon is now identified as Pre-Nectarian, prior to the
formation of the Nectaris basin (now understood to be
the oldest of the basins). Pre-Nectarian rocks are indeed
still present on the Moon, broken and mixed into the lunar
highlands. After this is the Nectarian, the geologically
brief period during which large impacts excavated many
of the Moon's major basins. The Imbrian period persists,
though now divided into Early and Late Imbrian. The Early
Imbrian period is defined as beginning with the formation
of the Imbrian basin and the remaining major basins. The
Late Imbrian period saw the filling of the basins by molten
lava rising up from the Moon's mantle. The longest geologic
period on the Moon is the Eratosthenian – essentially the
Moon's long middle age – during which the older, but
in many cases relatively well-preserved, craters were
excavated. There is also evidence of some volcanic activity
during this period, although not approaching the scale of
the Late Imbrian. Finally, the most recent features on the
Moon date from the Copernican. Beginning a little more
than a billion years ago, this period is mostly defined by
young, sharp-rimmed craters, the larger and more complex
containing well-defined central peaks, sloped walls and
outer ramparts. These craters are among the brightest

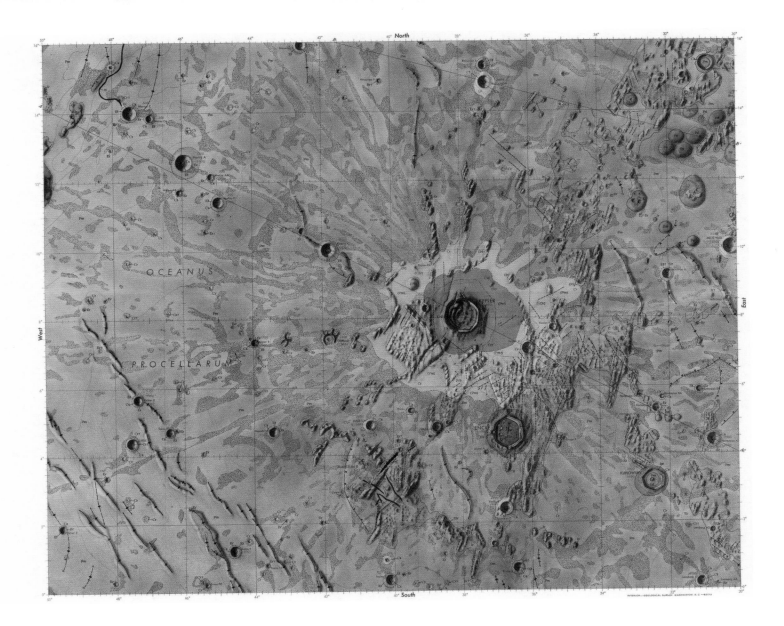

↑ Detail of the map of the Kepler region of the Moon, with colour-coded geological data based on the best imaging technology available at this time.

features on the Moon, and their equally bright rays of ejecta material overlie surrounding older terrain.[11]

Because the forty-four maps of the USGS Lunar Atlas were produced as lunar exploration proceeded and geological understanding evolved, there are inconsistencies between the maps that make it impossible to provide a universal key. Later maps are not only more detailed in the number and differentiation of lunar features, but also reflect more nuanced interpretations of the sequence of events that shaped the lunar surface. Nevertheless, it is still possible to see across all of these maps the features that helped geologists construct the timeline used today: the dark plains of the maria; the scarps and faults that ring the basins; the mountains and the rugged terrain of the lunar highlands; the degraded and deformed craters and walled plains; the craters of various sizes, complexities and brightness that speak to the majority of the Moon's history; and the bright ray systems created during the last billion years.

Preserved on the lunar surface, and presented here in these maps, is a history of violence in our solar system – the accretion of the planets was followed by a period of intense bombardment by large asteroids, followed by a more dispersed history of regular impacts that continues to shape the Moon's surface. Much of this history on Earth has been erased by the dynamic systems that recycle the Earth's crust, and by the liquid water that erodes and reshapes surface features over time. These same processes, and the fact that water is the dominant geologic force on Earth, are what keep our planet habitable. Their absence on the Moon is what makes it an effective chronicler of solar system events.

THE LUNAR GEOLOGIC TIMESCALE

BYA	ERA	GEOLOGICAL CHARACTERISTIC	REF
1.1	COPERNICAN	Craters with bright rays	
	ERATOSTHENIAN	Craters with faint rays	
3.16		Waning mare volcanism	
	LATE IMBRIAN	Younger mare basalts	
		Older mare basalts	
	EARLY IMBRIAN	Orientale basin	
3.85		Imbrium basin	
		Humboldtianum basin	
		Humorum basin	
		Crisium basin	
	NECTARIAN	Moscoviense basin	
		Hertzspring basin	
		Korolev basin	
3.92		Nectaris basin	
		Serenitatis basin	
		Nubium basin	
	PRE-NECTARIAN	Smythii basin	
4.6		South Pole-Aitken basin	
FORMATION OF THE MOON			

↑ As the geologic timescale reveals, after the formation of the Moon 4.51 billion years ago, the large basins were the first to form and the brightest craters the last. Many of the well defined craters without ray systems formed in the Eratosthenian period. The colours in the table relate to those featured on the maps.

↪ The original manilla envelopes for forty of the forty-four quadrangle maps. The maps were originally published individually in large format between 1962 and 1974 For distribution, each map was folded inside a manilla envelope identifying the author(s), quadrangle name, map number and scale.

INTRODUCTION— LUNAR NOMENCLATURE

	11 HERSCHEL 1-604	12 PLATO 1-701	13 ARISTOTELES 1-725			11-13
	23 RÜMKER 1-805	24 SINUS IRIDUM 1-602	25 CASSINI 1-666	26 EUDOXUS 1-705	27 GEMINUS 1-841	23-27
38 SELEUCUS 1-527	39 ARISTARCHUS 1-465	40 TIMOCHARIS 1-462	41 MONTES APENNINUS 1-463	42 MARE SERENITATIS 1-489	43 MACROBIUS 1-799	44 CLEO-MEDES 1-707 / 38-44
56 HEVELIUS 1-491	57 KEPLER 1-355	58 COPERNICUS 1-515	59 MARE VAPORUM 1-548	60 JULIUS CAESAR 1-510	61 TARUNTIUS 1-722	62 MARE UNDARUM 1-837 / 56-62
74 GRIMALDI 1-740	75 LETRONNE 1-385	76 RIPHAEUS MOUNTAINS 1-458	77 PTOLEMAEUS 1-566	78 THEOPHILUS 1-546	79 COLOMBO 1-714	80 LANGRENUS 1-739 / 74-80
92 BYRGIUS 1-755	93 MARE HUMORUM 1-495	94 PITATUS 1-485	95 PURBACH 1-822	96 RUPES ALTAI 1-690	97 FRACASTORIUS 1-720	98 PETAVIUS 1-794 / 92-98
	110 SCHICKARD 1-823	111 WILHELM 1-824	112 TYCHO 1-713	113 MAUROLYCUS 1-695	114 RHEITA 1-694	110-114
		125 SCHILLER 1-691	126 CLAVIUS 1-706	127 HOMMEL 1-702		125-127

INDEX MAP OF THE EARTHSIDE
HEMISPHERE OF THE MOON

Number of quadrangle name refers to lunar base map (LAC series);
additional number refers to published geologic map.

Lunar nomenclature has changed considerably since 1645 when Michel Florent van Langren (1600–1675) first attributed locations on the Moon with names taken from geographical locations on Earth, scientists, noblemen and monarchs. Since then observers at different periods have devised their own nomenclature systems as scientists' understanding of the Moon has evolved. More than sixty of Van Langren's original names have survived but only four remain attributed to their original locations. In 1647 John Hewelcke, or Hevelius (1611–1687), published three lunar maps using his own system of names based on terrestrial counterparts. Few of his proposed names have survived. By contrast, Joannes Baptisa Riccioli (1598–1671) was very successful. In 1651 he proposed around 200 lunar names, many of which have survived to this day in their original forms.

11
HERSCHEL
German/British
Astronomer
[1792–1871]
Named after Sir John
Frederick William Herschel.

12
PLATO
Greek Philosopher
[c. 427–347 BC]
Joannes Baptista Riccioli
suggested the name Plato
on his 1651 lunar maps.

13
ARISTOTELES
Greek Philosopher
[384–322 BC]
The International
Astronomical Union named
this after Aristotle in 1935.

23
RÜMKER
German Astronomers,
father and son
[1788–1862 / 1832–1900]
Named after Karl Rümker
and son Georg Rümker.

24
SINUS IRIDUM
['Bay of Rainbows']
The name survives
from Riccioli's original
suggestion. The Bay of
Rainbows is filled with lava.

25
CASSINI
Italian-French Astronomers,
father and son
[1625–1712 / 1677–1756]
Honours Giovanni Cassini
and son Jacques Cassini.

26
EUDOXUS
Greek Astronomer
[390–340 BC]
Named after Eudoxus of
Cnidus – mathematician,
and student of Plato.

27
GEMINUS
Greek Astronomer
[10–60 BC]
Named after Geminus
of Rhodes. The name
appeared on Riccioli's map.

38
SELEUCUS
Greek Astronomer
and Philosopher
[190–150 BC]
Named by Mary A. Blagg
and K. Muller in 1935.

39
ARISTARCHUS
Greek Astronomer
and Mathematician
[c. 310–230 BC]
This name is unchanged
from Riccioli's first map.

40
TIMOCHARIS
Greek Astronomer
and Philosopher
[320–260 BC]
Assigned by Riccioli and
has remained unchanged.

41
MONTES APENNINUS
[after the Apennine
Mountains, Italy]
One of the few Johannes
Hevelius names to survive
Riccioli's revisions.

42
MARE SERENITATIS
['Sea of Serenity']
Originally named by
Riccioli on his maps. The
Sea of Serenity is one of
the main dark lava seas.

43
MARCROBIUS
Roman writer of *Sea of
Scipio* and *Saturnalia*
[370–430]
Riccioli's creation to honour
Ambrosius Macrobius.

44
CLEOMEDES
Greek Astronomer
[c. 1 BC– AD 400]
Named by Riccioli and
remains unchanged in
today's lunar nomenclature.

56
HEVELIUS
Polish Astronomer
[1611–1687]
Originally appeared
as 'Hevelii' on Van
Langren's map of 1645.

57
KEPLER
German Astronomer
and Mathematician
[1571–1630]
Originally appeared as
'Keplerus' on Riccioli's map.

58
COPERNICUS
Prussian/Polish Astronomer
and Mathematician
[1473–1543]
Has survived in its original
form from Riccioli's map.

59
MARE VAPORUM
['Sea of Vapours']
Originally named by
Riccioli, the sea has kept
its name. Mare Vaporum
contains one crater.

60
JULIUS CAESAR
Roman General
[100–44 BC]
Named by Riccioli, Caesar
is one of the few military
figures with a lunar name.

61
TARUNTIUS
Roman Astrologer
and Mathematician
[unknown–86 BC]
Assigned by Riccioli to
honour Lucius Firmanus.

62
MARE UNDARUM
['Sea of Waves']
Named by J. H. Franz in
1906, Mare Undarum is
a shallow lunar mare and
contains five lunar domes.

74
GRIMALDI
Italian Mathematician
and Physicist
[1618–1663]
Riccioli named this for his
student Francesco Grimaldi.

75
LETRONNE
French Archaeologist
[1787–1848]
Named by Johann Heinrich
von Mädler to honour
Jean-Antoine Letronne.

76
MONTES RIPHAEUS
['Riphaeus Mountains']
Originally named by
Hevelius and slightly
amended to Montes
Riphaeus by Mädler.

77
PTOLEMAEUS
Greek Astrologer
and Mathematician
[c. AD 100–c. 170]
Named by Riccioli
to honour Ptolemaeus.

78
THEOPHILUS
Roman Emperor
[c. 385–412]
Named by Riccioli
to honour the Nicene
Pope of Alexandria.

79
COLOMBO
after Christopher
Columbus, Italian Explorer
[1451–1506]
Named after navigator
Christopher Columbus.

80
LANGRENUS
after Michel Florent
van Langren,
Flemish Astronomer
[1600–1675]
Modified from Langreni.

92
BYRGIUS
[1552–1632]
Named by Riccioli to
honour Swiss clockmaker
and mathematician Joost
Bürgi – logarithms inventor.

93
MARE HUMORUM
['Sea of Humours']
Originally called
Anticaspia by Gassendi,
Riccioli assigned the
mare its final name.

94
PITATUS
[unknown–c. 1550]
Riccioli named this after
Italian astronomer and
mathematician Pietro Pitati.
He authored several works.

95
PURBACH
[1423–1461]
Originally Purbachias
on Riccioli's map, in honour
of Austrian astronomer
Georg von Peuerbach.

96
RUPES ALTAI
[after the Altai
Mountains, Asia]
These were previously
named the Altai by
Blagg and Müller.

97
FRACASTORIUS
Italian Scholar
[1478–1553]
Riccioli named this after
Girolamo Fracastoro –
although spelt differently.

98
PETAVIUS
French Astronomer
[1583–1652]
Originally written by
Riccioli as Petavius Soc. I,
this name has survived.

110
SCHICKARD
German Astronomer
[1592–1635]
Originally spelt Schikardus
by Riccioli, this was named
for Wilhelm Schickard.

111
WILHELM
German Astronomer
[1532–1592]
One of Langren's surviving
names, after Wilhelm IV,
Landgrave of Hesse.

112
TYCHO
Danish Astronomer
[1546–1601]
Honoured by Riccioli for a
shared belief that the Earth
was the centre of the universe.

113
MAUROLYCUS
Sicilian Astronomer
[1494–1575]
Appeared on Riccioli's
map to honour
Francesco Maurolico.

114
RHEITA
Czech Astronomer
[1604–1660]
Riccioli originally wrote the
name as 'Reitha' – it now
appears as 'Rheita'.

125
SCHILLER
German Author
[1580–1627]
Originally spelt as
Schillerus on Riccioli's map,
it honours Julius Schiller.

126
CLAVIUS
German Astronomer
[1538–1612]
Appearing on Riccioli's
map, this was named
for Christopher Clavius.

127
HOMMEL
German Astronomer
[1518–1562]
Riccioli originally assigned
Homelius to a different
feature of the Moon.

INTRODUCTION—
MAPS 1–44

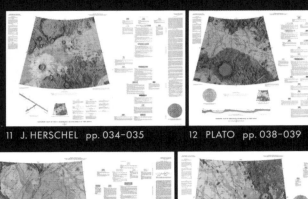

11 J. HERSCHEL pp. 034–035   12 PLATO pp. 038–039

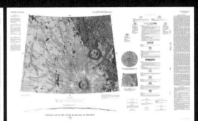

23 RÜMKER pp. 044–045   24 SINUS IRIDUM pp. 054–055   25 CASSINI pp. 056–057

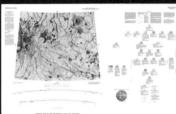
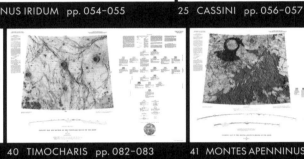
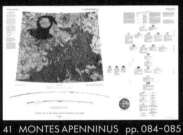

38 SELEUCUS pp. 074–075   39 ARISTARCHUS pp. 076–077   40 TIMOCHARIS pp. 082–083   41 MONTES APENNINUS pp. 084–085

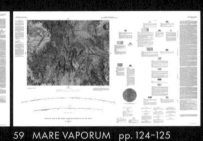

56 HEVELIUS pp. 104–105   57 KEPLER pp. 118–119   58 COPERNICUS pp. 120–121   59 MARE VAPORUM pp. 124–125

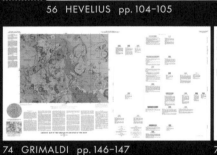

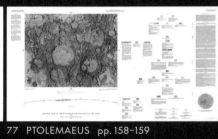

74 GRIMALDI pp. 146–147   75 LETRONNE pp. 148–149   76 RIPHAEUS MOUNTAINS pp. 156–157   77 PTOLEMAEUS pp. 158–159

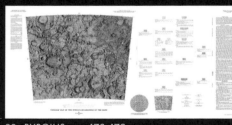
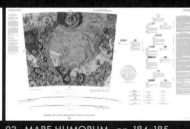

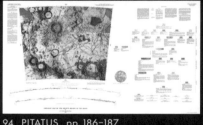

92 BYRGIUS pp. 178–179   93 MARE HUMORUM pp. 184–185   94 PITATUS pp. 186–187

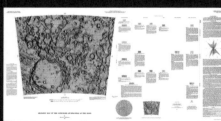
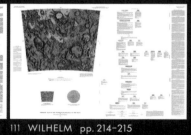

110 SCHICKARD pp. 212–213   111 WILHELM pp. 214–215

125 SCHILLER pp. 230–231

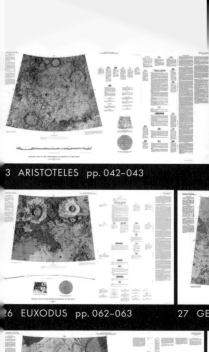

3 ARISTOTELES pp. 042–043

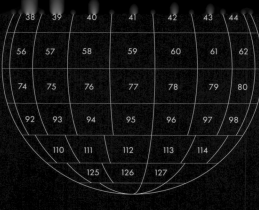

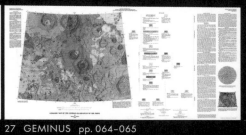

26 EUXODUS pp. 062–063

27 GEMINUS pp. 064–065

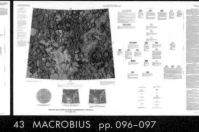

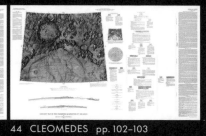

42 MARE SERENITATIS pp. 094–095

43 MACROBIUS pp. 096–097

44 CLEOMEDES pp. 102–103

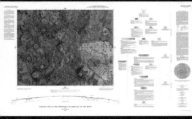

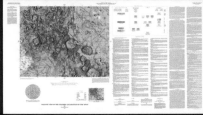

60 JULIUS CAESAR pp. 126–127

61 TARUNTIUS pp. 140–141

62 MARE UNDARUM pp. 142–143

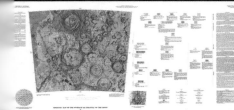

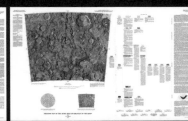

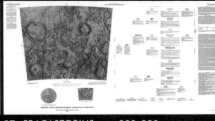

78 THEOPHILUS pp. 164–165

79 COLOMBO pp. 166–167

80 LANGRENUS pp. 176–177

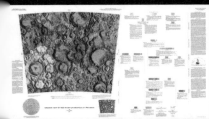

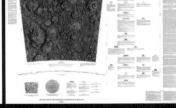

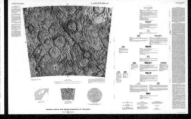

95 PURBACH pp. 194–195

96 RUPES ALTAI pp. 196–197

97 FRACASTORIUS pp. 202–203

98 PETAVIUS pp. 204–205

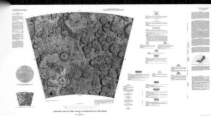

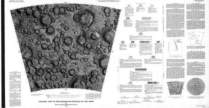

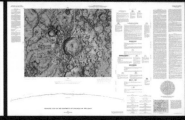

112 TYCHO pp. 218–219

113 MAUROLYCUS pp. 220–221

114 RHEITA pp. 228–229

THE PREHISTORIC LUNAR CALENDAR

B y definition, humanity's prehistoric ancestors left no written records of what they thought about their world or their interactions with it. Most of the 300,000 years humans have inhabited the Earth lie beyond our grasp – humankind's deep past is as tantalizing a mystery as its future. The story of who first noticed the movements or phases of the Moon, if such a claim can even be made, is buried irrecoverably beneath millennia of silence. But there is ample archaeological evidence to indicate that humans around the world – humans whose bodies and brains resemble those of the authors and readers of this book – watched the sky and made their own sense of the movements of the Sun, Moon and planets. They no doubt developed and communicated ideas about the Moon and its relation to their lives to family and community members, and across generations.

Prehistoric peoples viewed the night sky with their naked eyes alone. But they also saw the sky much more clearly, on a cloudless night, than can 21st-century people living in cities bathed in electric light. They watched an incredible theatre of lights unfold night after night, as the constellations moved, the planets wandered, and the Moon waxed and waned. On most nights, the Moon would emerge as the brightest and largest object in the sky. It went through the motions and phases familiar to skywatchers today. The added drama of an unpredicted lunar eclipse no doubt held our ancestors in thrall. It is no surprise, then, that the Moon is one of the earliest subjects of prehistoric records. Without written words, early humans began recording the changes they observed, and evidence of the attention they paid to the Moon is engraved in stone and in bone that has survived the ages.

Actually making sense of these records today – without imposing present-day assumptions and knowledge of the Moon's place in the solar system – is no easy feat. One of the oldest records that survives is a bone segment that archaeologists believe to be more than 30,000 years old. Found in the Blanchard cave in France's Dordogne valley in the early 20th century, the 'Blanchard bone' sat for years unappreciated in the collection of France's Musée des Antiquités. Then, in 1972, the same year that the last Apollo mission landed on the Moon, the American journalist and amateur anthropologist Alexander Marshack (1918–2004)

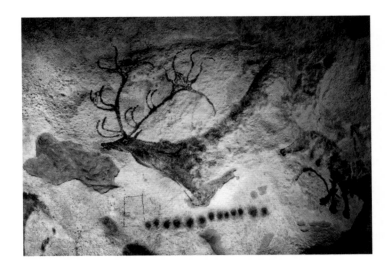

published the results of his own close analysis of the bone, performed with a hand lens and jeweller's loupe. He suggested that the sequential arrangement of dots on the bone corresponded with the phases of the Moon, each point representing one night. Star patterns and the Moon have also been found engraved in stone and painted on cave walls. Marshack argued that these were evidence of the beginning of scientific thought.[1]

Marshack's work was just one of many intellectual efforts attached to Apollo-era fascination with the Moon. His

book, *The Roots of Civilization: The Cognitive Beginnings of Man's First Art, Symbol and Notation*, in which he made his case for the prehistory of science, grew out of a NASA-sponsored project to provide historical context for the US space programme. Beginning in 1962, over the course of the same period in which NASA and the US Geological Survey compiled the Lunar Atlas, Marshack collaborated with NASA colleagues to construct a story of how civilization had reached a point of scientific and technological maturity, prepared at last to conquer the heavens. It was with this mission in mind that he saw order – scientific thought – in the seemingly random markings of the Blanchard bone and other younger specimens.[2] Many anthropologists disagreed with Marshack's interpretation, including his comparison of paleolithic timekeepers to the scientists and engineers planning the Apollo programme.[3] But his assertion that prehistoric people paid close attention to the Moon remains influential.

The Moon was bound to capture the attention of humanity's prehistoric ancestors. There is regularity to be found in its motions: like the Sun and planets, it moves from west to east against the backdrop of the stars. But, unlike any other object in the sky, its appearance changes from night to night. Over the course of twenty-nine nights, the Moon's face transforms, first igniting from a sliver of light into a full bright circle, then beginning to retreat into darkness, becoming at last fully dark – only to repeat this cycle anew.

What connections the ancient engravers of bone, or their communities, saw between these changes in the Moon's appearance and the world around them is impossible to say. They no doubt knew that a full Moon provided more visibility in the night-time, making nocturnal movement, work or hunting possible. Likewise, the dark of a new Moon might have been a time to stay protected by a communal fire or to move stealthily. As this bone calendar suggests, they may have found in the Moon's phases a convenient way to track the movement of time and, if they counted these lunar months, they may have been able to keep track of the seasons, predicting when the weather would grow colder or warmer. Their thoughts and inner lives remain inaccessible to their modern descendants, but it is certain that the world they inhabited – while in many ways familiar – was very different from ours.

↑
Anthropologists understand the thirteen dots painted beneath this representation of a deer to be half of the Moon's twenty-nine day cycle. This and other paintings, discovered inside the caves of Lascaux, France, in 1940, have been dated to 15,000 years ago. The caves appear to contain depictions of stars and constellations, suggesting that the people who painted them not only observed the night sky, but paid attention to its changes and their connections to the rhythms of nature. This knowledge helped them to survive and understand themselves.

↓
A serpentine sequence of sixty-nine engraved marks is discernable on the Blanchard bone, discovered by Abri Blanchard in a cave in Dordogne, France, in the early 20th century. In 1972 amateur anthropologist Alexander Marshack posited that the sequence of engraved notations represented the phases of the Moon over its twenty-nine day cycle. Dating from 30,000 years ago, during the Paleolithic era, the Blanchard bone is about 11 cm (4 1/2 in) in length with a flattened surface on which the marks are engraved.

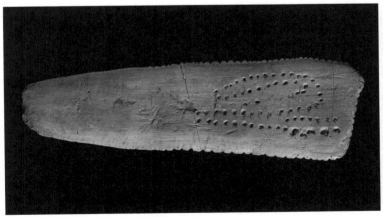

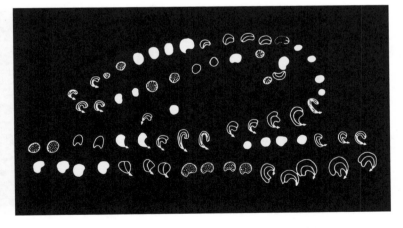

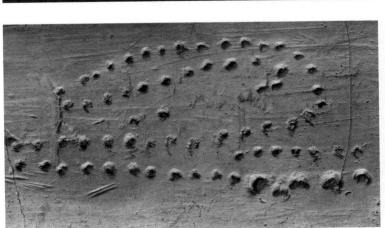

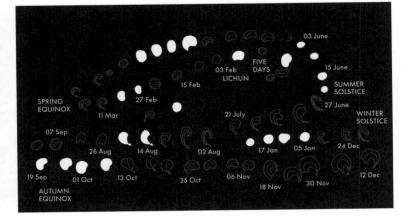

THE MOON IN ANCIENT EGYPTIAN CREATION MYTHS

↱
This relief from the Precinct of Amun-Re at the Karnak Temple Complex in Thebes (modern Luxor), dating back to the Nineteenth Dynasty (c. 1290–174 BC), portrays two Egyptian deities: one with the head of a lioness and the other with the head of an ibis. The ibis-headed deity is Thoth, associated with the Moon in Egyptian mythology. Here, Thoth fulfils his role as a divine scribe, as he records the duration of the king's reign. The goddess is identified in the hieroglyphs as Weret-hekau (meaning 'Great of Magic').

↳
The grand entrance to the temple of Khonsu at the Karnak Temple Complex is depicted in Maxime Du Camp's photograph *Propylone du Temple de Khons, Karnak*, 1849/51, printed in 1852. This imposing limestone gateway was erected during the reign of King Ptolemy III Euergetes in the 3rd century BC. The temple dedicated to Khonsu formed one element of the massive temple compound honouring the state deity, the divine creator god Amun-Re. Another complex, honouring his consort Mut, also resides in Karnak.

or millennia, the people of the Nile Valley gazed heavenward and, through keen observation of their natural surroundings, they conceived of a pantheon of deities who dwelled in the sky, but whose influence extended to the earthly realm of humankind. Many of these gods were associated with heavenly bodies including the Sun and Moon as well as individual stars and constellations. The belief that the sky served as the dwelling place of the gods appears in written texts as early as the Old Kingdom Pyramid Texts, funerary writings inscribed inside the burial chambers of royal pyramids beginning in the late Fifth Dynasty (c. 2360 BC).[1]

The all-pervading presence of the Sun in their predominantly desert environment probably accounted for the Egyptians' fascination with that celestial body, and a substantial portion of Egyptian religious belief is rooted in the consideration of the Sun's movement across the sky. However, the ancient Egyptians' attention also extended to its seemingly more mysterious nocturnal counterpart, the Moon. Depictions on the walls of tombs and temples often portray the Moon in its anthropomorphic divine form traversing the night sky in a sacred barque (boat), reminiscent of the river transport familiar to the Egyptians in their daily lives. Indeed, the characteristic crescent visible during the Moon's waxing and waning phases may have contributed to the idea that the Moon, like the Sun, sailed across the sky. In the ancient Egyptian language, the term for Moon was *iah* and this word incorporated a hieroglyphic sign that depicted a crescent Moon. Iah was also the name of a minor lunar deity.[2] Personal names such as Iah-mes (Ahmose), 'one born of the Moon', and Iah-hetep (Ahhotep), 'the Moon is at peace', were popular in certain periods of Egyptian history.

The gods Osiris and Min also held lunar associations,[3] and in later periods of ancient Egyptian history, when the country was ruled by Greeks and then Romans, the goddess Isis became linked to the Moon.[4] However, within the Egyptian pantheon, the gods Khonsu and Thoth are the most prominent lunar deities. Khonsu was the child of the mother goddess Mut and the great state god Amun, and he was influential in effecting the creation of new life in both animals and humans. Khonsu is credited with marking the passage of time and with healing abilities. He is typically shown in mummiform garb, and he carries the crook and flail – symbols of kingship. As a child god, he usually wears the 'sidelock of youth' hairstyle. He could also appear as a man with a falcon's head. His name, meaning 'traveller', may have alluded to the Moon's nightly journey across the sky. Khonsu appears in texts as early as the Old Kingdom, but he rose to greatest importance in the New Kingdom (c. 1550–1069 BC).[5] As with all lunar deities, he is shown wearing a crown consisting of a full Moon disc combined with the crescent Moon.

The ibis-headed god Thoth was also associated with the Moon. He was a god of wisdom and scribes and, as such, he was believed to have invented writing. He is often shown standing next to Re (the Sun-god) in his barque during Re's nightly journey through the underworld. Thoth was responsible for regulating the seasons, controlling the lunar phases and ordering the stars. He was associated with astronomy and mathematics and was the divine record-keeper. One of Thoth's sacred animals was the baboon, a nocturnal animal that goes to sleep only after greeting the new day.[6] Like Khonsu, Thoth often wears the combined full and crescent Moon crown. His key role was enshrined in the use of the name 'Thoth' for the first month of the Egyptian lunar calendar.

The ancient Egyptians also used celestial observations to organize their world, devising calendars and formulating timekeeping methods that served both sacred and secular purposes. Archaeological evidence in the Nile Valley suggests that the day and night sky were being observed as early as the 5th millennium BC. One of the oldest astronomical sites in the world is located at a site called Nabta Playa, about 100 kilometres (62 miles) west of Abu Simbel in southern Egypt. Here excavations have uncovered megalithic stone arrangements – the largest of which is a 3.6-metre (12-foot) circle of standing stones created about 6,500 years ago. This stone circle seems to have functioned as a form of early calendar, marking the onset of the summer solstice, which coincided with the arrival of the rainy season.[7]

The observable lunar phases were an essential timekeeping tool in ancient Egypt. The Egyptians devised a lunar calendar of 12 months of 30 days each, as defined by the duration of the lunar cycle. Each month began after the first observation of the new Moon. Important religious festivals were timed to coincide with the arrival of certain phases of the Moon. The months, known (in the Greek versions of

the Egyptian names from later textual sources) as Thoth, Phaophi, Athyr, Choiak, Tybi, Mekhir, Phamenoth, Pharmuthi, Pakhons, Payni, Epiphi and Mesore, incorporate the names of important gods and festivals. Because the lunar calendar was shorter than the solar year, a thirteenth lunar month was added every few years to keep the calendar coordinated with the agricultural seasons and their associated festivals. The yearly heliacal rising of the star Sothis (Sirius), visible on the eastern horizon just before dawn in midsummer (19 July of the Julian calendar), heralded New Year's Day and the arrival of the annual Nile flood, carrying water and silt to fertilize the fields. For the lunar calendar, the timing of this observation played a role in determining the need to employ an intercalary month. The Egyptians also observed a relationship between the Moon and the growth of plants, and noted that seeds were best sown at the time of a full Moon, so it is clear that lunar observations also served a practical function for this agrarian society.[8] The ancient lunar month names survived into the names used in the Coptic language, and were so engrained in agriculture that they are still used in modern Egypt.

The civil calendar, the earliest precursor to the modern Western calendar, was probably established for administrative and accounting purposes and emerged as early as the First Dynasty (c. 3100–2900 BC). The Egyptian civil year had three seasons: (1) Akhet (inundation); (2) Peret (emergence [of crops]); and (3) Shemu (harvest). Each season consisted of 4 months, further divided into 3 ten-day weeks, resulting in a fixed 360-day year and a systematic arrangement of 12 months. An additional 5 days, known as the epagomenal days ('days above the year'), were included, bringing the civil year to a total of 365 days. However, the Egyptians never introduced a leap year to the civil calendar, which always consisted of 365 and not 365 ¼ days. Consequently, they never addressed the disjunction between the solar year and the civil calendar, which continually shifted backwards by ¼ day every year. This defect in the civil calendar created complex, changing relationships between the civil and lunar calendars over the thousands of years of ancient Egyptian history.[9]

In addition to its divine associations and its role in religious and agricultural timekeeping, the cyclical nature of the Moon's phases paralleled the Egyptians' belief in life, death and rebirth and, therefore, mirrored their aspirations for an eternal afterlife.

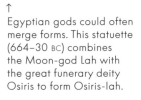

↑
Egyptian gods could often
merge forms. This statuette
(664–30 BC) combines
the Moon-god Iah with
the great funerary deity
Osiris to form Osiris-Iah.

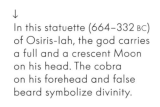

↓
In this statuette (664–332 BC)
of Osiris-Iah, the god carries
a full and a crescent Moon
on his head. The cobra
on his forehead and false
beard symbolize divinity.

↑
This 4th century BC faience
inlay depicts Thoth with
a feather that represents
the Egyptian concept
of *maat*, meaning order,
balance and harmony.

↓
An amulet (c. 664–332 BC)
of the lunar deity Thoth in
the form of an ibis-headed
man. Such objects were
popular personal talismans
in ancient Egypt.

↑
This inlay entitled *Horus of Gold* was created in the 4th century BC. It shows the falcon god Horus standing on the symbol for gold. Horas was a god of kingship.

↓
A bronze statue of the deity Horus (c. 664–525 BC). The ancient Egyptians equated Horus' left eye with the Moon and his right eye with the Sun.

↑
A gold pendant with a coloured inlay (c. 1295–1070 BC) of the falcon-headed god Khonsu. A full and a crescent Moon rests on top of his head.

↓
A bronze figurine of the Moon-god Khonsu (664–332 BC). Such figurines were often given as offerings to the gods in ancient Egyptian shrines and temples.

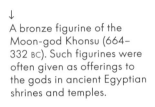

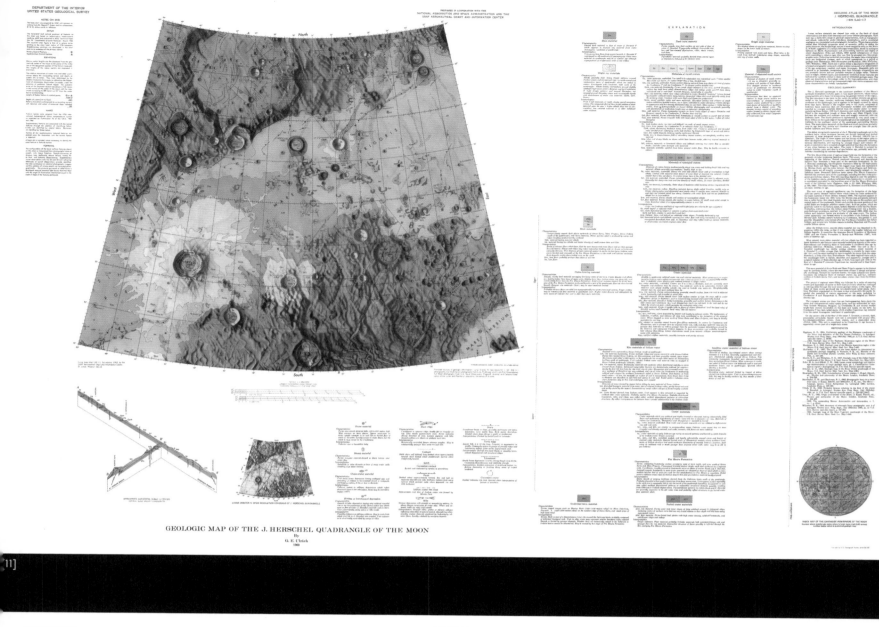

GEOLOGIC MAP OF THE J. HERSCHEL QUADRANGLE OF THE MOON

By
G. E. Ulrich
1969

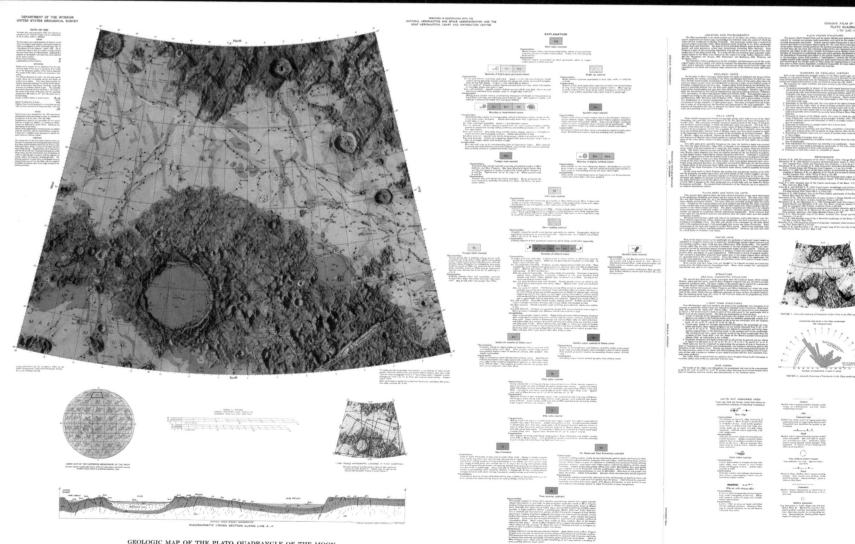

GEOLOGIC MAP OF THE PLATO QUADRANGLE OF THE MOON

By
John W. M'Gonigle and David Schleicher
1972

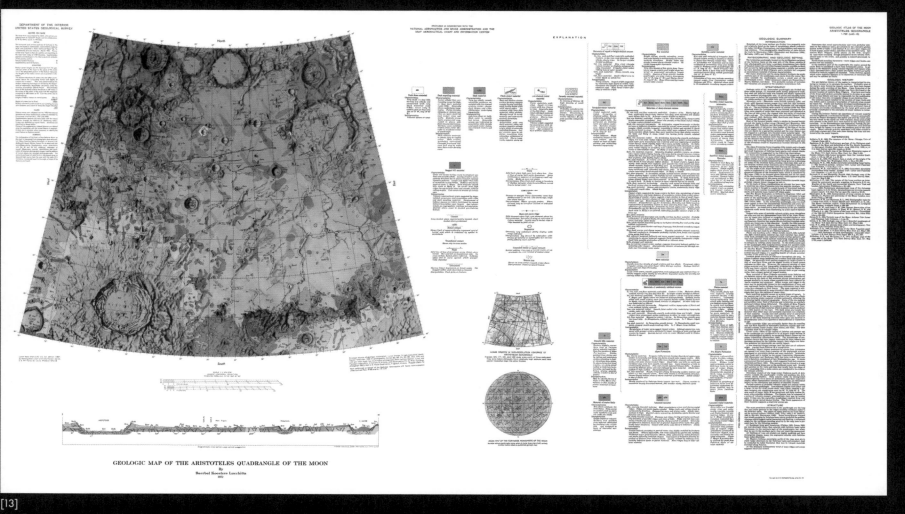

GEOLOGIC MAP OF THE ARISTOTELES QUADRANGLE OF THE MOON
By
Baerbel Koesters Lucchitta
1972

[13]

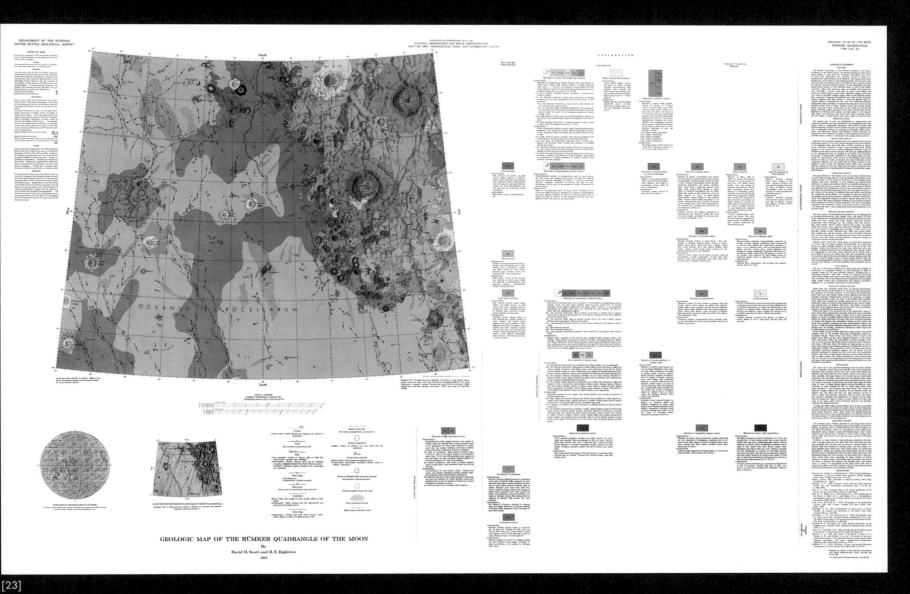

GEOLOGIC MAP OF THE RÜMKER QUADRANGLE OF THE MOON
By
David H. Scott and R. E. Eggleton
1973

[23]

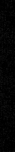

11
Facsimile:
Geologic Map of the J. Herschel Quadrangle of the Moon, 1969, by G. E. Ulrich

12
Facsimile:
Geologic Map of the Plato Quadrangle of the Moon, 1972, by John W. M'Gonigle and David Schleicher

13
Facsimile:
Geologic Map of the Aristoteles Quadrangle of the Moon, 1972, by Baerbel Koesters Lucchitta

23
Facsimile:
Geologic Map of the Rümker Quadrangle of the Moon, 1973, by David H. Scott and R. E. Eggleton

Karen ní Mheallaigh

THE MOON IN ANCIENT GREEK AND ROMAN MYTH

Myth

Lunar

↱
This marble relief (350–300 BC) depicts the lunar deity Selene. The Moon's phases are implied by the goddess's crescent crown, the circular object in her hand (full Moon) and her veil, with which she may cloak her face and 'disappear' (new Moon). Selene also wears starry earrings.

↱↱
Hecate, goddess of magic, night and the dead, was associated with the Moon and with crossroads, hence her triple-form depiction. 'Triple Hecate' also represented the three lunar phases (crescent, full, new) and, astronomically, the Moon's apparent three-part movement in space.

↱↱↱
This 2nd-century AD marble altar represents the night: Selene is flanked by Phosphoros – the morning star – and Hespheros, the evening star. Beneath Selene is the god of the sea, Poseidon/Neptune, as the Moon was thought to rise from the ocean in the east.

↳
This fragment of a sarcophagus panel (c. AD 210) shows Selene's encounter with Endymion, who reposes as Hypnos, god of sleep, as she hovers over him (far right). Common on coffins (or sarcophagi), the myth offers the comforting message that death is merely sleep, and that love endures even through death.

rom very ancient times the Greeks and Romans associated the Moon with the intimate rhythms of female life. Visually, its rotund waxing in the sky each month was thought to resemble the swelling of the pregnant womb, and menstruation was believed to be linked to its regular twenty-eight-day cycle of phases. The Moon was identified, consequently, with various goddesses, primarily the Greek Moon-goddess Selene herself (known to the Romans as Luna), who was believed to help women in childbirth.[1] But the Moon also wanes and disappears from view each month, and so, through this regular cycle of waxing and waning, it appears to undergo a microcosmic lifetime – of birth, flourishing and decline – over and over again. Greek and Roman myths, consequently, reflect deeply felt connections between birth, death and the Moon.

Arguably the most famous lunar myth is that of Selene and Endymion, which, appropriately, is first found in the writings of Sappho, a female poet of the 7th century BC.[2] In this myth the Moon-goddess falls in love with a handsome young hunter named Endymion. Endymion is cast into an enchanted sleep and hidden in a cave where Selene could visit periodically to enjoy his beauty in secret.

This myth, which has inspired so much later art (including famous paintings of the 17th and 18th centuries by Ubaldo Gandolfi, Nicolas Poussin and Anne-Louis Girodet), invites us to consider the Moon-goddess as an autonomous erotic agent. The focus is entirely on her desiring gaze, which is unusual in Greek and Roman myth, where the gaze is more usually commanded by the male, with the female as its more passive recipient. The Selene myth, therefore, challenged the usual protocols of gender-relations in ancient thought. Even the cave in which Endymion was sequestered for the goddess' pleasure provided an image of a private space where female desire might express itself without censure. The Moon, in this way, became a sort of mythological ally for women's private emotional worlds, particularly their experiences of desire.[3]

But Selene's regular visitations to Endymion's cave also offered a mythological explanation for why the Moon seems to vanish for a night or two each month during the new Moon phase. In fact, the myth reminds us that the Greeks and Romans associated the Moon with a powerful

visuality, not just in terms of female sexual desire, but in more naturalistic terms too, for the Moon was often thought of as a great eye in the sky gazing at the Sun, from whom it derived its light.[4] As early as the 5th century BC, the Greeks hit on the idea that the Moon's light was reflected, a theory known as 'heliophotism'. Our earliest witness to this idea survives in hauntingly beautiful lines of poetry by the Greek philosopher Parmenides (c. 500 BC), who describes the Moon as an 'alien light...gazing eternally at the rays of the Sun'.[5] Such visual associations, which are expressed so poignantly in the Selene myth, were intensified by the belief, widespread in antiquity, that the distinctive blotches on the lunar disc represented a *face*.[6] Through the myth of Selene and Endymion, therefore, we can see the ancient Greeks working out several core aspects of the Moon's nature, appearance and behaviour in the sky.

The Moon was also associated with death in ancient Greek and Roman thought; in fact, Endymion's eternal sleep was commonly used in ancient funerary art to allude to death.[7] Elsewhere, the idea is expressed through the Moon's association with the goddess Persephone (Proserpina to the Romans). In the well-known myth, Persephone was abducted and taken down to the underworld to be the bride of Hades, god of the dead. As her mother Demeter (Roman Ceres), the goddess of agriculture, searched desperately for her daughter, all the crops failed, for the goddess neglected her care of the fields. However, the two were eventually reunited: Persephone was granted permission to spend part of the year on the Earth's surface, and part among the dead below.[8] The myth, therefore, accounts for the cycle of the agricultural year: when mother and daughter are reunited, Demeter rejoices and the earth is fertile (spring, summer and the harvest); but when they are separated, as Persephone returns to the underworld, the vegetation dies as Demeter mourns (winter). In astronomical terms, this myth was also thought to represent the intimate relationship between the Earth and its satellite. Demeter represents the Earth, and the beautiful Persephone is the Moon, who cleaves closely to her mother but every so often is engulfed in Hadean shadow and separated from her mother's loving embrace, as the Moon undergoes eclipse and vanishes from the sky.[9]

This mythical association with Persephone, Queen of the Dead, also reflects the belief that the souls of the dead clustered around the Moon. In the tradition of Platonic philosophy, it was thought that the soul was separated from the body at death. The body decomposed into the Earth, but the incorporeal soul, newly liberated, ascended upwards to the Moon, where it underwent purification for its misdemeanours in life. Some very pure souls could then progress upwards to be united with the Sun, but most would spend time on the Moon until, suitably cleansed, they could be dispatched back down to Earth to serve out another lifetime and strive to improve.[10]

With this philosophical myth, we come full circle: the Moon is connected with both death and rebirth in a cosmic doctrine of reincarnation, mirroring once again the patterns of death and regeneration in its cyclical waxing and waning, which so fascinated the ancient Greeks.

This tauroctony relief was discovered in the mithraeum of Saint Stefano Rotondo at the end of the 3rd century AD. In the centre is the god Mithras sacrificing a bull. The luminaries in their flying chariots flank him: the Sun (Sol) left, the Moon (Luna) right. Mithras' golden face visually reminds us of his association with the Sun towards whom he directs his gaze. Such reliefs are commonly found in the grotto-like shrines devoted to the worship of Mithras, which were known as Mithraea. Since this was a mystery-cult, the meaning of this ritual scene is obscure to us and interpretations are legion. Astral motifs (Sun, Moon, stars) may point to an astrological meaning; other typical elements in the scene could represent constellations or zodiacal signs, for example the serpent (Hydra?) and scorpion (Scorpio?) beneath the bull, the dog (Canis Major/Minor?) and even the bull itself (Taurus?). The Moon was also often associated with a bull, so perhaps the scene somehow represents the triumph of the Sun over the Moon? We cannot be sure.

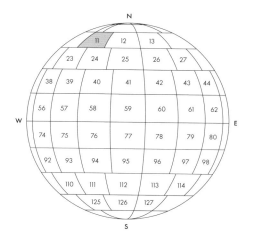

This quadrangle is named for the large remnant crater J. Herschel (diameter 154 kilometres, 95 ³/₄ miles) in the northwest corner. The crater is an ancient feature, whose structure has been noticeably altered by later impacts, volcanic events and tectonic forces. The walls of the crater have been reshaped into a ring of ridges. The sharper-edged craters in this quadrangle, such as Harpalus, are very young compared to Herschel. An impact crater, Harpalus has a sharp rim crest and is surrounded by rays of ejecta (material thrown out during the impact that formed it). That this material overlays the surrounding terrain indicates that the impact that created Harpalus (as well as that which created Bouguer to its east) happened after the nearby maria had filled with lava and became the smooth, dark features early astronomers described as lunar seas. This quadrangle also contains several small craters and crater chains that probably resulted from material ejected from larger impacts. In the south of the quadrangle is the Montes Jura mountain range (named for the Jura Mountains in France and Switzerland). In the mountains are the well-preserved crater Bianchini and the ruined crater Maupertuis, which has been almost erased by small impacts.

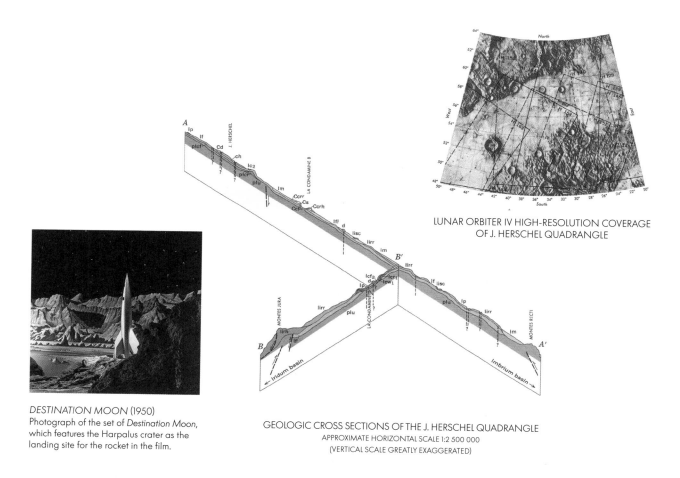

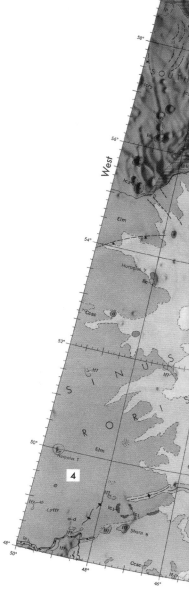

LUNAR ORBITER IV HIGH-RESOLUTION COVERAGE
OF J. HERSCHEL QUADRANGLE

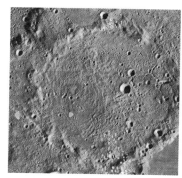

DESTINATION MOON (1950)
Photograph of the set of *Destination Moon*, which features the Harpalus crater as the landing site for the rocket in the film.

GEOLOGIC CROSS SECTIONS OF THE J. HERSCHEL QUADRANGLE
APPROXIMATE HORIZONTAL SCALE 1:2 500 000
(VERTICAL SCALE GREATLY EXAGGERATED)

J. HERSCHEL QUADRANGLE: KEY CHARACTERISTICS

1. J. HERSCHEL CRATER
Latitude: 62.0°N / Longitude: 42.0°W
Diameter: 154 km / 95 ³/₄ miles
Photograph: Lunar Orbiter IV (USA, 1967)

2. MARE FRIGORIS (SEA OF COLD)
Latitude: 56.0°N / Longitude: 1.4°E
Diameter: 1,446 km / 898 ½ miles
Photograph: unknown

3. HARPALUS CRATER
Latitude: 52.6°N / Longitude: 01.4°E
Diameter: 39 km / 24 ¼ miles
Depth 2.9 km / 1 ³/₄ miles
Photograph: Lunar Orbiter IV (USA, 1967)

4. SINUS RORIS (BAY OF DEW)
Latitude: 54.0°N / Longitude: 56.6°W
Diameter: 202 km / 125 ½ miles
Photograph: Lunar Reconnaissance Orbiter (USA, 2016)

[LAC-11]

I-604

J. Herschel Quadrangle

Maps

Lunar

035

5. FOUCAULT CRATER
Latitude: 50.4°N / Longitude: 39.7°W
Diameter: 23 km / 14 ¼ miles
Depth: 2.1 km / 1 ¼ miles
Photograph: Lunar Orbiter IV (USA, 1967)

6. BIANCHINI CRATER
Latitude: 48.7°N / Longitude: 34.3°W
Diameter: 38 km / 23 ½ miles
Depth: 3.1 km / 2 miles
Photograph: Lunar Orbiter IV (USA, 1967)

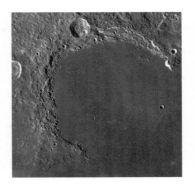

7. MONTES JURA
Latitude: 47.1°N / Longitude: 34°W
Diameter: 422 km / 262 ¼ miles
Max Height: 3.8 km / 2 ¼ miles
Photograph: LRO (USA, 2015)

8. MONTES RECTI
Latitude: 48.0°N / Longitude: 20.0°W
Length: 90 km / 56 miles
Max Height: 1.8 km / 1 mile
Photograph: Lunar Orbiter IV (USA, 1967)

'THE MOON IS MY MOTHER'— LAKOTA PERSPECTIVES

↱
The pictographic symbols of a black Moon and two red stars (top right) on the Lone Dog winter count (1851–52) symbolize the total solar eclipse that took place on 7 August 1869. Winter counts were used widely by Native Americans to record important events of the year.

↱↱
uŋȟčéla wílečhala, for Nathan Young (Waxing Crescent Peyote Moon) (2020) by Suzanne Kite depicts an upside-down crescent Moon above a cross.

akota people are as intricately connected to the cosmos as they are to the land. A non-metaphorical, complex relationship, including kinship, exists between and beyond the human and nonhuman entities of the sky, stars and Earth that populate the universe. The story of the separation of Hanwi (the Moon) from her partner Wi (the Sun) reflects an understanding of the mirroring of relations within the cosmos where, just as on Earth, relationships that form our constellations of existence are created and destroyed.

> When Hanwi arrived and saw Ite [a human] occupying her seat, she stood behind Wi with her robe over her head. When people saw this, they began to laugh at Hanwi. Inktomi [the trickster] laughed the loudest. Kanka [mother of Ite] sang a song of joy, but Wazi [father of Ite] was still afraid. Tate [Wind, husband of Ite] left the feast and returned to his own lodge, where he painted his face and the faces of his sons black in mourning over the loss of his wife.[1]

After Hanwi's face next to Wi was usurped by the beautiful Ite, Hanwe's face darkened with mourning, her joy eclipsed after the cruelty of her husband: she 'blackened her face and mourned with Tate and the people laughed at her no more'.[2] A story of love and betrayal, our night sky illuminates a vibrant cosmos, with knowledge and narratives that shift, adapt and change as frequently as the Moon, and as human relationships.

The Lakota 'felt a vivid relationship between the macrocosm, the star world and their microcosmic world on the plains,' as Ronald Goodman explains in *Lakota Star Knowledge*.[3] This relationship was understood in a metaphysical but also physical sense, manifested in the landscape:

> There was a constant mirroring of what is above by what is below. Indeed, the very shape of the earth was perceived as resembling the constellations. For example, the red clay valley which encircles the Black Hills looks like (and through Oral Tradition is correlated with) a Lakota constellation which consists of a large circle of stars.[4]

From at least 100 BC, the Lakota people calculated their decisions by the stars.[5] This relationship is documented

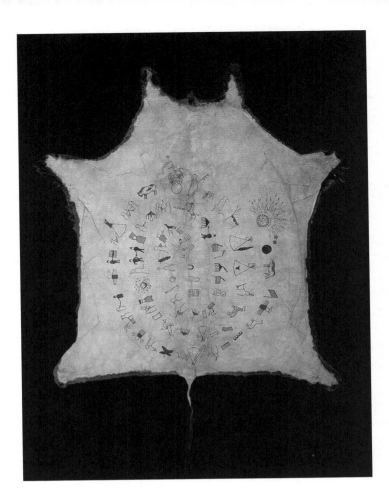

on two tanned animal hides, one mapping the Earth with buttes, ridges and rivers, and the other mapping the stars. Goodman writes: 'These two maps are the same...because what's on the earth is in the stars, and what's in the stars is on the earth'.[6] The intricate interplay between the cosmos and the land in Lakota cosmology underlines the profound interconnectedness between the celestial and terrestrial realms, shaping the very fabric of Lakota life and worldview.

However, when celestial bodies are treated as commodities to exploit and discard, they are devalued to mere objects, devoid of respect or autonomy. 'Object mastery and territorial possession are demonstrably part and parcel of the processes of genocide,' writes Dylan Rainforth.[7] Extending the ideology of dominance and ownership across the Earth and cosmos has resulted and will result in the subjugation of all entities within them, whether a people considers them 'beings' or not. A coordinated manipulation of knowledge attempts to contain places and beings to maps, trying to transform the very landscape into terra nullius.

Lakota art-making generates and sustains our intimacies with each other and the cosmos, where our environmental stewardship, Indigenous rights and decolonization efforts speak to the importance of critical engagement with our geographies and knowledge systems. In their debut poetry collection *Instructed by Haŋwí*, Autumn White Eyes explores being influenced by the Moon:

> we were searching for one another
> over and over
> like the moon follows the sun
> until they unite into eclipse,
> moments feel like life-times with you.[8]

Embedded within the diversity of cosmologies are principles of interconnectedness and respect for beings, guiding us towards sustainable practices that make clear our roles in engaging with other beings. Contemporary efforts to honour Indigenous sovereignty and reclaim ancestral lands echo the principles of our cosmology, fostering a deeper understanding of our interconnectedness with the land and each other.

In Lakota epistemology, there is a movement of knowledge through the act of art-making, and the transmission of knowledge is seen as interconnected decisions, situated in contexts, and reliant upon relationships with nonhuman beings who communicate in physical and nonphysical ways. To navigate the US Geological Survey (USGS) maps of the Earth and Moon is to navigate a history of human decisions. The history of dispossession of Lakota lands creates the past realities of my mothers and grandmothers, forming a liquid in and around USGS survey data of their burial grounds, all of which melts and reforms into a chart of stars, reflected in our human eyes. These maps serve not only as tools of navigation, but also as artefacts of colonialism, embodying the complex entanglement of power, knowledge and territory.

Maps chart our human desires for intimacy with the world, the cosmos and everything in between. USGS mappings of the lunar landscape also map space, time, geopolitical contexts, which, in turn, map the United States' desire to colonize, to own and to know. A desire to colonize celestial bodies reflects an ontology which does not consider nonhuman beings as worthy of beinghood. Eve Tuck writes, 'settler futurity is ensured through an understanding of Native-European relations as a thing of the past, and the inclusion of Native history as a past upon which a white future is ensured.'[9] 'Settler futurity' could encompass an epistemology which allows for Manifest Destiny, Pilgrim's Progress, and an American destiny of ownership over an entire continent, and beyond, into the stars. This colonial mentality extends beyond earthly realms, perpetuating systems of domination and exploitation even within the cosmic domain, reinforcing hierarchies that privilege human interests above all else.

Where does knowledge come from, how is knowledge communicated and what do humans do with that knowledge? Human desire impels us to become more intimately knowledgeable about the Moon; this can reflect our choices and values. Lakota methods of astronomy and storytelling make clear our connections, relationships, contexts and, most importantly, ethics.

As we look up to the skies from our planet, which often darkens its face in light of the suffering of its human and nonhuman beings, let these words guide our humble position in a relational cosmos: 'The sun is my father. The moon is my mother. Let no one have power over them.'[10]

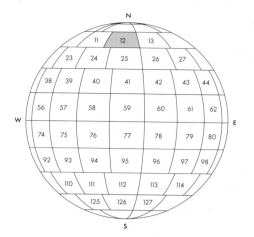

The Plato quadrangle is in the north-central part of the Moon's near side. It contains the northern edge of the ancient Mare Imbrium (Sea of Showers) lava plain. The smooth terrain of Mare Imbrium and Mare Frigoris (Sea of Cold) to its north is interrupted by rugged features, including part of the Montes Alpes mountain range (named for the European Alps), which forms the northeastern edge of Mare Imbrium and whose peaks reach as high as 2.4 kilometres (over 1 ½ miles). Within Mare Frigoris, to the west of the Montes Alpes and against the 'shore' of Mare Imbrium, is the large impact crater Plato (stretching 101 kilometres, or nearly 63 miles, across), the feature that gives this quadrangle its name. The crater's floor was filled in by lava, giving it a smooth surface with a dark appearance. Compare this to the seemingly unaltered, sharp-rimmed impact crater Archytas to the northeast, with its interior peaks and rings of rough material. At the north of this quadrangle from west to east are the impact craters Fontenelle, which marks the northern edge of Mare Frigoris, and Timaeus, part of the southern wall of the W. Bond plain. Between these two lay the resurfaced remnants of the crater Birmingham.

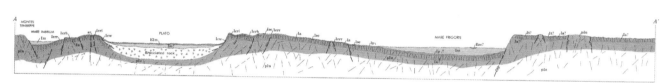

GEOLOGIC CROSS SECTION OF THE PLATO QUADRANGLE
APPROXIMATE HORIZONTAL SCALE 1:1 000 000 (VERTICAL SCALE GREATLY EXAGGERATED)

AZIMUTH FREQUENCY OF LINEAMENTS IN FIVE AREAS IN THE PLATO QUADRANGLE

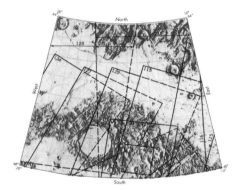

LUNAR ORBITER PHOTOGRAPHIC COVERAGE OF PLATO QUADRANGLE

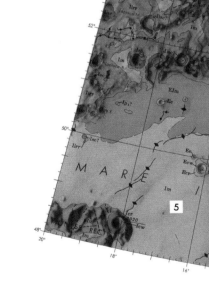

PLATO QUADRANGLE: KEY CHARACTERISTICS

1. PLATO
Latitude: 51.6°N / Longitude: 9.3°W
Diameter: 101 km / 62 ¾ miles
Depth: 1,468 m / 4,816 ft, 3 in
Photograph: LRO (USA, 2014)

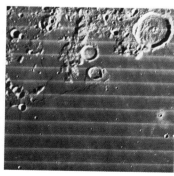

2. FONTENELLE CRATER
Latitude: 63.4°N / Longitude: 18.9°W
Diameter: 38 km / 23 ½ miles
Depth: 1.8 km / 1 mile
Photograph: Lunar Orbiter IV (USA, 1967)

3. ARCHYTAS CRATER
Latitude: 58.7°N / Longitude: 5°E
Diameter: 32 km / 20 miles
Depth: 2.4 km / 1 ½ miles
Photograph: Lunar Orbiter IV (USA, 1967)

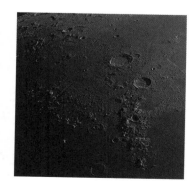

4. MARE FRIGORIS
Latitude: 56°N / Longitude: 1.4°E
Diameter: 1,446 km / 898 ½ miles
Photographed from Earth, 2017

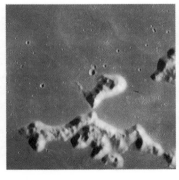

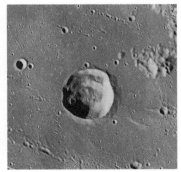

5. MARE IMBRIUM (SEA OF SHOWERS)
Latitude: 32.8°N / Longitude: 15.6°W
Diameter: 1,146 km / 712 miles
Photograph: LRO (USA, 2014)

6. MONTES TENERIFE
Latitude: 47.89°N / Longitude: 13.19°W
Diameter: 112 km / 69 ½ miles
Max Height: 2.4 km / 1 ½ miles
Photograph: Lunar Orbiter IV (USA, 1967)

7. MONTES ALPS
Latitude: 48.36°N / Longitude: 58°W
Diameter: 334 km / 207 ½ miles
Max Height: 2.4 km / 1 ½ miles
Photograph: Lunar Orbiter IV (USA, 1967)

8. PROTAGORAS CRATER
Latitude: 56.0°N / Longitude: 7.3°E
Diameter: 22 km / 13 ¾ miles
Depth: 2.1 km / 1 ½ miles
Photograph: LRO (USA, 2011)

Samantha Thompson

Matter

Lunar

THE MOON AND A GEOCENTRIC UNIVERSE

↱ Semi-Tychonic representation of the world's construction according to 16th-century astronomer Paul Wittich, from Andreas Cellarius' *Harmonia macrocosmica* (1660). Tycho Brahe's cosmos was a compromise between geocentric and heliocentric models of the universe. The planets orbited the Sun and the Sun and Moon circled a stationary Earth.

→ The planisphere of Ptolemy, also from *Harmonia macrocosmica*. Ptolemy's cosmos placed Earth nested at the centre with the Sun, Moon and planets in concentric spheres around it. Ptolemy's predictions of thepositions of the planets in the night sky were so effective that they were used for astronomical charts for over a thousand years.

↳ Diagram of the cosmos, from Matfre Ermengaud's 14th-century *Breviari d'Amor* (Book of Hours of Love). The map depicts Aristotle's geocentric cosmos, where everything within the sublunar sphere is corruptible and mutable, and the realm beyond the Moon is perfect.

reek philosopher Aristotle (384–322 BC), in his treatise *De Caelo* (On the Heavens), posited a geocentric model of the universe, with the Sun, Moon, planets and stars arranged in perfect circles in a heavenly aether. A motionless Earth sat at the centre. In this model, the Moon moved in the boundary between the imperfect terrestrial sphere, composed of the changeable elements earth, water, air and fire, and the perfect celestial sphere, made of a fifth substance, quintessence.[1] The Moon's spherical shape was a key point of support for Aristotle's geocentric model, but its visible imperfections needed to be explained away. For centuries most ancient Greek, ancient Roman and medieval philosophers held the belief that the Moon, as a heavenly body, must be spherical and perfectly smooth. Any models of the universe put forward by philosophers had somehow to resolve observations of the Moon's actual motion, and the shadowing and features on its surface, with the assertion that it was innately a perfect, heavenly body.

The concept of the geocentric model of the universe seems to have entered Greek astronomy in the 6th century BC with the philosopher Anaximander (died after 547 BC). At the Milesian school in Ionia, Anaximander conceived a cosmology that placed a cylindrical-shaped Earth at the centre.[2] In this universe, the Earth floated freely, without the support of water or pillars, a unique supposition at the time.[3] The Earth was surrounded by rings, each filled with fire; this fire was visible through holes that appeared as the Moon, Sun and stars. In an attempt to explain the phases of the Moon, he suggested that the holes in the Moon's ring changed shape. Though he placed these bodies in the wrong order, with the stars closest to the Earth and the Moon furthest away, his suggestion that they moved in circles remained in subsequent understandings of the cosmos. And though Anaximander used non-mythological explanations for the motions of the Moon, Sun and stars, he still viewed the celestial bodies as deities.

By the 5th century BC, philosophers had largely accepted that the Earth and Moon were spherical in shape. Anaximander's pupil Pythagoras (active 6th century BC) had combined empirical evidence for this claim with an aesthetic rationale. As the Moon changed phases, Pythagoras observed the curvature of the terminator, the line separating the light and dark areas of the Moon. He posited that this curvature could only be produced on a spherical object. Once he accepted that the Moon was a sphere, it followed that the Earth and all other heavenly bodies must be round as well. This fit in well with the preconceived idea that the sphere was a perfect shape and worthy of the heavens. Philolaus (c. 470–390 BC), a philosopher of the Pythagorean school, took this idea a step further and offered a view of the universe in which the central object was neither the Earth nor the Sun, but an unseen Central Fire that shone upon everything, including the Moon. It was this Central Fire that produced the Moon's changing phases. By not placing the Earth at the centre, the Pythagorean astronomical system offered a worldview that suggested that the Earth on which observers were standing was itself in motion, but this was still based upon religion and aesthetics rather than science. The idea lay dormant until cited by Nicolaus Copernicus (1473–1543) nearly 2,000 years later.

The Ionian philosopher Anaxagoras (born c. 500 BC) questioned the divinity and perfection of the Moon and offered a physical, rather than spiritual, explanation of its appearance and nature. He observed mountains on its surface and posited that it was made of rock, like the Earth. He also theorized that the Sun, too, was made of rock, but rock that burned with a great light that shone on

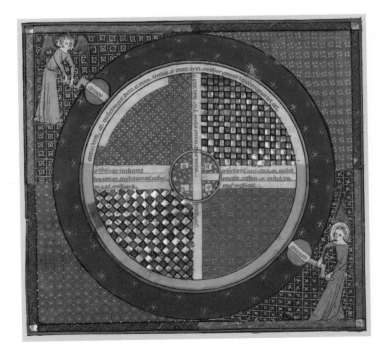

the Moon, lighting up its face.[4] While others had thought that the Moon reflected light, Anaxagoras was one of the first to suggest that this light came from the Sun. Only fragments of his writing survive, but it is possible to see how he may have relied on geometry and observation, rather than religious belief, in developing his geocentric model of the universe. When the Moon is visible during the day, the illuminated side can always be observed as facing the Sun. Anaxagoras' model explained the phases of the Moon but also lunar eclipses, an apparent darkening of the Moon when the Earth passes between the Moon and Sun, and solar eclipses, when the Moon passes in front of the Sun and casts a shadow on the Earth.[5] Breaking with tradition and religion, and offering explanations based more consistently on observation, the Ionian philosopher was an early proponent of what would eventually become natural philosophy, and then science. His ideas, however, were not widely accepted at the time and he was charged with impiety and exiled from Athens for suggesting that the Moon and Sun were not deities.

The notion that the Moon and other celestial objects were perfect, heavenly bodies remained the prevailing belief in later variations of the geocentric model of the universe. For Plato (c. 429–347 BC), the Earth stood still at the centre and everything else moved at a uniform rate, in perfect circles. Aristotle held that 'the shape of the heavens is of necessity spherical'.[6] The perfect, spherical Moon only looked imperfect because substances from Earth tainted it, in its close proximity to Earth's realm. Ptolemy of Alexandria (2nd century AD) added epicycles to this model to account for the retrograde motion observed for some of the planets. For these Greek philosophers, the observed spherical shape of the Moon aligned with their mystical understanding of the heavens, but the features visible on its face needed to be explained with non-empirical evidence.

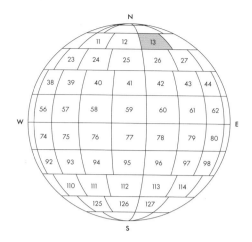

A belt of mountains formed from the degraded partial rims of old, lava-filled craters occupies the northern portion of this quadrangle. Rilles and scarps speak to the deformation of this terrain and its features over time. South of the mountains, the terrain becomes smoother as it joins the dark lava plains of Mare Frigoris (Sea of Cold). South of the mare, rugged terrain resumes, including the large impact crater Aristoteles after which this quadrangle is named. This area was probably once more heavily cratered, although much of this was erased by lava floods and in-filling by ejecta blanket material. Some time in the middle of the Moon's geologic history, Aristoteles crater was excavated, as was the crater Hercules, the edge of which is visible in the southeast corner of this map. These craters, as well as much of the rest of the terrain shown on this map, appear to have been filled with layers of ejecta material.

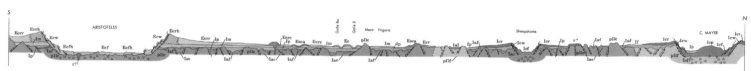

GEOLOGIC CROSS SECTION OF THE ARISTOTELES QUADRANGLE
APPROXIMATE HORIZONTAL SCALE 1:2 500 000 (VERTICAL SCALE GREATLY EXAGGERATED)

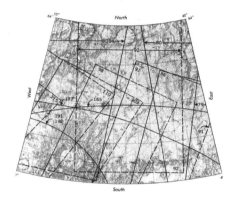

LUNAR ORBITER IV HIGH-RESOLUTION COVERAGE
OF ARISTOTELES QUADRANGLE

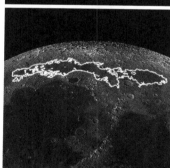

MOSAIC OF THE MOON
Multiple images taken by NASA's (LRO) in 2019 form a composite image (above) of the near side of the Moon. A zoomed-in frame of the mosaic (below) with Mare Frigoris highlit at the top and the large Aristoteles crater visible centre left.

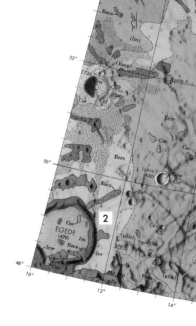

ARISTOTELES QUADRANGLE: KEY CHARACTERISTICS

1. ARISTOTELES CRATER
Latitude: 50.2°N / Longitude: 17.4°E
Diameter: 87 km / 54 miles
Depth: 3.3 km / 2 miles
Photograph: Lunar Orbiter IV (USA, 1967)

2. EGEDE CRATER
Latitude: 48.7°N / Longitude: 10.6°E
Diameter: 37 km / 23 miles
Photograph: Lunar Orbiter IV (USA, 1967)

3. C. MAYER CRATER
Latitude: 63.2°N / Longitude: 17.3°E
Diameter: 38 km / 23 ½ miles
Photograph: SMART-1 (EU, 2006)

4. KANE CRATER
Latitude: 63.1°N / Longitude: 26.1°E
Diameter: 55 km / 34 ¼ miles
Depth: 3.3 km / 2 miles
Photograph: Lunar Orbiter IV (USA, 1967)

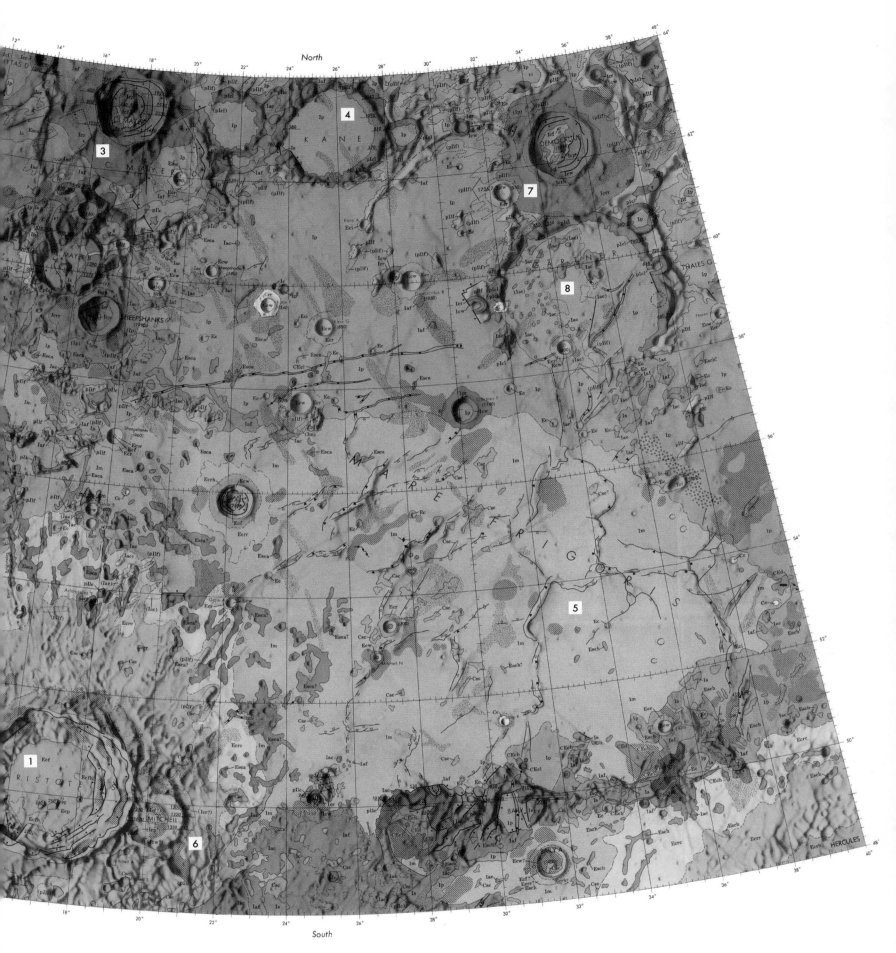

North

South

5. MARE FRIGORIS (SEA OF COLD)
Latitude: 56°N / Longitude: 1.4°E
Diameter: 1,446 km / 898 ½ miles
Photograph: Lunar Orbiter IV (USA, 1967)

6. MITCHELL CRATER
Latitude: 50.2°N / Longitude: 17.4°E
Diameter: 87 km / 54 miles
Depth: 3.3 km / 2 miles
Photograph: Lunar Orbiter IV (USA, 1967)

7. DEMOCRITUS CRATER
Latitude: 62.3°N / Longitude: 35°E
Diameter: 39 km / 24 ¼ miles
Depth: 2 km / 1 ¼ miles
Photograph: Lunar Orbiter IV (USA, 1967)

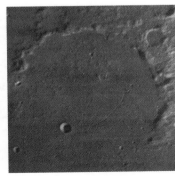

8. GÄRTNER CRATER
Latitude: 59.1°N / Longitude: 34.6°E
Diameter: 102 km / 63 ½ miles
Depth: 1,300 m / 4,265 ft
Photograph: Lunar Orbiter IV (USA, 1967)

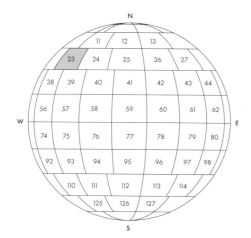

The ancient history of this region was obscured by the ejecta blanket produced by the impact that excavated the Iridum basin. Volcanic formations are present, the most notable being the Rümker Hills (sometimes Mons Rümker) that interrupt the smooth plains of Oceanus Procellarum (Ocean of Storms). Twenty-two volcanic domes comprise this feature, the results of lava extruding from vents in the plains. Lava flows have also flooded crater floors and depressions in the terra. Basalt lava filled the maria of Oceanus Procellarum, which dominate this quadrangle. Only a handful of young, bright craters appear in this region. In 2019, Israel's Beresheet spacecraft crashed catastrophically in Mare Serenitatis (Sea of Serenity). On 1 December 2020, the Chinese Chang'e-5 lander successfully set down at a site christened Statio Tianchuan. The spacecraft collected samples of lunar rock and soil and sent them back to Earth for analysis.

CHANG'E 5	ROCKET	Long March 5
	CREW	N/A
	LAUNCH DATE	20:30:12, 23 November 2020 (UTC)
	LAUNCH SITE	Wenchang Space Launch Center
	PAYLOAD	Lunar Mineralogical Spectrometer, Camera
	LUNAR ORBIT INSERTION	28 November 2020
	LANDING DATE	15:13:00, 01 December 2020 (UTC)
	LANDING SITE	Mons Rümker
	LANDING COORDINATES	43.00°N, 52.00°W Ⓐ

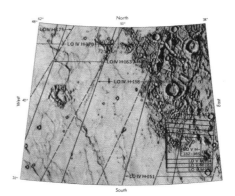

LUNAR ORBITER IV HIGH-RESOLUTION COVERAGE OF RÜMKER QUADRANGLE

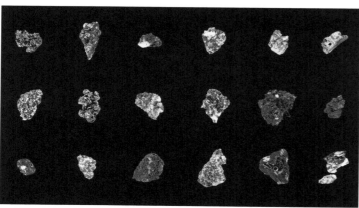

Agglutinate 0.088mg	Breccia 0.361mg	Breccia 0.109mg	Breccia 0.211mg	Basalt 0.164mg	Glass 0.138mg
Breccia 0.273mg	Agglutinate 0.131mg	Basalt 0.330mg	Agglutinate 0.186mg	Glass 0.732mg	Glass 0.216mg
Breccia 0.134mg	Basalt 0.182mg	Glass 0.402mg	Basalt 0.663mg	Glass 0.703mg	Basalt 0.429mg

LUNAR SOIL PARTICLES
Micrographs of lunar soil particles that were retrieved by the Chang'e-5 probe. The mission collected samples weighing almost 1.731 g in total.

RÜMKER QUADRANGLE: KEY CHARACTERISTICS

1. RÜMKER HILLS
Latitude: 40.8°N / Longitude: 58.1°W
Diameter: 70 km / 43 ½ miles
Max Height: 1,300 m / 4,265 ft 1 in
Photograph: Lunar Orbiter IV (USA, 1967)

2. RÜMKER E CRATER
Latitude: 38.62°N / Longitude: 57°W
Diameter: 6.96 km / 4 ¼ miles
Photograph: LRO (USA, 2013)

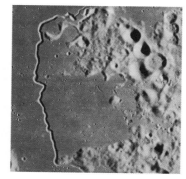

3. RIMA SHARP
Latitude: 46.02°N / Longitude: 50.36°W
Length: 107 km / 66 ½ miles
Depth: 50–300 m / 164–984 ft 3 in
Photograph: Lunar Orbiter IV (USA, 1967)

4. SHARP CRATER
Latitude: 45.7°N / Longitude: 40.2°W
Diameter: 40 km / 24 ¾ miles
Depth: 3.2 km / 2 miles
Photograph: LRO (USA, 2010)

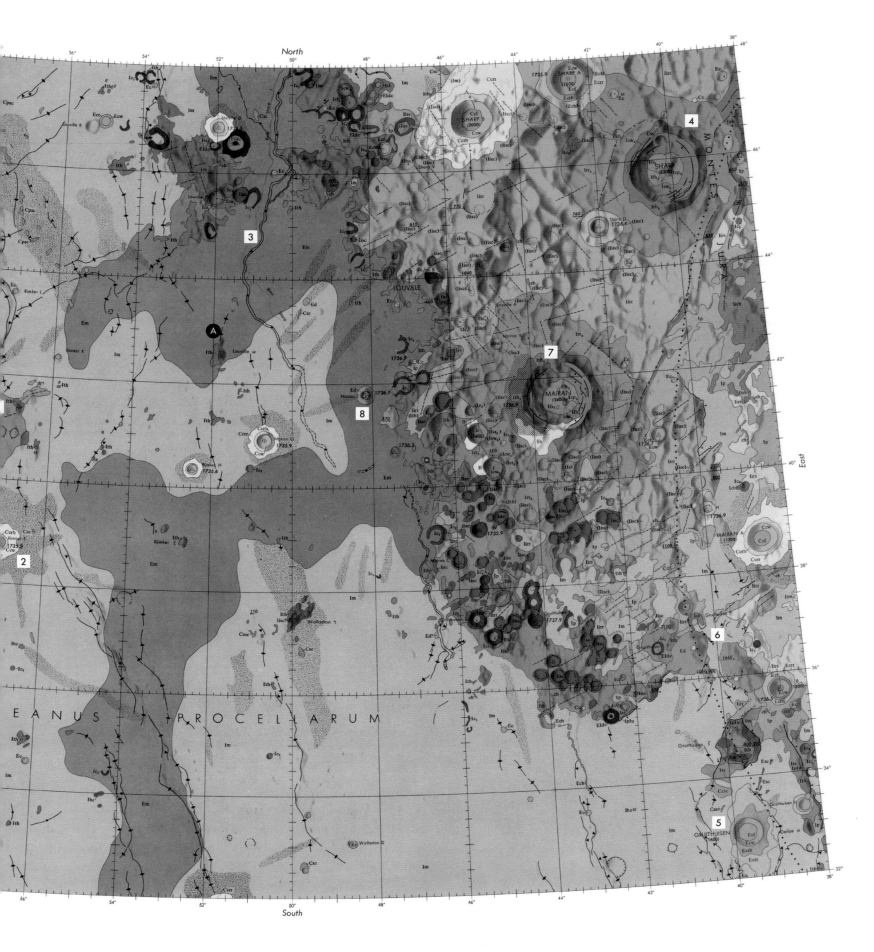

North

South

East

OCEANUS PROCELLARUM

5. GRUITHUISEN CRATER
Latitude: 32.9°N / Longitude: 39.7°W
Diameter: 16 km / 10 miles
Depth 1.9 km / 1 ¼ miles
Photograph: LRO (USA, 2019)

6. GRUITHUISEN DOMES
Latitude: 36.3°N / Longitude: 40°W
Diameter: 55 km / 34 ¼ miles
Max Height 1.8 km / 1 mile
Photograph: LRO (USA, 2022)

7. MAIRAN CRATER
Latitude: 41.6°N / Longitude: 43.4°W
Diameter: 40 km / 24 ¾ miles
Depth 3.4 km / 2 miles
Photograph: Lunar Orbiter IV (USA, 1967)

8. MAIRAN T VOLCANO
Latitude: 41.79°N / Longitude: 48.3°W
Diameter: 6 km / 3 ¾ miles
Height 600 m / 1968 ft 6 in
Photograph: LRO (USA, 2022)

→
This sketch from Leonardo da Vinci's *Codex Leicester* (1504–8) depicts the thin crescent Moon with a faint glow distinguishing the rest of the disc from the background sky. Da Vinci correctly attributed this glow to the effect of sunlight reflected from Earth illuminating the Moon's night-time surface, supplanting theories that the Moon had its own internal luminosity.

↱↱
This plate from Andreas Cellarius' *Harmonia Macrocosmica* (1660) provides a detailed explanation for the phases of the Moon as a result of changing solar illumination on the Moon's Earth-facing hemisphere. Notably, this diagram implies an Earth-centred cosmology, although elsewhere the atlas also covers the increasingly influential Copernican or Sun-centred model of the universe.

↳↳
An Indian manuscript of Zakariyya al-Qazwini's *Wonders of Creation* (c. 1650–1700). In a comprehensive survey of marvels originally completed around AD 1280, Persian geographer al-Qazwini describes how phases arise due to the relative positions of Sun and Moon in the sky, and also how eclipses may occur at times of the new and full Moon.

↳
Twelve plates from Robert Stawell Ball's *An Atlas of Astronomy* (1892). For his popular Victorian star atlas Ball compiled a series of detailed charts with accompanying keys identifying prominent features on the Moon's face in various phases.

he Moon's phases are probably the most obvious lunar phenomenon of all, and the changing shape of Earth's satellite has inspired countless myths and legends as well as providing insights into its true nature and humanity's place in the wider cosmos. The cycle of phases sees the Moon first appear in the evening sky after sunset as a slender crescent, moving slowly eastwards from night to night as it grows or 'waxes' in size. After passing through its semicircular 'first quarter' phase it develops a bulging gibbous shape, growing to the brilliant disc of full Moon. Thereafter, it begins to shrink, passing through 'waning gibbous' and last quarter before eventually narrowing to a crescent as it approaches the Sun from the west in the pre-dawn sky. Technically, the cycle begins and ends with the lunar disc being completely invisible, but because this moment marks the point where the Moon moves past the Sun in the sky, aside from during solar eclipses the true 'new Moon' is always lost in daylight.

The cycle of phases, known as a synodic month, lasts 29.53 days on average. In many ancient cultures this cycle was seen to have great astrological and mythological significance and often associated with stories of death and rebirth (see pp. 30–31). On a more practical level it was also the basis for most early calendar systems – for instance Babylonian calendars of the 21st century BC used a system of twelve lunar months in a year, each beginning at the first sighting of the evening crescent Moon, and with an additional 'intercalary' month inserted to keep the lunar and solar calendars roughly in step. From around 499 BC, meanwhile, the insertion of intercalary months began to be standardized, with the recognition that 19 solar years are almost exactly equal to 235 lunar months (the so-called Metonic cycle).

The lunar phases arise through the changing angles of the Sun and Moon as seen from Earth. One half of the lunar surface is always illuminated by sunlight, but the amount of the lit hemisphere visible from Earth changes as the Moon circles the Earth and the Sun's direction in the sky shifts more slowly throughout the month. The earliest recorded proponent of this model was Anaxagoras, a philosopher who taught in Athens in the first half of the 5th century BC. Fragments of his writings quoted by later scholars show he believed both the Sun and Moon to be large balls of material ejected from the Earth. The Sun, he thought, was a glowing sphere of fiery stone, while the Moon was made of colder rocks that shone only by reflecting its light.

While the basic cause of the phases went unquestioned in the Western world (and was also independently established by stargazers in China and India), not everyone was certain that it was the whole story. Many philosophers, following the influential ideas of Aristotle (384–322 BC), attributed the Moon's cyclical changes to its position at the boundary between the turbulent, changeable elements of Earth and the perfect, perpetual celestial realm.

Nevertheless, careful study of the phases also helped early astronomers learn other secrets about the nature of the Moon. For example the changing curve of the terminator (the line between the Moon's lit and unlit sides) was convincing evidence that the Moon had a spherical shape. Another lingering concern was the dim glow seen from the dark side of the lunar sphere when the crescent Moon is at its thinnest, sometimes known as 'the old Moon in the new Moon's arms'. Some took this glow as evidence that the Moon was itself slightly luminous, and it was not until

the early 16th century AD that Leonardo da Vinci correctly described how the Moon's dark face is faintly illuminated by sunlight reflecting off Earth itself.

Around the 3rd century BC, meanwhile, Greek mathematician Aristarchus of Samos made use of the phases in his work *On the Sizes and Distances of the Sun and Moon*. Reasoning that first and last quarters occur when the Sun is directly overhead from the Moon's western and eastern limbs, he envisaged a right-angled triangle with Earth, Moon and Sun at its three corners. In such a triangle, the relative distances of the Sun and Moon could be calculated by measuring the angular separation of the two bodies in Earth's sky at the quarter phases.[1]

Aristarchus' theory was faultless, but a limited ability to measure angles and determine precisely when the Moon was half illuminated (in an age long before telescopes) led him to the severe underestimate that the Sun is 20 times further from Earth than the Moon (in fact it is about 400 times more distant). Despite this, the fact that the Sun was apparently so large encouraged Aristarchus to become an early advocate for a heliocentric (Sun-centred) model of the universe.

The desire for a more precise measurement of lunar phases became particularly significant with the foundation of Islam in the early 7th century. Islamic months begin with the sighting of the crescent Moon or *hilal* in the evening sky after sunset, and although the 29.5-day cycle might seem to make this calculation a trivial exercise, the reality is that the slightly elliptical shapes of both the lunar orbit and Earth's orbit around the Sun complicate matters considerably by causing variations in the speed at which Sun and Moon appear to move across the sky. Classical astronomers (most notably Ptolemy of Alexandria in the 2nd century AD) developed models that attempted to describe these 'inequalities' in motion, but Islamic astronomers took them much further. For instance Habash al-Hasib al-Marwazi, working in Baghdad during the 9th century, developed sophisticated mathematical tools for taking the inequalities into account, calculating the true positions of the Sun and Moon and, therefore, predicting the first visibility of the new crescent Moon at the start of the month.[2]

Islamic and later medieval European astronomers remained hampered, however, by the underlying flaws of Ptolemy's model – most significantly the positioning of Earth at its centre and the insistence on uniform circular motion. More accurate predictions of lunar phases remained impossible until the 17th century, when the lunar theory of Isaac Newton[3] for the first time took into account the competing gravitational influences of the Earth and Sun on the motion of the Moon.

DAY 3

DAY 4

DAY 7

DAY 8

DAY 11

DAY 12

DAY 5

DAY 6

DAY 9

DAY 10

DAY 13

DAY 14

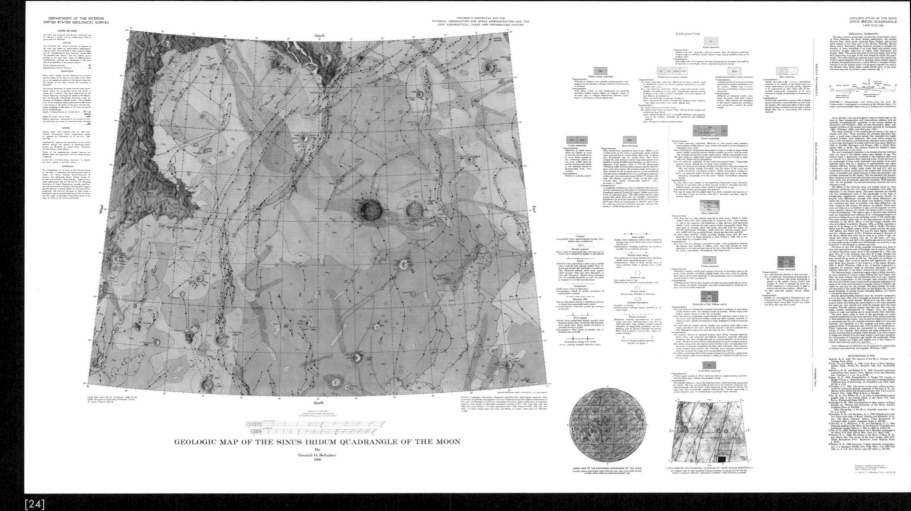

GEOLOGIC MAP OF THE SINUS IRIDUM QUADRANGLE OF THE MOON
By
Gerald G. Schaber
1969

[24]

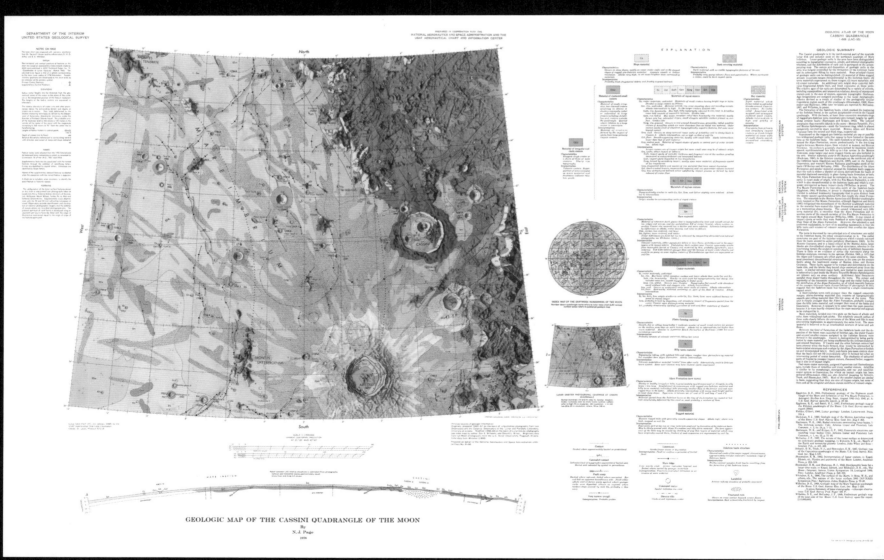

GEOLOGIC MAP OF THE CASSINI QUADRANGLE OF THE MOON
By
N.J. Page
1970

[25]

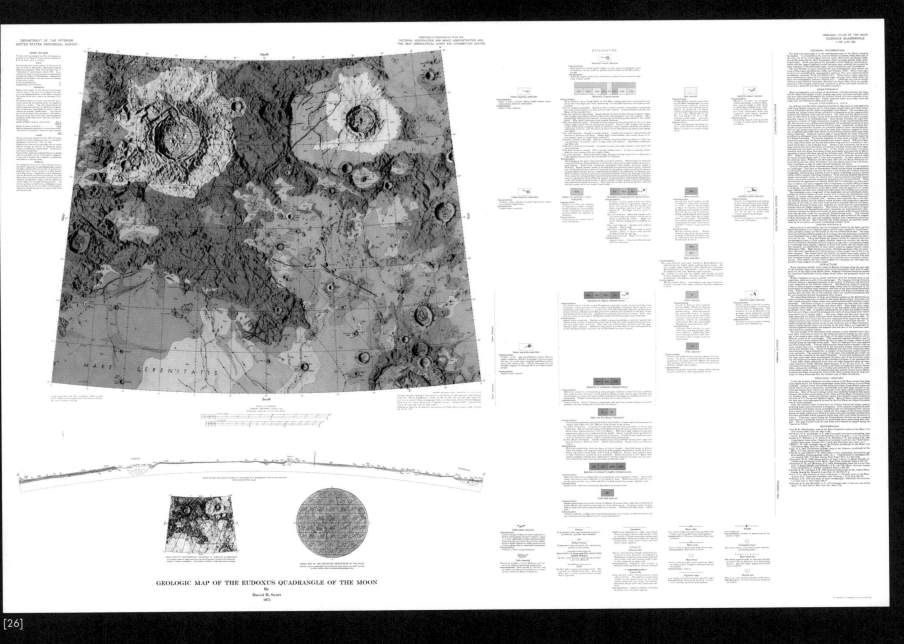

[26]

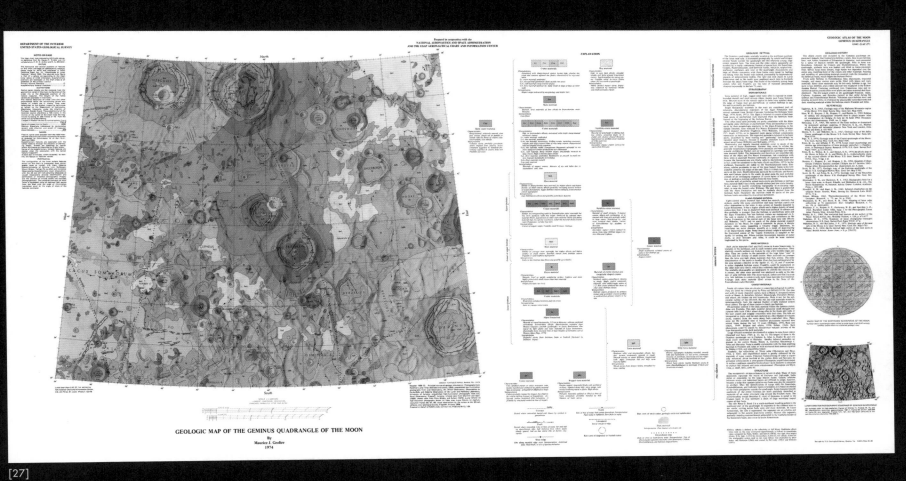

[27]

24
Facsimile:
*Geologic Map of the Sinus Iridum
Quadrangle of the Moon*, 1969,
by G. Schaber

25
Facsimile:
*Geologic Map of the Cassini
Quadrangle of the Moon*, 1970,
by N. J. Page

26
Facsimile:
*Geologic Map of the Eudoxus
Quadrangle of the Moon*, 1972,
by David H. Scott

27
Facsimile:
*Geologic Map of the Geminus
Quadrangle of the Moon*, 1974,
by Maurice J. Grolier

THE FEMALE CYCLE AND THE MOON

he association of menstruation with the Moon dates back to antiquity. Before modern scientific methods and technologies, it was common for humans to look to the natural world to make sense of the hidden inner workings of their bodies; as early as the 5th century BC, it was recognized that menstrual rhythms followed a similar month-long cycle to the Moon.[1] According to medical men in antiquity, the relationship between the menstrual and lunar cycles was evidence that human bodies (the microcosm) were connected to the wider universe (the macrocosm). Aristotle (384–322 BC), the Greek philosopher, claimed that healthy women menstruated during the coldest time of the month, when the Moon waned, because their bodies were colder than men's.[2] For the next two centuries, the works of ancient authorities, including Aristotle, laid the foundation for natural philosophical and medical knowledge. By the early modern period (AD 1500–1750), when astrological ideas and practices reached a zenith, the notion of menstrual synchrony with the Moon was an established part of the Western worldview. In fact, the word 'menstruation', which first appeared in English texts in the 1680s, is derived from the Greek *mene* ('Moon').[3]

AN ALMANACK OR PROGNOSTICATION for the year of our LORD 1660. Being the third after Bissextile or Leap-year. Calculated for the Meridian of London, and may without exception serve for England, Scotland, and Ireland.

By SARAH JINNER, Student in Astrology. London, Printed for the Company of Stationers.

← The title page from Sarah Jinner's 1658 almanac. It is one of the only surviving astrological texts from the early modern period that was written by a woman for female readers. Alongside a lunar calendar, Jinner's almanacs contain advice on sex and reproduction.

↳ Robert Fludd's woodcut from *History of the Two Worlds* (1617–18) depicts the universe as a complex, harmonious system. The image demonstrates the interconnected nature of all things, from the celestial realms to the earthly sphere. Drawing on the ancient Greek theory of the macrocosm and the microcosm, Fludd suggested that human bodies (microcosm) were made from the same elemental qualities as the wider universe (macrocosm).

↱ Two illustrations from a 1720 edition of *Aristotle's Master-piece Completed*. First published in 1684, the book was written by an unknown author claiming to be Aristotle, who was widely viewed at the time as an expert on sex and reproduction matters. Both a sex manual and a book containing practical advice on pregnancy and childbirth, this third version of the book is divided into two parts: the first describes male and female sex organs, matters concerning virginity and sexual intercourse; and the second part covers pregnancy, the foetus, infertility and midwifery. In addition to providing advice to women on how to become pregnant and how to determine the sex of the child, the author also warns against having sex with a beast for fear of giving birth to a monstrous child.

The Effigies of a monstrous Child, by reason of the excess of the material Seed.

The Effigies of a Monster half Man and half Dogg.

It is generally presumed that very little is known about menstruation in the pre-modern world. While it is true that historians have only recently turned their attention towards menstruation in the past, it is mainly due to cultural taboo in the present. Not only have humans always had a lot to say about menstruation, but they left behind their ideas and experiences in writing. Another commonly held presumption is that, as in many cultures today, menstruation was not discussed openly. As one would expect, there is evidence to suggest that menstrual knowledge was shared orally among pre-modern European women, but it was not confined to this tradition. From the 1500s, European printing presses published works that discussed menstruation in vernacular languages to make their contents accessible to broader swathes of the (literate) population.

One such text was written by the Dutch physician Levinus Lemnius (1505–1568). In 1559, Lemnius printed *The Secret Miracles of Nature*, a Latin 'book of secrets' that revealed occult (meaning 'hidden from the senses') knowledge for human manipulation. Almost a century later, in 1658, *Secret Miracles* was printed in English and became a foundational text for other best-selling medical works, including the famous sex manual *Aristotle's Masterpiece* (1684).[4] In an oft-repeated passage, Lemnius detailed the relationship between menstruation and the Moon. Near the basin of the Rhine, he claimed, there was a community of sailors' wives who could only give birth to monsters. When their husbands returned from months at sea, they were eager to lie in their marital beds, without considering whether the timing was astrologically auspicious. If a couple copulated during the waning Moon, when the woman had her menses, they would conceive a *manekind* ('Moon birth'). Unlike healthy children, *manekinds* were born with physical deformities, including scales and pointed tails.[5] Although this story appears bizarre to modern sensibilities, it shows how the Moon had a dual purpose to early modern Europeans: as both a calendar, and a source of occult influence.

For thousands of years, humans have used the changing lunar phases to mark the passing of time. One of the most important tools in the early modern household was the almanac, a cheap pamphlet containing calendars and information for using planetary motions to time everyday activities, from sowing seeds to letting blood. Women had been using farmers' and sailors' almanacs to manage their menstruating bodies for more than a century but, from the 1650s, they began to write almanacs for their own use. Much like modern-day wall calendars, these pamphlets were cheaply made and easily discarded, but a few copies survive. Sarah Jinner's (fl. 1658–1664) almanacs contained lunar calendars, as well as recipes for treating menstrual disorders using the Moon's power.[6] Although these texts were popular, by the 1660s, Jinner was censored by men

publishing satirical editions under her name.[7] Almanacs were found in most early modern households but, even without one, it was still possible for women to track their menstrual rhythm using the Moon; Lemnius suggested that couples could find an appropriate time for sexual activity by simply observing the night sky.[8]

Aside from its use as a calendar, early modern people also assumed that the Moon had the power to instigate menstruation. Since Aristotle's cold-air theory, several men had proposed explanations for the connection. For Lemnius, the answer was found in the Moon's relationship to moisture. The sailors' wives lived close to the Rhine, a large body of water that held great attraction to the Moon's power. As the luminary moved the river's waters towards the sea, it also drew the fluid blood from women's bodies. On the surface, his story could suggest that the Moon's influence over menstruation was harmful; after all, the monsters were called *manekinds* ('Moon births'). However, the physician stressed that the Moon was not entirely to blame. Instead, it was the toxic combination of sexual intercourse with menstruation.[9] In the 16th century, when Lemnius was writing, two key sources of menstrual knowledge were the Bible (Leviticus 15) and Pliny the Elder's (AD 23–79) *Natural History*. Leviticus warned of the dangers of having sex with a menstruating woman while Pliny claimed that the poisonous qualities of menstrual blood were amplified under the Moon.[10] Combining these two ideas, Lemnius argued that sex forced menses back into the womb where it putrefied and turned poisonous. Under normal circumstances, however, the Moon's influence over menstruation was beneficial as it ensured women's menstrual rhythms were regular and their bodies kept healthy for procreation.[11]

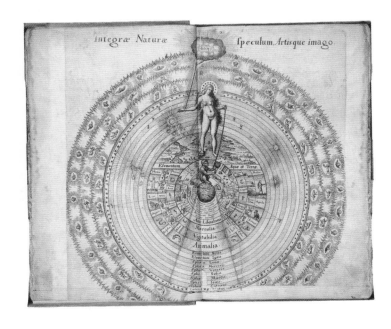

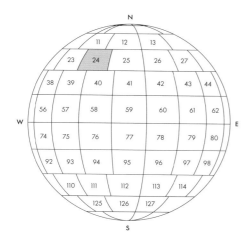

The feature that gives this quadrangle its name is the Sinus Iridum (Bay of Rainbows) lava plain, which is an extension of the northwest rim of Mare Imbrium (Sea of Showers). The Montes Jura mountain range that borders Iridum gives the feature its bay-like appearance, even though Iridum is in fact the remains of an impact crater and the mountains were once a ring formed when the crater was excavated. Over time, floods of lava flowed over the southern rim of the crater and filled the basin, giving it its smooth, flat appearance. In Giovanni Cassini's 1679 map of the lunar surface, Promontorium Heraclides, located at the southwest edge of Montes Jura, appeared as a siren's head, reinforcing the idea that the maria were filled with water. This quadrangle also includes several terra (highland) 'islands' (isolated mountains) dotting the mare, including Montes Teneriffe and Montes Recti. On 17 November 1970, the Soviet Luna 17 lander delivered Lunokhod 1, the first robotic rover to explore the surface of another world, to the lunar surface at coordinates 38.24°N, 35.00°W (not marked). In 2013, China's Chang'e-3 mission landed at 44.12°N, 19.51°W in northern Mare Imbrium.

LUNA 17	ROCKET	Proton-K + Blok D
	CREW	N/A
	LAUNCH DATE	14:44:01, 10 November 1970 (UTC)
	LAUNCH SITE	Tyuratam (Baikonur Cosmodrome)
	PAYLOAD	Lunokhod 1 Rover
	LUNAR ORBIT INSERTION	15 November 1970
	LANDING DATE	03:46:50, 17 November 1970 (UTC)
	LANDING SITE	Mare Imbrium
	LANDING COORDINATES	38.24°N, 35.0°W Ⓐ

CHANG'E 3	ROCKET	Long March 3B Y-23
	CREW	N/A
	LAUNCH DATE	17:30:00, 01 December 2013 (UTC)
	LAUNCH SITE	Xichang Satellite Launch Centre
	PAYLOAD	Yutu Moon Rover
	LUNAR ORBIT INSERTION	06 December 2013
	LANDING DATE	13:11:00, 14 December 2013 (UTC)
	LANDING SITE	Mare Imbrium
	LANDING COORDINATES	44.12°N, 19.51°W Ⓑ

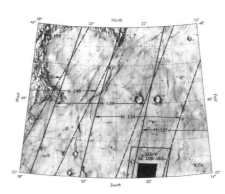

LUNAR ORBITER IV HIGH-RESOLUTION COVERAGE OF SINUS IRIDUM QUADRANGLE

YUTU MOON ROVER
A segment of the 360° timelapse panorama taken by the landform camera on the Chang'e-3 lander, 22 December 2013. In this frame a Yutu rover heads south.

SINUS IRIDUM QUADRANGLE: KEY CHARACTERISTICS

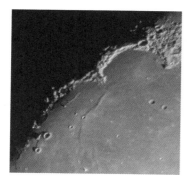

1. MONTES JURA
Latitude: 47.1°N / Longitude: 34°W
Diameter: 422 km / 262 ¼ miles
Max Height: 3.8 km / 2 ¼ miles
Photographed from Earth (USA, 2021)

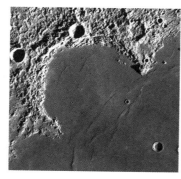

2. SINUS IRIDUM
Latitude: 32.8°N / Longitude: 15.6°W
Diameter: 1,146 km / 712 miles
Photograph: LRO (USA, 2014)

3. PROMONTORIUM LAPLACE
Latitude: 46.8°N / Longitude: 25.5°W
Height: 2.6 km / 1 ½ miles
Photographed from Earth

4. PROMONTORIUM HERACLIDES
Latitude: 40.3°N / Longitude: 33.2°W
Diameter: 50 km / 31 miles
Photographed from Earth

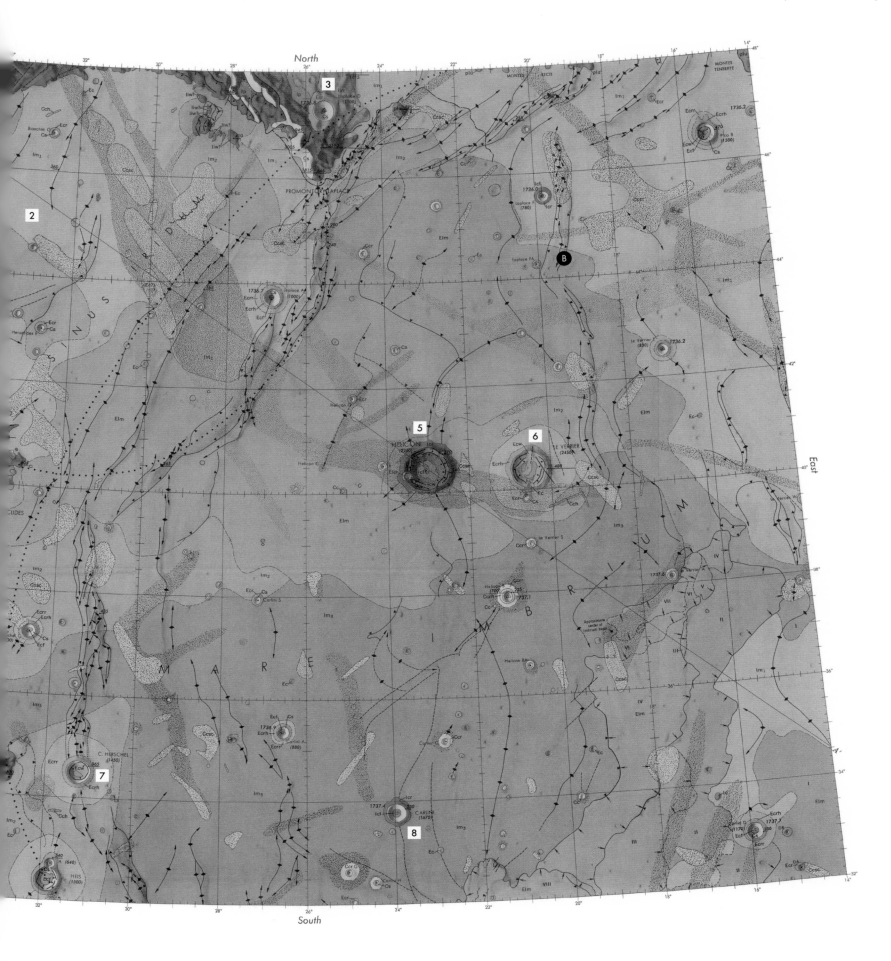

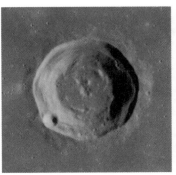

5. HELICON CRATER
Latitude: 40.4°N / Longitude: 23.1°W
Diameter: 25 km / 15 ½ miles
Depth: 500 m / 1,640 ft 5 in
Photograph: Lunar Orbiter IV (USA, 1967)

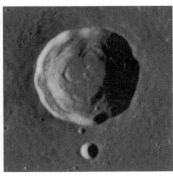

6. LE VERRIER CRATER
Latitude: 40.3°N / Longitude: 20.6°W
Diameter: 20 km / 12 ½ miles
Depth: 2.1 km / 1 ¼ miles
Photograph: Lunar Orbiter IV (USA, 1967)

7. C. HERSCHEL CRATER
Latitude: 34.5°N / Longitude: 31.2°W
Diameter: 13.4 km / 8 ¼ miles
Depth: 1.9 km / 1 ¼ miles
Photograph: LRO (USA, 2013)

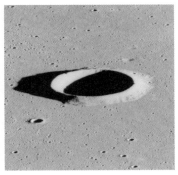

8. CARLINI CRATER
Latitude: 33.7°N / Longitude: 24.1°W
Diameter: 11 km / 6 ¾ miles
Depth: 2 km / 1 ¼ miles
Photograph: Apollo 15 (USA, 1971)

[LAC-25]

I-666

Cassini Quadrangle

Maps

Lunar

056

GEOLOGIC ATLAS OF THE MOON

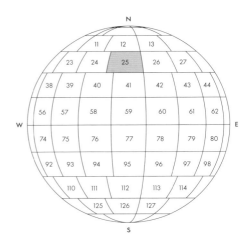

Just southwest of the Montes Alpes mountain range is Cassini crater. Filled with lava and marked by a number of smaller craters in its interior and on its rim and ramparts, this crater is probably one of the oldest features shown in this quadrangle. The western edge of this quadrangle is characterized by rough terrain, terra with rolling hills and ridges, as well as the mountains of the Montes Alpes and Montes Caucasus ranges. In the south, in the Mare Imbrium (Sea of Showers) lava plain, is the large rayed crater Aristillus, with the smaller crater Theaetetus to its northeast. Aristillus, based on the integrity of its rim, terraced wall and central peaks, is probably very young, excavated during the latest period in the Moon's geologic history. Amid the ridges that texture the smooth mare, geologists also traced the outlines of the concealed craters and basins beneath the plains.

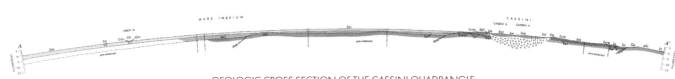

GEOLOGIC CROSS SECTION OF THE CASSINI QUADRANGLE
APPROXIMATE HORIZONTAL SCALE 1:1 000 000 (VERTICAL AND HORIZONTAL SCALES APPROXIMATELY EQUAL)

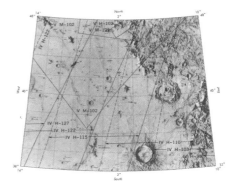

LUNAR ORBITER PHOTOGRAPHIC COVERAGE
OF CASSINI QUADRANGLE

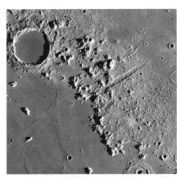

MONTES ALPES
Photograph showing the Montes Alpes, taken by a LRO in 2015. Paired with (below) a satellite photograph of the Alps, after which the Montes Alpes were named, taken in 2002.

CASSINI QUADRANGLE: KEY CHARACTERISTICS

1. CASSINI CRATER
Latitude: 40.2°N / Longitude: 4.6°E
Diameter: 57 km / 35 ½ miles
Depth: 1,200 m / 3,937 ft
Photograph: LRO (USA, 2015)

2. ARISTILLUS CRATER
Latitude: 33.9°N / Longitude: 1.2°E
Diameter: 55 km / 34 ¼ miles
Depth: 3.6 km / 2 ¼ miles
Photographed from Earth (USA, 2019)

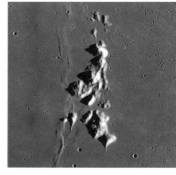

3. MONTES SPITZBERGENSIS
Latitude: 34.42°N / Longitude: 5.22°E
Max Height: 1,400 m / 4,593 ft 2 in
Photograph: LRO (USA, 2014)

4. MONT BLANC
Latitude: 45.41°N / Longitude: 0.44°E
Width: 25 km / 15 ½ miles
Max Height: 3.8 km / 2 ¼ miles
Photograph: Lunar Orbiter IV (USA, 1967)

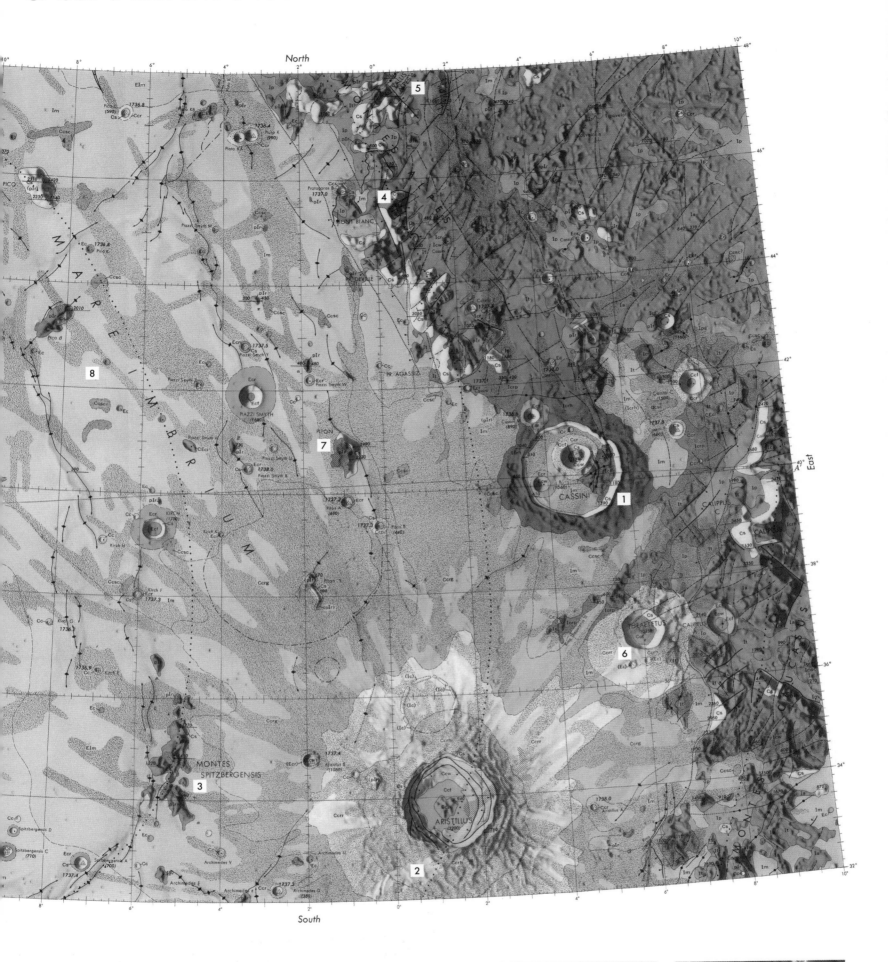

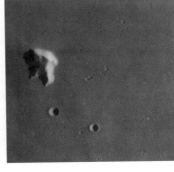

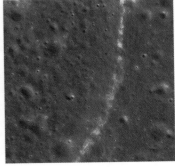

5. VALLIS ALPES
Latitude: 49.21°N / Longitude: 3.63°E
Length: 166 km / 103 ¼ miles
Max Width: 10 km / 6 ¼ miles
Photographed from Earth (USA, 2006)

6. THEAETETUS CRATER
Latitude: 37°N / Longitude: 6°E
Diameter: 25 km / 15 ½ miles
Depth: 2.8 km / 1 ¾ miles
Photograph: LRO (USA, 2011)

7. MONS PITON
Latitude: 40.6°N / Longitude: 1.1°W
Width: 25 km / 15 ½ miles
Height: 2.25 km / 1 ½ miles
Photograph: Lunar Orbiter IV (USA, 1967)

8. MARE IMBRIUM
Latitude: 32.8°N / Longitude: 15.6°W
Diameter: 1,146 km / 712 miles
Photograph: LRO (USA, 2012)

Gerardo Aldana

MAYAN LUNAR ASTROLOGY

↱
The Moon-goddess occupies the cartouche on the extreme left of the skyband. Here, she wears a flower ear-spool ornament and a symbol of vital energy on her nose. The hieroglyph UJ, for Moon, fills the right third of the cartouche, while a rabbit clings to her upper arm in the centre.

↱↱
The Sun-god, K'inich Ajaw, is identifiable on the Skyband Bench by his distinctive nose, chiselled tooth, and the flower markings on his forehead, back and upper arms. Skybands much more commonly include symbolic representations of celestial entities, with the flower symbol of the Sun substituted here for this more elaborate personification.

↱↱↱
This reclining figure has the face of the Sun-god, but the markings on his upper arm and upper thigh are clearly Ak'bal symbols for darkness, leading some scholars to identify him as the Night Sun.

↱↱↱↱
The hieroglyph in the centre composed of two circles, held in place by what looks like a stylized and inverted number three, represents EK' for 'star' or, more generally, 'celestial body'. Here EK' is held by the arm of a personified celestial being with a scorpion tail, which can be seen on the left-hand side of the cartouche.

→
During the late 8th century AD, a member of the Copán elite commissioned the so-called 'Skyband Bench' for the largest room (Structure 66C in Group 8N-11) of the residential complex. Celestial 'Moan' bird heads cap each end of the bench, while primordial deities known as Pawahtuns are depicted holding up 'the sky' on two of its four legs. The bench is currently preserved in the sculpture museum at the archaeological site.

hroughout the civilizations of ancient Mesoamerica (the region encompassing northern to southern Mexico, Guatemala, Belize, El Salvador and northern Honduras), artists and scientists found inspiration in the celestial realm. A rich body of examples comes from the hieroglyphic inscriptions and iconography of the Classic Mayan period (AD 250–900).

During the reign of the sixteenth *k'uhulajaw* ('ruler') of Copán, Yax Pahsaj Chan Yopat, at the very end of the Classic period, c. AD 800, Mayan artists and scribes broke significantly from traditional practice. Previously, stone monuments had been carved strictly for the *k'uhulajaw* in the city centre, but now artists produced elaborate stone artwork for themselves and for others of the nobility.[1] A stone bench carved with a hieroglyphic inscription naming the artist, Mak Chanil, is one well-known (and beautiful) example. Another is the 'Skyband Bench', located in the same elite neighbourhood on the banks of the Copán River, a kilometre (over half a mile) east of the centre of the city of Copán.

Scholars refer to it as the 'Skyband Bench' because its front edge is carved with a standard rectangular iconographic motif including representations of celestial entities. In Classic Mayan art the skyband usually represents the body of the animated Milky Way, known as the 'Celestial Dragon'. On the Skyband Bench at Copán, there is no inscription naming the owner of the bench or the structure it is housed within, nor is there a date specifying when the monument was dedicated.[2] What it lacks in historical detail, though, it makes up for in artistry. Here are shown four celestial entities in embellished form: K'inich Ajaw (theSun); an animated star with a scorpion's tail; darkness inanthropomorphic form; and the Moon. The Moonhieroglyph is accompanied by the Moon-goddess, whocarriesa rabbit in the crook of her arm. The rabbit seen on the face of the Moon is the Mesoamerican equivalent of the Western'Man in the Moon' and appears widely throughout the region.

Classic Mayan period scribes used an elaborate calendric system for their historical, mythological and astronomical records.[3] The 260-Day Count (made up of 13 numbers and 20 day signs) along with the 365-Day Count (18 months of 20 days each, supplemented by 5 'nameless days') were used by daykeepers throughout Mesoamerica in a 'Calendar Round' that repeated every 52 years. To this, Mayan scribes added a Long Count, which simply counted days forward from a 'zero date' anchored around 3100 BC. This Long Count allowed for the unique placement of dates historically, parallel to the Western Julian Day Number system. Most public monumental inscriptions began with a Long Count date, followed by the corresponding 260-Day Count date and often, when space allowed, a 'Supplementary Series' of additional information, including Lunar Series records, and then finally the 365-Day Count date.

Lunar Series records are made up of a set of five or six hieroglyphs. In its Classic period form, the first component of a Lunar Series record was a number attached to the glyph for the verb *jul*, 'to arrive'. This tells the reader how many days 'ago' the New Moon 'arrived', i.e. it provides the Moon Age.[4] The final three glyphs in a Lunar Series record

state the number of days in the current lunar month – either 29 or 30 – in order to accommodate the observational synodic period of 29.5306 days. This number was prefaced by the phrase *u ch'ok k'aaba*, or 'it is the youthful name of'. The beginning and the end phrases of a Lunar Series record, then, gave the observational characteristics of the current Moon and assigned it a name.

The perhaps more curious part of a Lunar Series record was the glyph block right in the middle. Referred to as Glyph C of the Supplementary Series, it was made up of four parts:

- ► the hieroglyphic verb *k'al*, known as the 'flat hand glyph'
- ► a coefficient ranging from 1 to 6
- ► one of three different deity heads (Ixim, Yax B'ahlam Ajaw or Kimil) known as the 'lunar patron'
- ► the hieroglyph for the Moon, *UJ*

K'al reads 'to enclose' and refers to the completion of space-time ceremonial activity. This verb was used for astronomical and calendric records, such as solar and Venus periods, as well as the completion of cycles within the Long Count.[5] In the Lunar Series, *k'al* refers to the 'completion' of synodic periods of the Moon, with five or six periods allocated to each of the three patrons. A full sequence for one lunar patron featured six Moons over a 177-day period, yielding an average lunar month of 29.5 days.

As a whole, the Lunar Series provided a couplet description of the Moon. For example: 'Twenty-seven days ago, the New Moon of the 2-Kimil-Moon-enclosing arrived. [Glyph X] is the name of the 29 days'.

The Moon Age was based on observation, but the components of Glyph C were not. Each of the numbered periods of a given patron Moon began and ended with a synodic period, but the numbers and patrons assigned to any given Moon were arbitrary. As a result, scribes of different cities often assigned the same observed lunar period to lunar patrons in different sequences. This pattern temporarily changed, however, at a critical political point during the Late Classic period, c. AD 700. After the defeat of the most powerful dynasty of Calakmul by the army of

Tikal, for a period of over forty years, all Lunar Series records at all sites across the entire region followed the same sequence.[6]

By the Late Classic, around the time that artists at Copán were carving the Skyband Bench, scribes at Xultun painted a sequence of 162 Moons on a wall within a room attached to the back of a pyramidal structure, recognized as an 'astronomer's workshop'.[7] This sequence – painted alongside other astronumerological tables – tracked the incorporation of 'extra' 30-day periods into a larger period of 4,784 days to generate an approximate synodic lunar month of 29.5309 days. This approximation matches the one utilized at the Mayan city of Palenque decades earlier, under the patronage of K'inich Kan B'ahlam, suggesting the collegiality of the Mayan scientific community.

Three lunar patrons, each governing six Moons, permit the tracking of eighteen-Moon sets, each of which corresponds to an eclipse season of sorts. When a lunar or solar eclipse occurs, a further solar or lunar eclipse may follow six and/or twelve Moons later. An eighteen-Moon count would make it straightforward to establish 'eclipse warning' dates. Versions of this periodicity were recorded in the final step of tracking eclipse cycles, which can be found in the Eclipse Table in the Dresden Codex (a Mayan

← Jaina-style figurine depicting the Moon-goddess and her rabbit companion (AD 600–900). Jaina Island off the coast of Campeche, Mexico, was a site for elite burials, which were often accompanied by life-like ceramic figurines. These figurines portrayed people from all sections of society, as well as deities and mythological figures.

↵ Late Classic period (c. AD 700) scribes painted cylindrical polychrome drinking vessels known as *k'ib*, which in many cases depicted historical and mythological scenes. This rollout photograph was taken by Justin Kerr, who created an extensive database of images of Mayan ceramics, enriching modern scholarship's understanding of Classic Mayan culture.

↳ Flat stone monument Stela 3 from Piedras Negras (c. AD late 7th century) and author illustration of the historical date, which includes Long Count, Calendar Round and Supplementary Series components. Below the date and hieroglyphic text is a depiction of Ix Winikhaab Ajaw, a noblewoman from the city of 'Namaan', and wife of the Piedras Negras *k'uhulajaw* (ruler), along with their daughter.

↓↓ Pages 55 through 58 of the Dresden Codex (1200–50) – the last four pages of the Eclipse Table. Each page is divided into an upper and a lower half, so that the full table runs from the top of page 51 to the top of page 58, and then continues from the bottom of page 51 to the bottom of page 58.

hieroglyphic manuscript containing ceremonial almanacs and astronomical calculations). Although eclipse periods are recognizable in the Codex, scholars have yet to match them up with any historical records, suggesting that there is still much to learn about this table and about Mayan approaches to lunar records.[8]

While the astronomical records are fascinating and provide extensive material for further investigation, there are also striking literary and artistic renderings. A Late Classic period ceramic vessel owned by K'ahk' Tiliw Chan Chaak, *k'uhulajaw* of the city of Naranjo, was painted with a scene including a rabbit, apparently the one who eventually resided on the Moon. Although the entire narrative is lost, it appears that a rabbit served as the royal scribe to the Lords of the Underworld. According to Mayan myth, the Lords were vanquished by the Hero Twins, and the rabbit was charged with taking possession of their finery. On this pot, the sun deity, K'inich Ajaw, protects the rabbit from the highest Lord of the Underworld, who is trying to get back his apparel.[9] Whether the rabbit's long-term refuge from the Lord was a new home in the Moon, the archaeological record does not yet tell us. Like the rendering of the Moon on the Skyband Bench at Copán, though, this painted vessel demonstrates that astronomy was part of Mayan art and literature as well as fodder for heady Mayan mathematicians.

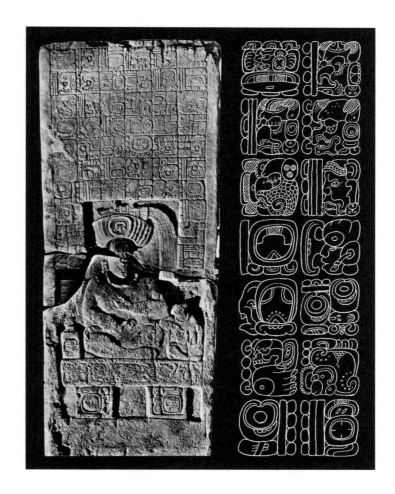

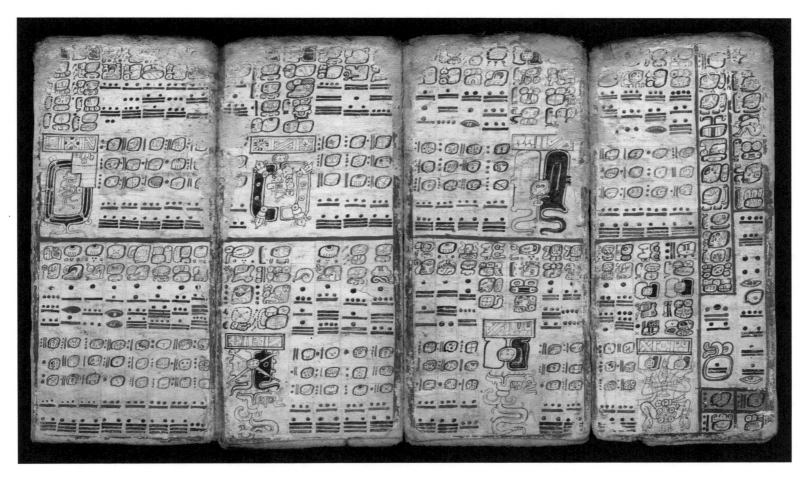

[LAC-26]

I-705

Eudoxus Quadrangle

Maps

Lunar

062

GEOLOGIC ATLAS OF THE MOON

The lower portion of this quadrangle is dominated by the volcanic basalt plains of Mare Serenitatis (Sea of Serenity) and Lacus Somniorum (Lake of Dreams). More plains continue in the northeast corner of the quadrangle in Lacus Mortis (Sea of Death). In the west are more rugged highlands, such as the Montes Caucasus mountain range and the remnant wall of the crater that formed the basin of Lacus Mortis. The large craters Eudoxus and Bürg are the youngest prominent features in this quadrangle. Their sharp rims, terraced walls, ramparts and ray systems indicate that both craters formed during the Moon's most recent geologic period, the Copernican. Many of the oldest features of this region may no longer be visible, having been covered by material related to the larger nearby impacts that formed the Serenitatis basin, Lacus Mortis crater and other features.

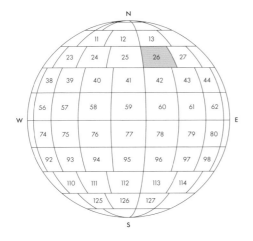

BERESHEET	ROCKET	Falcon 9 B5
	CREW	N/A
	LAUNCH DATE	01:45:00, 22 February 2019 (UTC)
	LAUNCH SITE	Cape Canaveral Air Force Station
	PAYLOAD	Magnetometer, Laser Retroreflector
	LUNAR ORBIT INSERTION	04 April 2019
	LANDING DATE	13:11:00, 11 April 2019 (UTC)
	IMPACT SITE	Mare Imbrium
	LANDING COORDINATES	32.60° N, 19.34° E Ⓐ

LUNAR ORBITER PHOTOGRAPHIC COVERAGE
OF EUDOXUS QUADRANGLE

BERESHEET IMPACT SITE
Two photographs of Mare Serenitatis. The first was taken on 16 December 2016 before Beresheet impacted the Moon's surface. The second was taken of the same section of Mare Serenitatis after the spacecraft crashed onto the surface on 22 April 2019.

EUDOXUS QUADRANGLE: KEY CHARACTERISTICS

1. EUDOXUS CRATER
Latitude: 44.3°N / Longitude: 16.3°E
Diameter: 67 km / 41 ¾ miles
Depth: 3.4 km / 2 miles
Photograph: Lunar Orbiter IV (USA, 1967)

2. BÜRG CRATER
Latitude: 45°N / Longitude: 28.2°E
Diameter: 40 km / 24 ¾ miles
Depth: 1.8 km / 1 mile
Photographed from Earth (USA, 2006)

3. MARE SERENITATIS
Latitude: 28°N / Longitude: 17.5°E
Diameter: 674 km / 418 ¾ miles
Photographed from Earth (Belgium, 2020)

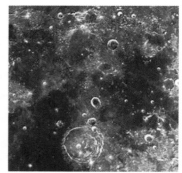

4. LACUS SOMNIORUM
Latitude: 37.56°N / Longitude: 30.8°E
Diameter: 424.76 km / 264 miles
Photograph: Clementine Orbiter (USA, 1994)

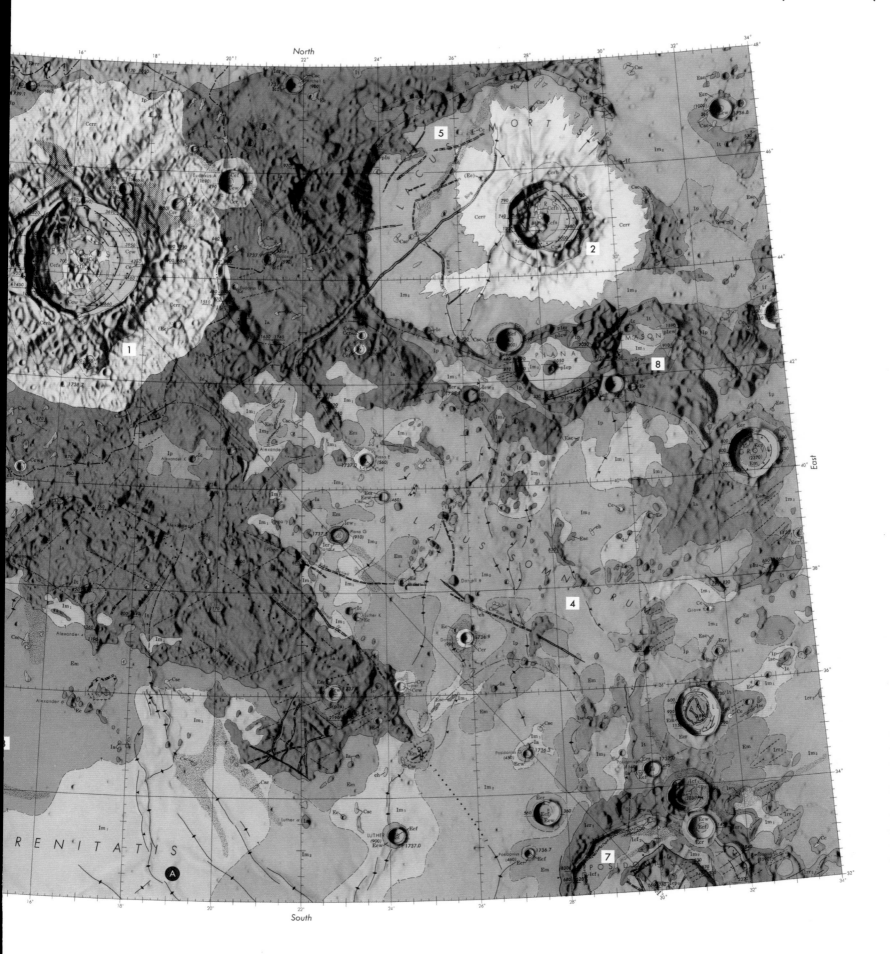

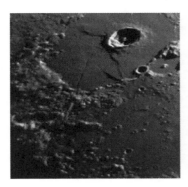

5. LACUS MORTIS
Latitude: 45.13°N / Longitude: 27.32°E
Diameter: 158.78 km / 98 ¾ miles
Photographed from Earth (Greece, 1997)

6. MONTES CAUCASUS
Latitude: 38.4°N / Longitude: 10°E
Diameter: 445 km / 276 ½ miles
Max Height: 6 km / 3 ¾ miles
Photograph: Apollo 15 (USA, 1971)

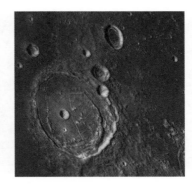

7. POSIDONIUS CRATER
Latitude: 31.88°N / Longitude: 29.99°E
Diameter: 95 km / 59 miles
Depth: 2.3 km / 1 ½ miles
Photographed from Earth (Germany, 2005)

8. MASON CRATER
Latitude: 42.6°N / Longitude: 30.5°E
Diameter: 43 km / 26 ¾ miles
Depth: 1.9 km / 1 ¼ miles
Photograph: Lunar Orbiter IV (USA, 1967)

[LAC–27]

I–841

Geminus Quadrangle

I–841

Maps

Lunar

064

GEOLOGIC ATLAS OF THE MOON

11	12	13		
23	24	25	26	27
38	39	40	41	42	43	44
56	57	58	59	60	61	62
74	75	76	77	78	79	80
92	93	94	95	96	97	98
110	111	112	113	114
125	126	127

This quadrangle is home to ancient craters that have been distorted multiple times by faults created by the large impacts that formed the Moon's large basins. These events also deposited ejecta material onto the quadrangle and the surrounding region. Volcanic plains and mare materials have filled the depressions and structural troughs of the region. Cratering throughout the later periods of the Moon's geologic history continued to shape the region. The largest craters in the quadrangle—Franklin, Atlas, Cepheus, Geminus, and Hercules—formed in this order over the course of the Moon's middle geologic history. The eponymous Geminus crater is worn but still well preserved, with its features muted by space weathering and erosion. The youngest craters are small simple bowl-shaped craters dotted across the quadrangle.

HAKUTO-R M1	ROCKET	Falcon 9
	CREW	N/A
	LAUNCH DATE	07:38:13, 11 December 2022 (UTC)
	LAUNCH SITE	Cape Canaveral Air Force Station
	PAYLOAD	*Rashid*, Japanese Lunar Excursion Vehicle
	LUNAR ORBIT INSERTION	21 March 2023
	LANDING DATE	Unknown
	IMPACT SITE	Close to Atlas Crater
	LANDING COORDINATES	47.58 N, 44.09 E Ⓐ

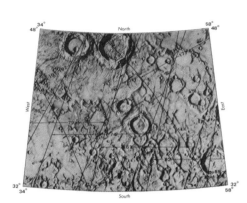

LUNAR ORBITER PHOTOGRAPHIC COVERAGE OF GEMINUS QUADRANGLE

GEMINUS QUADRANGLE: KEY CHARACTERISTICS

1. GEMINUS CRATER
Latitude: 34.5°N / Longitude: 56.7°E
Diameter: 86 km / 53 ½ miles
Depth: 5.4 km / 3 ¼ miles
Photograph: LRO (USA, 2017)

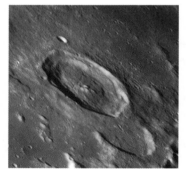

2. FRANKLIN CRATER
Latitude: 38.8°N / Longitude: 47.7°E
Diameter: 56 km / 34 ¾ miles
Depth: 2.7 km / 1 ¾ miles
Photographed from Earth (USA, 2005)

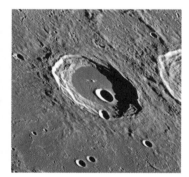

3. HERCULES CRATER
Latitude: 46.82°N / Longitude: 39.21°E
Diameter: 68.32 km / 42 ½ miles
Depth: 3.2 km / 2 miles
Photograph: LRO (USA, 2011)

4. HUMBOLDTIANUM BASIN RING
Latitude: 56.8°N / Longitude: 81.5°E
Diameter: 650 km / 404 miles
Photograph: LRO (USA, 2011)

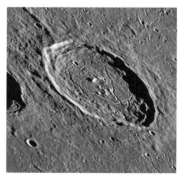

5. ATLAS CRATER
Latitude: 46.7°N / Longitude: 44.4°E
Diameter: 87 km / 54 miles
Depth: 2 km / 1 ¼ miles
Photograph: LRO (USA, 2014)

6. BERZELIUS CRATER
Latitude: 36.55°N / Longitude: 50.95°E
Diameter: 49 km / 30 ½ miles
Depth: 1.7 km / 1 mile
Photograph: LRO (USA, 2014)

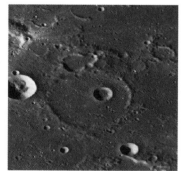

7. CHEVALLIER CRATER
Latitude: 45.01°N / Longitude: 51.57°E
Diameter: 51.83 km / 32 ¼ miles
Depth: 1,200 m / 3,937 ft
Photograph: LRO (USA, 2016)

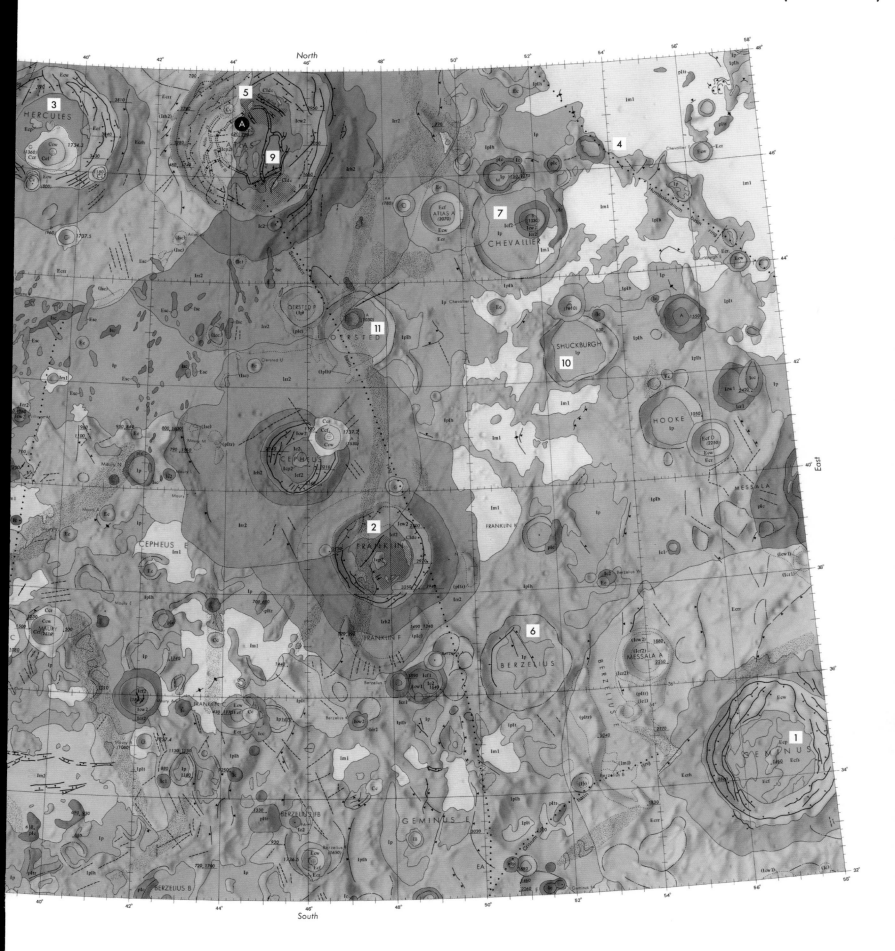

North

East

South

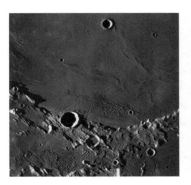

8. SERENITATIS BASIN
Latitude: 28°N / Longitude: 17.5°E
Diameter: 920 km / 571 ¾ miles
Photograph: LRO (USA, 2018)

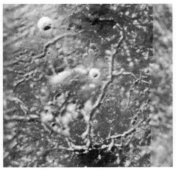

9. RIMA ATLAS
Latitude: 47.5°N / Longitude: 43.6°E
Length: 60 km / 37 ¼ miles
Photograph: Clementine Orbiter
(USA, 1994)

10. SHUCKBURGH CRATER
Latitude: 42.6°N / Longitude: 52.8°E
Diameter: 38 km / 23 ½ miles
Depth: 1,300 m / 4,265 ft 1 in
Photograph: Lunar Orbiter (USA, 1967)

11. OERSTED CRATER
Latitude: 43.1°N / Longitude: 47.2°E
Diameter: 42 km / 26 miles
Photograph: LRO (USA, 2010)

THE MOON AND MEDIEVAL NATURAL PHILOSOPHY

'And as the soul within your dust is shared by different organs, each most suited to a different potency, so does that Mind unfold and multiply its bounty through the varied heavens, though that Intellect itself revolves upon its unity.'

Dante Alighieri, *Paradiso*, Canto II: lines 133–38[1]

↑
Medieval manuscripts often contained illustrations of figures contemplating the heavenly spheres. The spheres represented God's unerring handiwork in constructing an entire world system through which to enact His plan. This illustration is found in a 15th-century French translation of the 4th-century AD Christian theologian and philosopher Augustine of Hippo's *De civitate Dei* (The City of God). Its depiction of Philosophy holding the planetary spheres accompanied Augustine's argument to the Romans of the truth of Christianity. It also reflects the medieval conviction that the Earth-centred cosmos aligned with the Christian view of creation.

n *Paradiso* (Paradise), the third book of his *La Divina Commedia* (Divine Comedy), 14th-century Italian poet Dante Alighieri (1265–1321) ascends through the heavens with his angelic guide, Beatrice, a journey that allows Dante to present each of the spheres of heaven as a physical and religious experience. His first stop on this journey is the Moon. Here he asks his guide why the Moon has spots – the dark features long seen as a face (the 'Man in the Moon') or an animal by human onlookers. Is it, he asks, that the popular legend of Cain making his way to the Moon is true? Or is it, he wonders, an inconsistency in the density of the Moon's substance? Beatrice assures him that neither of these explanations is correct, and that the answer lies in the purpose of the universe. The heavens hold the source of all difference on Earth. To Dante and his contemporaries, each of the heavenly spheres exerted its own influence. But, as Beatrice states, each was also part of one whole, like the organs in the human body – a *machina mundi*, or world machine, through which God's will was done. This was the source and purpose of all difference in the heavens and on Earth.

The idea that the heavens were an interacting system responsible for enacting God's will on Earth was widely shared among medieval Jewish, Muslim and Christian natural philosophers. This was imagined as a somewhat mechanical system, but more so as a tapestry of subtle connections and relationships devised by the Creator. The Moon's sphere was the closest to Earth, and its position denoted the boundary between the sublunar world – composed of the four corruptible elements of earth, water, air and fire – and the realm of the heavens. The heavens, including the Moon, were made of a fifth incorruptible element: the quintessence, or aether.

For some natural philosophers, such as Albertus Magnus (c. 1200–1280), the Sun and the Moon together acted to govern the heavens – the Sun, residing king-like in the centre of the heavens, would gather the influences of all the bodies of heaven, then transmit them to the Moon to be distributed to the sublunar world below. To Magnus, the Moon was 'queen of the heavens'.[2] This relationship between the Sun and Moon gave special significance to events such as lunar and solar eclipses, during which the collective influence of the heavens was understood to be especially great.

The medieval natural philosopher would have read Ptolemy of Alexandria's 2nd-century AD study of astrology, the *Tetrabiblos*, and would have known that the Sun and Moon

That the Sun and the Moon could so readily be seen to effect changes upon the Earth and to be so connected to seasonal changes and the marking of time was one of Ptolemy's primary pieces of evidence that all the planets and stars would have their own influences and, as the Sun and Moon differed from each other, so too would each of the planets. Ptolemy went on to define the Moon's power as being fed by the 'moist exhalations' from the Earth – the Moon in turn acted upon the waters, it humidified parts of the world, and it acted 'to soften and cause putrefaction in bodies'.[4]

In Beatrice's explanation above, her mention of the organs of the body speaks to another aspect of medieval natural philosophy. The universe was created to be an assemblage of parts and subtle interconnections, and so too the human body. The universe was the macrocosm, while the body was the microcosm. The functioning of the body was analogous to the movements of the Sun, Moon, planets and stars, and it was also under their influence. In place of the four elements, the body contained four humours – black bile, yellow bile, phlegm and blood. The humours had qualities analogous to the elements, combinations of either hot or cold with dry or wet.

The medieval physician understood illness to be the result of imbalances in the humours. Among the tools the physician relied upon to restore balance to the humours was a knowledge of the influence of the heavens, which could affect not only the body but the efficacy of the medicines and techniques at his disposal. The Moon was the most significant heavenly body in determining health and illness. The humours were understood to be fluids and, like the waters of the Earth, under the Moon's influence.[5] Blood-letting, for example, was a common 'treatment', and in order to determine the proper time to do it, the physician needed knowledge of the Moon's phase and position in the zodiac.

were understood to be the two most powerful heavenly bodies – the Sun because of its size, and the Moon because of its proximity to Earth. After first explaining that the Sun is always affecting everything on Earth, Ptolemy proclaimed that the Moon:

> as the heavenly body nearest the earth, bestows her effluence most abundantly upon mundane things, for most of them, animate or inanimate, are sympathetic to her and change in company with her; the rivers increase and diminish their streams with her light, the seas turn their own tides with her rising and setting, and plants and animals in whole or in some part wax and wane with her.[3]

↑
Dante's guide through hell and purgatory was the 1st-century BC Roman poet Virgil, whose *Aeneid* also depicted a journey through the underworld. This illustration from a 14th-century Italian manuscript depicts Dante meeting Beatrice in the Garden of Eden in Canto XXX of *Purgatorio*, when it comes time for him to ascend into the heavens. An embodiment of faith and love, Beatrice acts as Dante's guide and teacher from this point on as they ascend toward God's Empyrean – the final and most divine heavenly sphere.

↱
In Greek mythology, King Agamemnon must sacrifice his daughter Iphigenia to the goddess Artemis, who is withholding the winds needed to reach Troy. While in the sphere of the Moon, Beatrice uses this and other examples to impress upon Dante that, while vows are sacred, there are times in which keeping a vow will do great evil.

→
Beatrice and Dante arrive at the Moon's sphere in Canto II of *Paradiso*. 'Direct your mind to God in gratefulness,' she tells him, 'He has brought us to the first star.' Here Dante meets the souls who took vows they did not keep.

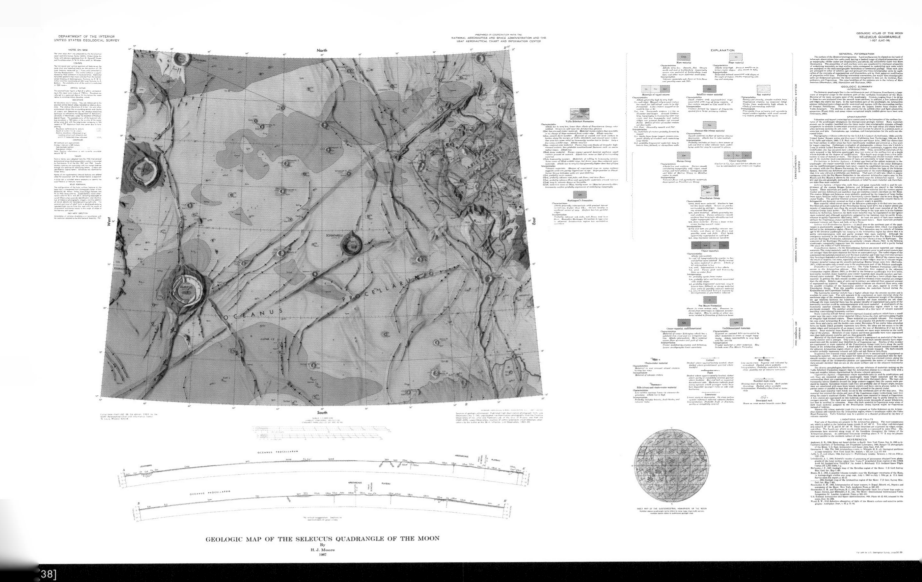

GEOLOGIC MAP OF THE SELEUCUS QUADRANGLE OF THE MOON

By

H. J. Moore

1967

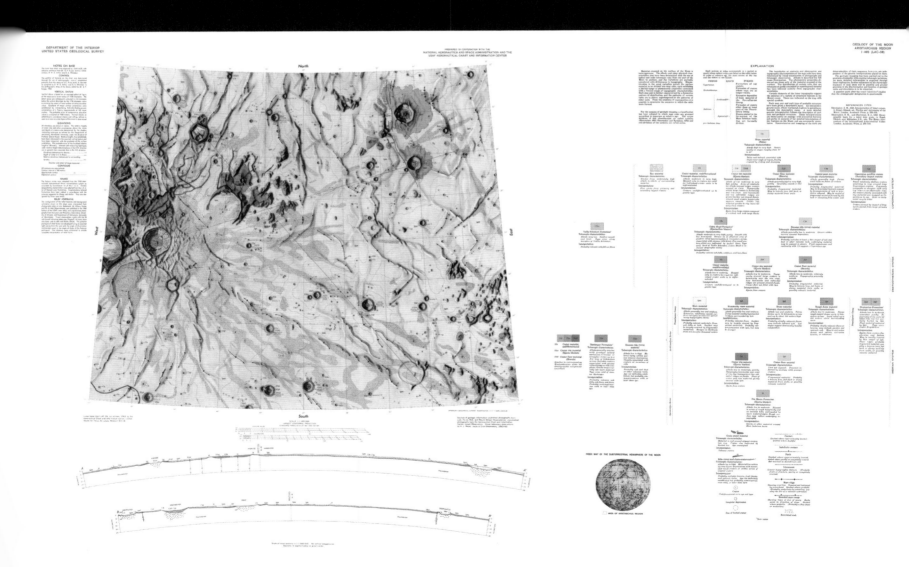

GEOLOGIC MAP OF THE ARISTARCHUS REGION OF THE MOON

By

H. J. Moore

1965

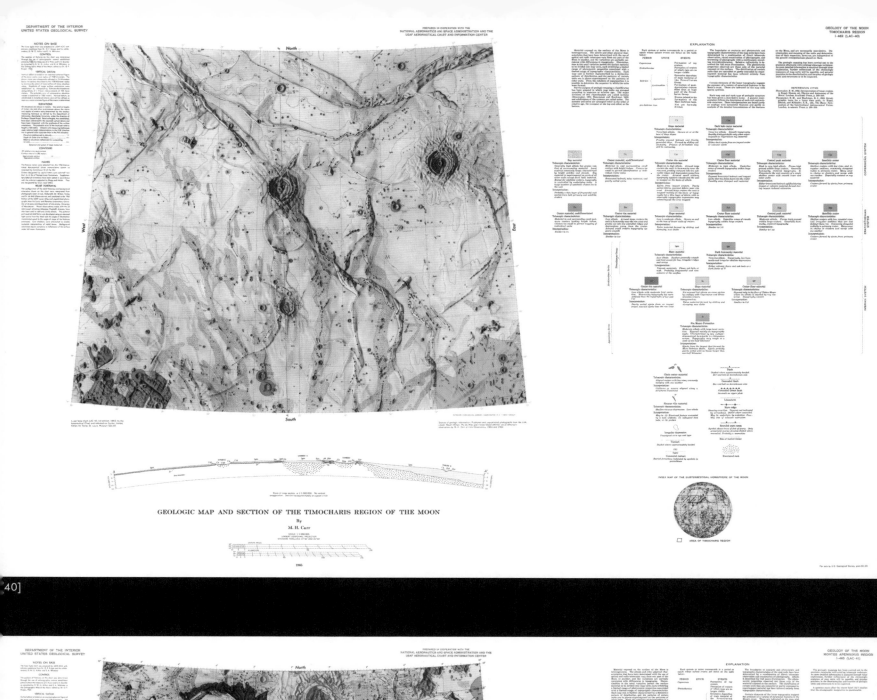

GEOLOGIC MAP AND SECTION OF THE TIMOCHARIS REGION OF THE MOON

By

M. H. Carr

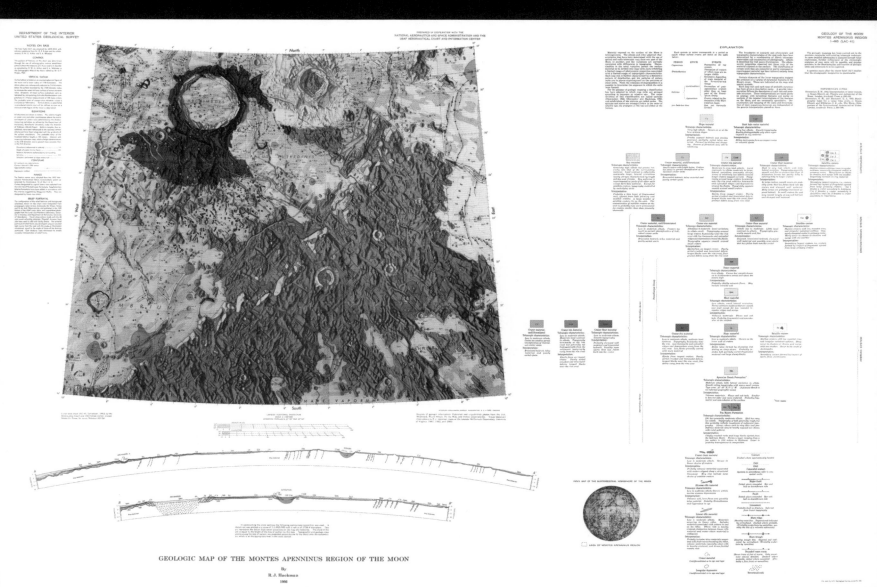

GEOLOGIC MAP OF THE MONTES APENNINUS REGION OF THE MOON

By

R. J. Hackman

'MOONSTRUCK' – LUNACY AND THE FULL MOON

↑
Albrecht Dürer's *Melencolia I* (1514) depicts a gloomy winged female figure – the personification of melancholia.

→
L'influance de la lune sur la teste des femmes (The Influence of the Moon on Women's Minds) (1652) depicts five women dancing beneath a full Moon. Above the head of each woman rests a crescent Moon with an eye, symbolizing the influence of the Moon on women's minds.

↱↗
A werewolf terrorizes children in this woodcut by Lucas Cranach the Elder, entitled *The Werewolf or The Cannibal* (c. 1510–12).

↳↓
Weird Sisters; Ministers of Darkness; Minions of the Moon (1791) by caricaturist James Gillray parodies Henry Fuseli's painting of the three witches in Shakespeare's *Macbeth*. Three ministers gaze at Queen Charlotte depicted as the crescent Moon, concerned about the future after the pronouncement of the madness of King George III, portrayed as the dark side of the Moon.

he Moon's influence on our planet is undeniable. Its gravity pulls the waters of our oceans, creating the tides. Its light cues the migration of birds,[1] the foraging of wildebeests,[2] and the synchronized mating of corals.[3] It is not surprising, then, that since ancient times, humans have wondered if the cycle of the Moon might exert similar power over our minds.

Some of the earliest records of the purported link between the Moon and madness can be found in ancient Greek and Roman texts. An oft-repeated saying, attributed to the 5th-century BC physician Hippocrates, claims that 'One who is seized with terror, fright, and madness during the night is being visited by the Goddess of the Moon'. Pliny the Elder, a Roman author and natural philosopher, wrote that the Moon, 'with her winding and turning in many and sundry shapes, hath troubled much the wits of the beholders'.[4] He conjectured that the full Moon caused more dew to form, leading to increased moisture in the brain.[5] The very word 'lunatic' comes from Latin *lunaticus*, from the Latin word for the Moon, *luna*. People with mental illnesses and such neurological disorders as epilepsy were said to be 'moonstruck', rendered mad by lunar influence.[6]

The belief continued through the medieval and early modern periods in Europe. At the turn of the 17th century, Shakespeare wrote in *Othello*: 'It is the very error of the moon; She comes more near the earth than she was wont, And makes men mad.'[7] And the Moon's phases in particular were referenced by the English legal expert William Blackstone in the 18th century: 'A lunatic, or *non compos mentis*, is properly one who hath lucid intervals, sometimes

enjoying his senses and sometimes not and that frequently depending upon the changes of the moon.'[8] Mental health patients at London's Bethlehem Hospital (also known as Bedlam) were chained and beaten at certain phases of the Moon to prevent violence, a practice that went on (according to a 1943 article in the *American Journal of Psychiatry*) until 1808.[9]

While medieval and early modern Europe had plenty of werewolf lore, a transformation triggered by the full Moon is missing from most of these stories.[10] One of the few stories to mention the Moon at all, from 13th-century writer Gervase of Tilbury, claimed that the werewolf Chaucevaire transformed at the new Moon. The full Moon trope did not really take off until the film *Frankenstein Meets the Wolf Man* was released in 1943. In it, the titular werewolf, who had been killed in the original 1941 *The Wolf Man*, is resurrected when the light of a full Moon hits his long-dead body; he transforms into a wolfish creature at the next full Moon. Countless books, movies, TV shows and video games released since then have included full-Moon werewolf transformations. It even crops up in works outside the traditional horror genre, such as the *Harry Potter* book and film series and the music video to Michael Jackson's 'Thriller' (1982).

Contemporary vampire stories have largely lost their historic connections to the Moon, but the first major such tale to be published in English, John William Polidori's *The Vampyre* (1819), includes the revival of a vampire when he is 'exposed to the first cold ray of the Moon that rose after his death'.[11]

Rumours about the power of the full Moon have persisted into the 21st century. Anecdotal evidence about monthly spikes in crime, unruly classrooms, and busy hospital emergency departments abound on such internet forums as Reddit, with threads on teacher forums titled 'The FULL MOON is making my kids crazy!!!'[12] and security guards asking each other, 'You guys dealing with full moon bs too?'[13]

In recent decades, countless scientists have attempted to pinpoint a connection between the Moon and unruly or otherwise strange behaviour. In 1992, a paper published in the journal *Psychological Reports* analysed twenty previous studies examining a link between the cycle of the Moon and suicidal behaviour. The authors concluded: 'Most studies indicated no relation between lunar phase and the measures of suicide. The positive findings conflicted, have not been replicated, or were confounded with variables such as season, weekday, weather, or holidays. ...there is insufficient evidence for assuming a relationship between the synodic lunar cycle and completed or attempted suicide.'[14] Similarly, a 2010 study in the *Journal of Criminal Justice* examined police data over a five-year period; it found no link between criminal activity and the phases of the Moon.[15] In a 2009 article for *Scientific American*, psychologists Dr Hal Arkowitz and Dr Scott Lilienfeld summarized their analysis of studies on the Moon's effects on the human psyche thus: 'The lunar lunacy effect appears to be no better supported than is the idea that the moon is made of green cheese.'[16]

However, some researchers insist that there is more to the story.

A 2021 study published in the journal *Science Advances* demonstrated a correlation between the lunar cycle and the circadian rhythms that govern wakefulness. The connection appeared in remote communities in Argentina that lacked electric lights, as well as in urban Seattle (though the effects were stronger in Argentina, where the study participants lost up to an hour and a half of sleep on nights preceding the full Moon).[17] Since lack of sleep can exacerbate some mental health conditions, including bipolar disorder, some have argued that the full Moon's brightness might have been a contributor to some episodes of supposed 'lunacy', especially in historical accounts before artificial light pollution began outshining the Moon's gentle gleam.[18]

Other researchers have spotted correlations between the Moon and the mood cycles of people with bipolar disorder. In 2018, Dr Thomas Wehr published a study on seventeen individuals with rapid cycling bipolar disorder, a pattern of four or more manic or depressive episodes within one year.[19] Over the course of an average of 1.6 years' worth of observations of each patient, he found that all of these patients' mood cycles were synchronous, at some point, with subtle and complex aspects of the Moon's cycle, including its effects on the tides. However, there was no singular lunar phase or position that affected all of the patients – the subjects did not, for example, all experience mania around the full Moon. (The mechanism by which people's brains 'sense' these lunar cycles remains unknown, as does the reason why the different patients seemed to sync with different aspects of lunar behaviour.)

While the bulk of scientific evidence opposes a link between the full Moon and 'lunacy', the debate is far from settled. If nothing else, the Moon has demonstrated its power over the human mind through our never-ending stories and legends about it.

⇈

Légendes rustiques (Rustic Legends) (1858) is a collection of ghost stories from the region of Berry, France, retold by George Sand, who spent her childhood there. It is illustrated with twelve plates by her son Maurice Sand. This plate, entitled *Les Laveuses de nuit ou Lavandières* (Night Washers or Washerwomen), depicts the spectres of bad mothers condemned to wash the swaddling clothes of their victims forever.

↑

This plate from *Légendes rustiques* is entitled *Les Trois Hommes de Pierre* (The Three Stone Men) and illustrates a story told by those living in the villages of the Creuse area in lower Berry. Three gigantic rocks of the region come to life at night and take the form of three gigantic figures. They walk at speed along the riverbank, even though they have no legs, waving their arms and terrifying anyone they encounter, asking them if they want arms.

↱

Maurice Sand's illustration of *Le Lupeux* (The Lupus) depicts a man walking in moonlight towards a row of topped trees, lining a pond. He is being tormented by the voice of a demon known as the Lupus, which tries to engage him in conversation. With promises of intriguing adventures and romantic trysts, the Lupus attempts to lure his victim to the edge of the pond. If he succeeds, the demon will then appear, tell him to look into the pond and push him into the icy waters.

↳

Entitled *Les Lupins* (The Wolves) this plate from *Légendes Rustiques* depicts a row of wolf-like creatures, which stand on their hind legs and lean against the wall of a cemetery under a crescent Moon. Their eyes shine like fiery blood and their breath stinks. However, they are sorrowful and timid creatures, the result of some past humiliation to one of their kind, and if someone walks past them, they run away, shouting, 'Robert is dead, Robert is dead.'

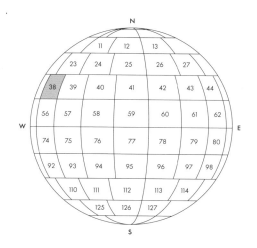

This quadrangle is mostly defined by the smooth, ridged terrain of Oceanus Procellarum's (Ocean of Storms) lava plains. This terrain is interrupted by craters, including the Seleucus crater that gives this quadrangle its name. Unlike Briggs crater to its northwest and Schiaparelli crater to its northeast, Seleucus remains well preserved with an intact rim, a terraced wall and a rampart that meets the mare abruptly. Southeast of Seleucus crater, marked with the words 'LUNA 13' at the coordinates 18.52°N, 62.04°W, is the landing site of the Soviet robotic spacecraft Luna 13, the third spacecraft to land successfully on the Moon. The northeast corner of this quadrangle holds a portion of the Aristarchus Plateau that continues into the Aristarchus Quadrangle. This plateau includes a segment of the Vallis Schröteri, a sinuous rille that almost resembles a channel. This rille – the largest on the Moon – was probably formed by a lava flow.

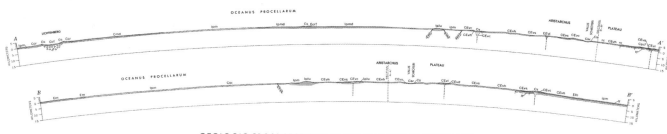

GEOLOGIC CROSS SECTIONS OF THE SELEUCUS QUADRANGLE
APPROXIMATE HORIZONTAL SCALE 1:1 000 000 (NO VERTICAL EXAGGERATION)

LUNA 13	ROCKET	Molniya-M
	CREW	N/A
	LAUNCH DATE	10:17:08, 21 December 1966 (UTC)
	LAUNCH SITE	Tyuratam (Baikonur Cosmodrome)
	PAYLOAD	Panoramic Television Camera
	LUNAR ORBIT INSERTION	N/A
	LANDING DATE	18:01:00, 24 December 1966 (UTC)
	LANDING SITE	Oceanus Procellarum
	LANDING COORDINATES	18.52°N, 62.04°W (A)

LUNA 13 PANORAMA
Section of the lunar surface panorama returned by Luna 13 on 26 December 1966, showing the highly cratered floor. The shadow of the lander with its set of four antennas is visible in this section.

SELEUCUS QUADRANGLE: KEY CHARACTERISTICS

1. SELEUCUS CRATER
Latitude: 21°N / Longitude: 66.6°W
Diameter: 61 km / 38 miles
Depth: 3 km / 1 ¾ miles
Photograph: LRO (USA, 2015)

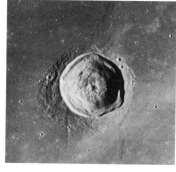

2. SCHIAPARELLI CRATER
Latitude: 23.4°N / Longitude: 58.8°W
Diameter: 24 km / 15 miles
Depth: 2.1 km / 1 ¼ miles
Photograph: Lunar Orbiter IV (USA, 1967)

3. BRIGGS CRATER
Latitude: 26.5°N / Longitude: 69.1°W
Diameter: 37 km / 15 miles
Depth: 1,200 m / 3,937 ft
Photograph: Lunar Orbiter IV (USA, 1967)

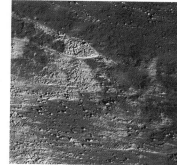

4. ARISTARCHUS PLATEAU
Latitude: 26.4°N / Longitude: 50.4°W
Diameter: 230 km / 143 miles
Photograph: LRO (USA, 2010)

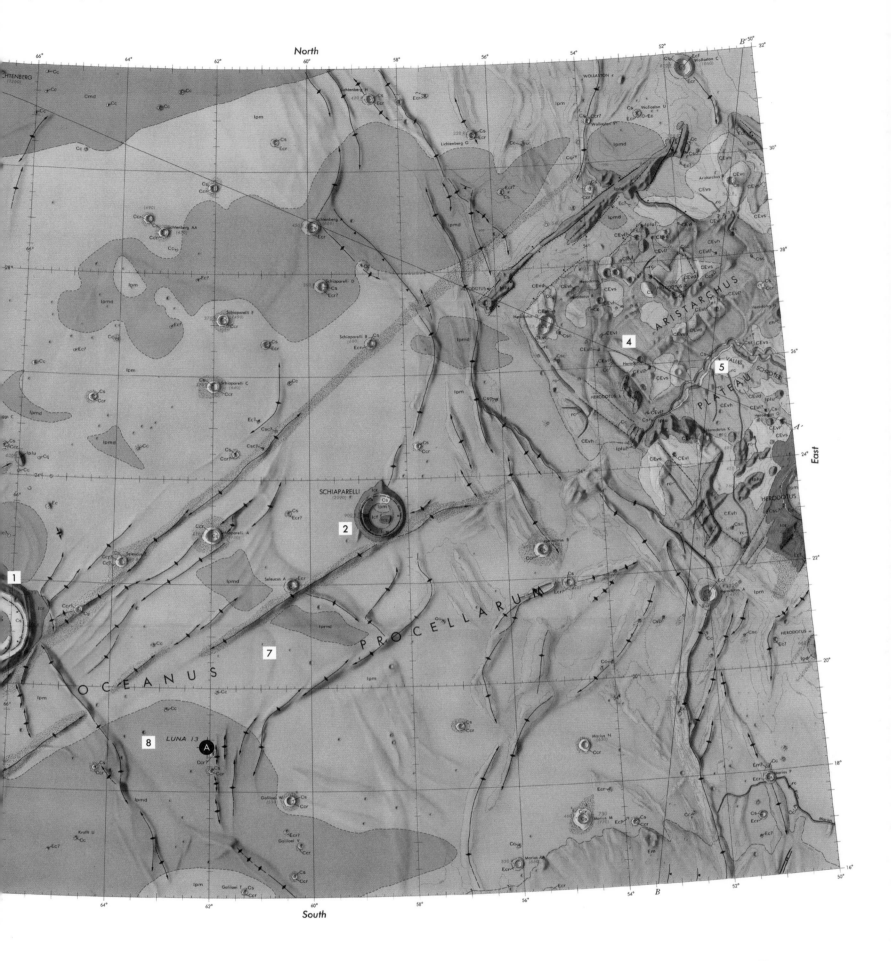

5. VALLIS SCHRÖTERI
Latitude: 26.2°N / Longitude: 50.8°W
Max Diameter: 168 km / 104 ½ miles
Photograph: Apollo 15 (USA, 1971)

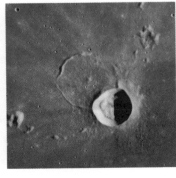

6. LICHTENBERG CRATER
Latitude: 31.8°N / Longitude: 67.7°W
Diameter: 20 km / 12 ½ miles
Depth: 1,200 m / 3,937 ft
Photograph: Lunar Orbiter IV (USA, 1967)

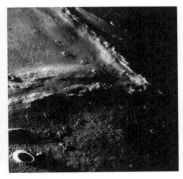

7. OCEANUS PROCELLARUM
Latitude: 18.4°N / Longitude: 57.4°W
Diameter: 2,592 km / 1,610 ½ miles
Photograph: Apollo 15 (USA, 1971)

8. LUNA 13 PANORAMA
Latitude: 18.52°N / Longitude: 62.04°W
Landing Date: 24 December 1966
Photograph: Luna 13 (USSR, 1966)

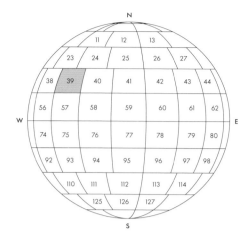

The impact crater Aristarchus is one of the brightest features on the Moon's near side. Its size, exceptional brightness, and its location on the Aristarchus Plateau surrounded by the dark lava plains of Oceanus Procellarum (Ocean of Storms) make it an easy feature to spot from Earth. The brightness of this crater corresponds to its age, as it is one of the youngest craters on the Moon. The integrity of its walls and central peak also correspond to its young age, as does the bright ejecta blanket. Nearby craters Herodotus and Prinz are markedly different from Aristarchus. They have been flooded with lava, giving them a dark appearance, and, in the case of Prinz, much of its rim has been degraded. The nearby Montes Harbinger mountains mark the boundaries of Oceanus Procellarum and Mare Imbrium (Sea of Showers). Several small craters also populate this quadrangle, many of which are secondary craters created by material ejected by the impact that excavated Aristarchus. Also visible are sinuous rilles and ridges associated with tectonic activity and the flow of lava. In the northeast of the quadrangle are the craters Diophantus and Delisle, both of which are older than Aristarchus but younger than the mare.

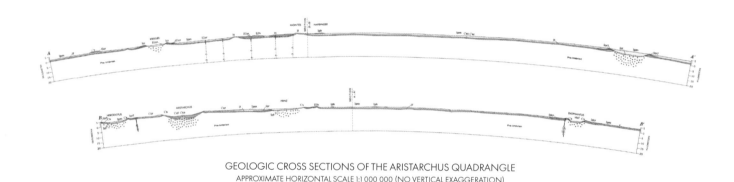

GEOLOGIC CROSS SECTIONS OF THE ARISTARCHUS QUADRANGLE
APPROXIMATE HORIZONTAL SCALE 1:1 000 000 (NO VERTICAL EXAGGERATION)

ARISTARCHUS QUADRANGLE: KEY CHARACTERISTICS

1. ARISTARCHUS CRATER
Latitude: 23.4°N / Longitude: 47.4°W
Diameter: 40 km / 24 ¾ miles
Depth: 2.7 km / 1 ¾ miles
Photograph: Apollo 15 (USA, 1971)

2. HERODOTUS CRATER
Latitude: 23.2°N / Longitude: 49.7°W
Diameter: 35 km / 21 ¾ miles
Depth: 1.5 km / 1 mile
Photograph: Apollo 15 (USA, 1971)

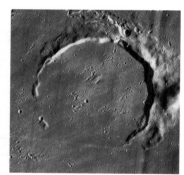

3. PRINZ CRATER
Latitude: 25.5°N / Longitude: 44.1°W
Diameter: 46 km / 28 ½ miles
Photograph: Luna Orbiter V (USA, 1967)

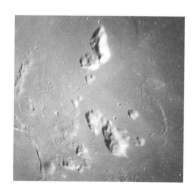

4. MONTES HARBINGER
Latitude: 26.9°N / Longitude: 41.3°W
Diameter: 93 km / 57 ¾ miles
Max Height: 2.5 km / 1 ½ miles
Photograph: Apollo 15 (USA, 1971)

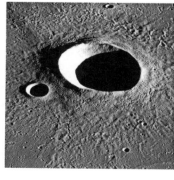

5. DIOPHANTUS CRATER
Latitude: 27.6°N / Longitude: 34.3°W
Diameter: 19 km / 11 ¾ miles
Depth: 3 km / 1 ¾ miles
Photograph: Apollo 17 (USA, 1972)

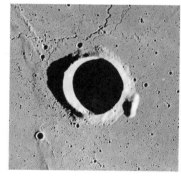

6. BRAYLEY CRATER
Latitude: 20.9°N / Longitude: 36.9°W
Diameter: 14.5 km / 9 miles
Depth: 2.8 km / 1 ¾ miles
Photograph: Apollo 17 (USA, 1972)

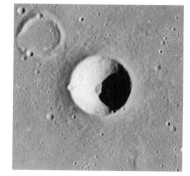

7. ÅNGSTRÖM CRATER
Latitude: 29.9°N / Longitude: 41.6°W
Diameter: 9.55 km / 6 miles
Depth: 2 km / 1 ¼ miles
Photograph: Apollo 15 (USA, 1971)

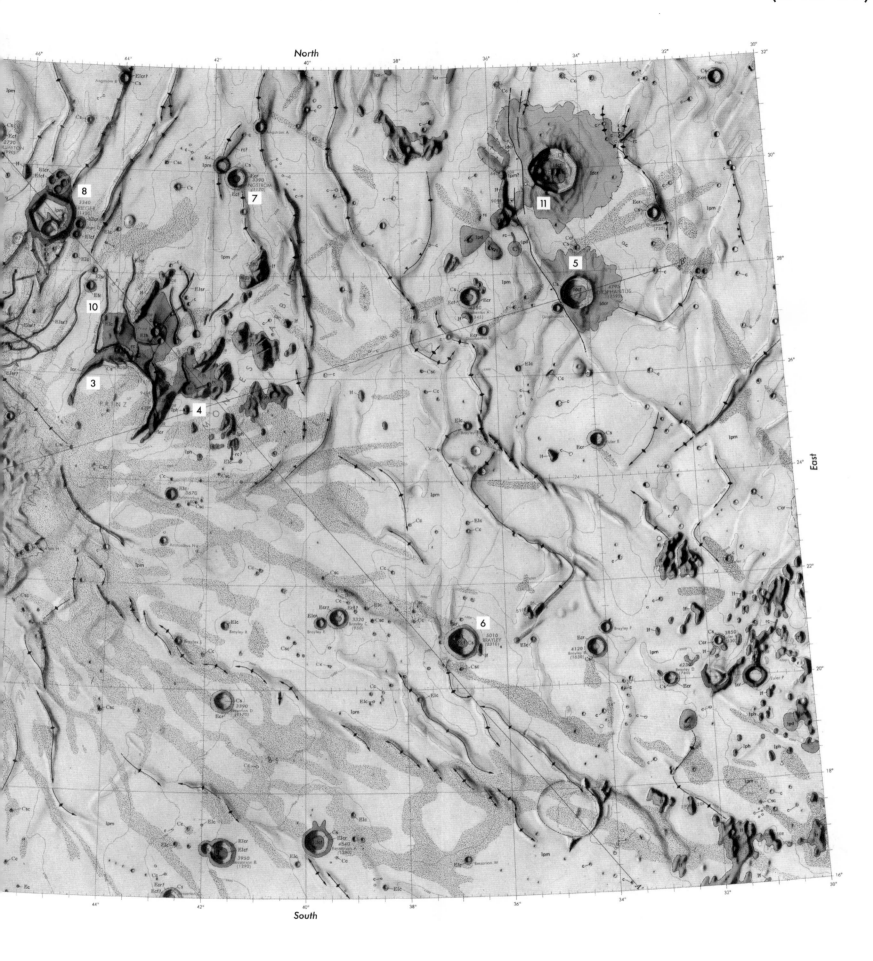

8. KRIEGER CRATER
Latitude: 29.02°N / Longitude: 45.61°W
Diameter: 22.87 km / 14 ¼ miles
Depth: 1,100 m / 3,608 ft 11 in
Photograph: Apollo 15 (USA, 1971)

9. VALLIS SCHRÖTERI
Latitude: 26.2°N / Longitude: 50.8°W
Max Diameter: 168 km / 104 ½ miles
Photograph: Apollo 15 (USA, 1971)

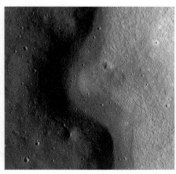

10. RIMAE PRINZ REGION
Latitude: 27.32°N / Longitude: 43.54°W
Length: 65 km / 40 ½ miles
Photograph: LRO (USA, 2010)

11. DELISLE CRATER
Latitude: 29.9°N / Longitude: 34.6°W
Diameter: 25 km / 15 ½ miles
Depth: 2.6 km / 1 ½ miles
Photograph: Apollo 15 (USA, 1971)

THE SYMBOLISM OF THE MOON IN ASTROLOGY

Myth

Lunar

↱
Two folios from the Persian *Mu'nis al-ahrar fi daqa'iq al-ash'ar* (The Free Man's Companion to the Subtleties of Poems) by Jajarmi, 1340–41, show the Moon visiting the houses of Gemini (top), Cancer (centre) and Leo (bottom).

↱↱
Two 15th-century German manuscript illustrations that feature the Moon. The first (left) shows the Moon's rulership of Monday. The twenty-four divisions in the diagram represent the hours of the day, and the Moon's creature, the crab, is shown at the bottom. In the second (right), the Moon is represented as a woman on a horse, perhaps indicating an independent spirit.

↳
This clay seal, dating to the Third Dynasty of Ur (now in southern Iraq) (c. 2100 BC), shows Sin, the Moon-god, conferring his authority on the king, Ur-Nammu, as he provides order and advice to the three figures on the left.

↳↳
This portrayal of a 'zodiac man' from Heinrich Laufenberg's *Regimen* (1400) shows the signs of the zodiac over the parts of the body they rule, for example the Cancerian crab rules the breasts and stomach.

←
This tarot card from the *Tarot dit de Charles V* (King of France from 1380–1420) deck shows the Moon's importance as a measure of the count of calendar days: the man on the left holds a pair of compasses to measure the Moon's position in the sky, while the man on the right relates this to tables in a book.

he Moon has a central role in timekeeping, mythology, religion and astrology. The last of these has been a necessary guide in many (if not all) traditional cultures in establishing the most propitious moments for performing important activities. The twenty-nine day cycle from one full Moon to the next is one of the defining features of culture: humanity's earliest ancestors could see to travel or hunt during the light of the full Moon, but at the dark of the new Moon they had to shelter from unseen threats. And so the thought emerged that different kinds of activity were favoured at different phases of the Moon.

In ancient Babylon, where modern astrology has its origins, the Moon-god, Sin, counted the days, keeping the calendar. According to a text on one clay tablet, if the Moon 'becomes late at an inappropriate time and does not become visible', it warns of 'attack of a ruling city'; if it was observed on the 16th of the month it favoured the king of Subartu (now northern Iraq); or, if it was surrounded by a halo through which another planet could be seen, then 'robbers will rage'.[1] If lunar omens for the king were particularly bad then prayers could be said to the Sun, and slaves were sacrificed to appease the gods.[2] Eclipses were not all bad, and when the mighty Assyrian emperor Sargon II (721–705 BC) took an astrologer on his successful eighth military campaign, 'Magur ["the boat", a name for the Moon], lord of the tiara [made an eclipse]': after consulting the Sun for reassurance, Sargon then assembled his army, ready for battle.[3] It is easy to imagine the scene as the eclipse ended and the Sargon's huge force began to move.

From Babylon, astrology moved west to Egypt and Greece, before taking an eastward turn and travelling to India. In Egypt an important lunar deity was Thoth, who, like Sin, counted the days, measuring time, but also invented every one of the arts and sciences (see p. 24 [Ancient Egypt]).

The Greeks adopted Babylonian astrology from the 3rd century BC onwards, adapting and expanding it so that it became a complete system for understanding the universe and acting within it. The Greeks made much of the Moon's

rapid journey through the twelve zodiac signs, over a period of just over twenty-seven days. For reasons that are no longer clear, they declared that the Moon 'ruled' Cancer, where it was particularly strong, and was 'exalted' in Taurus, where it was also favourable. It was best not to plan events when the Moon was in one of the opposite two signs, Capricorn and Scorpio, which could arouse unforeseen problems.

For the Greeks, the Moon was female, having long been an embodiment of the goddess Selene (probably from the word *selas*, meaning 'brightness' or 'light'), but acquired a range of other associations that allowed astrologers to analyse the destinies of individuals and even whole societies. The 2nd-century AD writer Vettius Valens, who lived in Antioch, compiled a list of its associations, some of which are specifically gendered, including conception, birth, mothers, breasts, the 'mistress of the house', queens and goddesses.[4] There was also a list of less gender-specific connections, such as older brothers, homes and property, crowds, ships, the left eye, the stomach, silver and silver objects, and anything white. Above all, the Moon was reflective, having no light of its own, but always responding to its bright partner, the Sun. Valens' contemporary, Ptolemy of Alexandria, argued that, like the tides, everything ebbs and flows in accord with lunar cycles.[5] Ptolemy also introduced psychology into astrology eighteen hundred years before Sigmund Freud, arguing that the Moon's position indicates

our emotional tendencies. If, for example, the Moon was in a good relationship with Mercury, the planet of mental processes, a person's emotional and intellectual selves would work together; but, if there were no relationship between them, the person's heart would take them in one direction, but their mind in another, precipitating stress.

Later astrologers implicated the Moon in calculating, or 'electing', the timing of events in order to secure the best results. The aim was to have the Moon placed in a sympathetic zodiac sign, ideally Cancer or Taurus. They also allocated meanings to the planets. Saturn became obstructive because it was slow, Mars dangerous because of its reddish tinge, while Venus was benevolent because of its brightness, and Jupiter was benign on account of its connection to the king of the gods. The careful astrologer, therefore, advised action when the Moon was linked to Venus or Jupiter, for the best chance of success, but to avoid at all costs periods when the Moon was linked to Mars or Saturn.

Astrologers could answer questions from clients on the same basis, and a positive answer to any inquiry was pretty much certain if the Moon was in Cancer or linked to Venus or Jupiter. And if there was a problem one could aways resort to magical ritual. This is a Greek cure for gout, appealing to the Moon in the making of a magical talisman:

> Another amulet for the foot of the gouty man: You should write these names on a strip / of silver or tin. You should put in on a deer skin and bind it to the foot of the man named, on his two feet: 'THEMBARATHEM OUREMBRENOUTIPE / AIOXTHOU SEMMARATHEMMOU NAIOOU, let NN, whom NN bore, recover from every pain which is in his knees and two feet'. You do it when the Moon is [in the constellation].[6]

In India, the Moon was – and still is – known as Chandra, a male god. The Indians imported Greek astrology in the 2nd century AD but retained the Moon's male gender, and adapted the entire system to the Hindu deities. Indian astrology, or *jyotish* (the 'science of light') has the same methods for timing as astrology in the West, but places much more emphasis on rituals designed to secure divine blessings. For example, if an individual has a personal problem of a lunar nature, which might include family and marital issues, they may help resolve it by fasting on Monday (the Moon's day), wearing white clothes, wearing a pearl ring, drinking milk, using silver utensils, and respecting their mother, as well as all older women. And, if there is a temple containing an image of Chandra nearby, then they can visit it to make offerings and chant a lunar mantra.

Western astrology long ago abandoned the planetary deities, but has retained the classical qualities described by Valens and Ptolemy. Margaret Hone's *Modern Textbook of Astrology*, which was the major teaching manual in English for thirty years from 1951, bases the personality profile of the lunar person on the Moon's reflective qualities.[7] The person born under a strong Moon, Hone wrote, shapes their behaviour according to influences from family, school, friends, society and work, never quite being themselves, but always being who they think they are expected to be.

Modern astrology has also retained the classical associations of the zodiac signs, but largely given up the idea that some signs are inherently more fortunate than others. For current astrologers the Moon symbolically represents emotional expression, and is linked to family life and marriage. The person born when the Moon is in Aries, it is said, has forceful emotions and a volatile family life. Taurus, meanwhile, provides a yearning for stability, Gemini for cerebral relationships and Cancer for security. Moon-in-Leo types need to be the centre of attention, Virgos prioritize practical issues, Libras seek balance and harmony, Scorpios are intense, Sagittarians crave their freedom, Capricorns respect tradition and conservative values, Aquarians need to be respected as individuals, and Pisceans are sensitive romantics.

←
These paintings of the Moon and Cancer from Cristoforo de Predis's 15th-century illuminated manuscript of *De Sphaera* (On the Sphere) present the Moon's female nature and command of all things relating to water.

1
A 12th-century miniature from *The Marvels of Creation* by al-Qazwini.

2
The illustrated horoscope of Prince Iskandar created in 1411 in Shiraz, Iran.

3
A *mappa mundi* (map of the world), from *The Catalan Atlas*, Barcelona, 1375.

4
This 16th-century chart of the zodiac signs includes Venus, Cupid and a Bishop Saint.

5
Celestial map from Ottoman manuscript *Zübdet'üt Tevarih* (1598) by Seyyid Lokma.

6
This astrolabe (c. 1552–57) includes a marker to show the position of the Moon.

[LAC-40]

I-462

Timocharis Quadrangle

Maps

Lunar

082

Timocharis is the largest impact crater in this quadrangle, located near the northeast corner. If this crater once had a central peak, it has since been obliterated by the smaller, more recent impact that excavated a crater at its centre. Lambert crater, to the west of Timocharis, has terraced walls and a rough interior, and also has a small crater at its centre. Just to the south of Lambert, and partially covered by the ramparts extending from Lambert's walls, is Lambert R, a crater that is only just visible to telescopic observers as it has been almost completely covered by the lava that formed Mare Imbrium (Sea of Showers). Further south is the crater Pytheas, with a sharp rim and ray system indicating that it is relatively young. In the west is the crater Euler, with a low central peak, terraced walls and a small ray system. The surrounding lava plains of the maria, which make up the bulk of this quadrangle, are marked by ridges and chain crater systems. A portion of the Montes Carpatus mountain range that forms the southern edge of Mare Imbrium is visible in the southwest corner of this quadrangle.

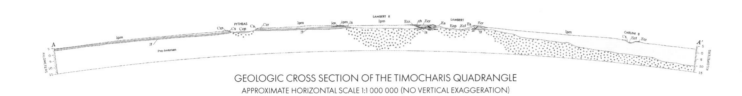

GEOLOGIC CROSS SECTION OF THE TIMOCHARIS QUADRANGLE
APPROXIMATE HORIZONTAL SCALE 1:1 000 000 (NO VERTICAL EXAGGERATION)

MONTES CARPATUS
Photograph of the Montes Carpatus taken by an LRO in 2014 (far left). The mountain range was named after the Carpathian Mountains which are shown in a satellite photograph (left), taken on 14 March 2007.

TIMOCHARIS QUADRANGLE: KEY CHARACTERISTICS

1. TIMOCHARIS CRATER
Latitude: 26.7°N / Longitude: 13.1°W
Diameter: 34 km / 21 ¼ miles
Depth: 3.1 km / 2 miles
Photograph: Apollo 15 (USA, 1971)

2. TIMOCHARIS CRATER WALL
Latitude: 26.7°N / Longitude: 13.1°W
Diameter: 34 km / 21 ¼ miles
Depth: 3.1 km / 2 miles
Photograph: Apollo 15 (USA, 1971)

3. LAMBERT CRATER
Latitude: 25.8°N / Longitude: 21°W
Diameter: 30 km / 18 ¾ miles
Depth: 2.7 km / 1 ¼ miles
Photograph: Apollo 15 (USA, 1971)

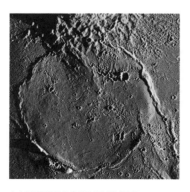

4. LAMBERT R GHOST CRATER
Latitude: 23.88°N / Longitude: 20.66°W
Diameter: 55 km / 34 ¼ miles
Photograph: Apollo 15 (USA, 1971)

5. MONTES CARPATUS
Latitude: 14.5°N / Longitude: 24.4°W
Diameter: 361 km / 224 ¼ miles
Max Height: 2.4 km / 1 ½ miles
Photograph: LRO (USA, 2014)

6. LA HIRE
Latitude: 27.66°N / Longitude: 25.51°W
Max Diameter: 25 km / 15 ½ miles
Height: 1.5 km / 1 mile
Photograph: Apollo 15 (USA, 1971)

7. MARE IMBRIUM
Latitude: 32.8°N / Longitude: 15.6°W
Diameter: 1,146 km / 712 miles
Photograph: Lunar Orbiter IV (USA, 1967)

[LAC–40]

1–462

Timocharis Quadrangle

Maps

Lunar

083

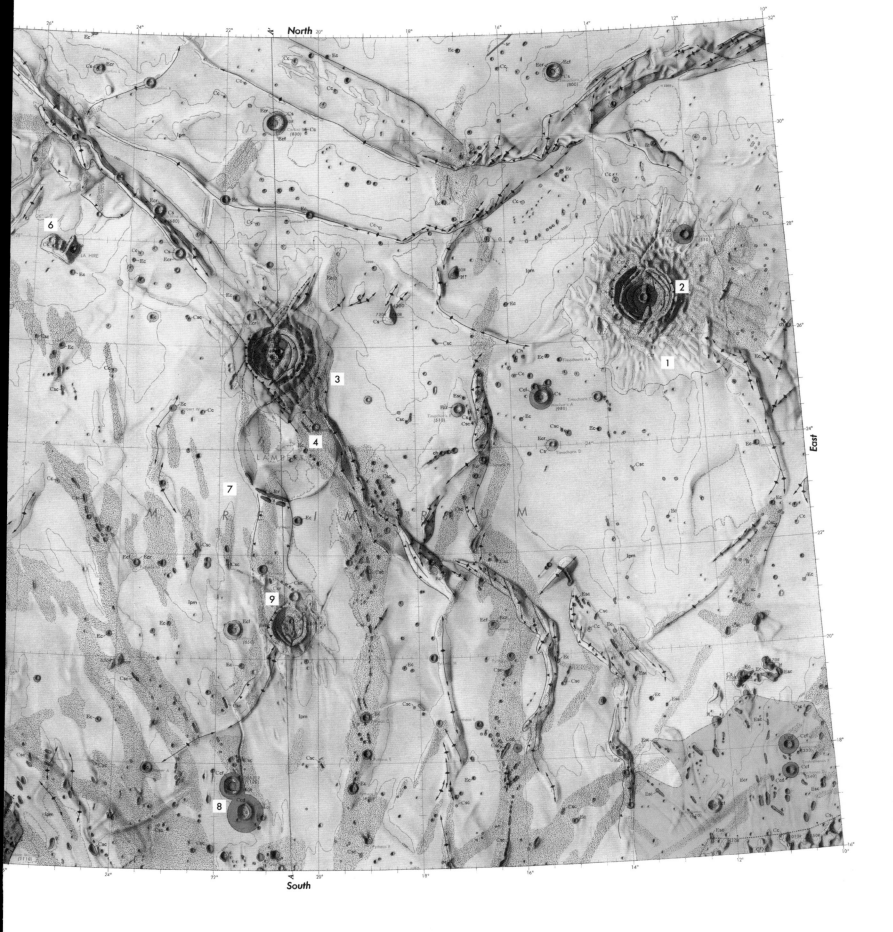

North

South

East

8. DRAPER CRATER
Latitude: 17.6°N / Longitude: 21.7°W
Diameter: 9 km / 5 ½ miles
Depth: 1.7 km / 1 mile
Photograph: Lunar Orbiter IV (USA, 1967)

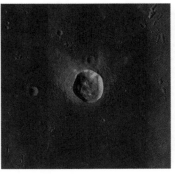

9. PYTHEAS CRATER
Latitude: 20.5°N / Longitude: 20.6°W
Diameter: 20 km / 12 ½ miles
Depth: 2.5 km / 1 ½ miles
Photograph: LRO (USA, 2012)

10. TOBIAS MAYER CRATER
Latitude: 15.6°N / Longitude: 29.1°W
Diameter: 33 km / 20 ½ miles
Depth: 2.9 km / 1 ¾ miles
Photograph: Lunar Orbiter IV (USA, 1967)

11. EULER CRATER
Latitude: 23.3°N / Longitude: 29.2°W
Diameter: 27 km / 16 ¾ miles
Depth: 2.2 km / 1 ¼ miles
Photograph: Apollo 17 (USA, 1972)

[LAC-41]

I-463

Montes Apenninus Quadrangle

Maps

Lunar

084

GEOLOGIC ATLAS OF THE MOON

The Montes Apenninus mountain range marks the meeting of the southeastern edge of Mare Imbrium (Sea of Showers) and the Terra Nivium highlands. The rugged terrain of these highlands features many named mountains. These include the massif Mons Hadley and the Mons Hadley Delta. The Mons Hadley Delta was visited by the Apollo 15 astronauts, who landed their Lunar Module in the valley between these two features at coordinates 26.13°N, 3.63°E (this landing site is not marked as it occurred after this map was drawn). Years earlier, on 13 September 1959, the Soviet Union's robotic Luna 2 became the first spacecraft to reach the Moon when it crashed near the crater Archimedes, the largest crater on Mare Imbrium. This crater has a complete rim, significant outer ramparts and terraced interior walls, but has a smooth floor that is the result of in-filling by lava. To the east is the crater Autolycus, which is half the size of Archimedes and significantly younger. It has a complete rim and a ray system that overlays Archimedes. Between these craters and Montes Apenninus is the smooth lava plain Palus Putredinis (Marsh of Decay). Crater chains and rilles weave through this quadrangle.

LUNA 2	ROCKET	R-7 + Blok E
	CREW	N/A
	LAUNCH DATE	06:39:40, 12 September 1959 (UTC)
	LAUNCH SITE	Tyuratam (Baikonur Cosmodrome)
	PAYLOAD	Radiation Detectors, Magnetometer
	LUNAR ORBIT INSERTION	N/A
	LANDING DATE	23:02:23, 14 September 1959 (UTC)
	IMPACT SITE	near Autolycus Crater
	LANDING COORDINATES	30°N, 0°W (A)

APOLLO 15	ROCKET	Saturn V AS-510
	CREW	D. R Scott, A. M. Wordenm J. B. Irwin
	LAUNCH DATE	14:32:00, 26 July 1971 (UTC)
	LAUNCH SITE	Cape Kennedy Air Force Station
	PAYLOAD	Endeavor (CM-112) Falcon (LM-10)
	LUNAR ORBIT INSERTION	29 July 1971
	LANDING DATE	23:16:00, 29 July 1971 (UTC)
	LANDING SITE	Hadley Plains
	LANDING COORDINATES	26.13°N, 3.63°E (B)

ENDEAVOUR
View of the SIM bay of the service module from the window of the command module.

APOLLO 15 LUNAR LANDING: 29 JULY 1971

LUNAR ROVING VEHICLE
Photograph of James Irwin standing beside a Lunar Roving Vehicle (LRV), taken by David R. Scott during the first Extra Vehicular Activity (EVA) on 31 July 1971.

VIEW FROM ENDEAVOUR WINDOW
View of craters Krieger (left) and Prinz (right) from the Endeavour window, capturing the reflection of either David R. Scott or James Irwin from inside the cabin.

EXAMINING A LUNAR BOULDER
David R. Scott reaches for a hammer at Station 9a on the Hadley Rille rim during EVA 3 on 2 August 1971. James Irwin's reflection is visible in his visor.

REPEATING GALILEO'S EXPERIMENT
The hammer and feather dropped by David R. Scott to prove that there is gravity atsurface level on the Moon, captured for a televised audience during EVA 3.

MONTES APENNINUS QUADRANGLE: KEY CHARACTERISTICS

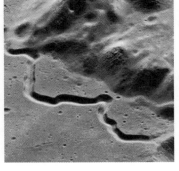

1. MONTES APENNINUS
Latitude: 20°N / Longitude: 3°W
Length: 600 km / 372 ¾ miles
Max Height: 5 km / 3 miles
Photograph: Lunar Orbiter IV (USA, 1967)

2. MONS HADLEY
Latitude: 26.69°N / Longitude: 4.12°E
Diameter: 25 km / 15 ½ miles
Height: 4.5 km / 2 ¾ miles
Photographed from Earth, 2014

3. RIMA HADLEY
Latitude: 25.72°N / Longitude: 3.15°W
Length: 80 km / 49 ¾ miles
Max Width: 2 km / 1 ¼ miles
Photograph: Apollo 15 (USA, 1971)

4. ARCHIMEDES CRATER
Latitude: 39.7°N / Longitude: 4.2°W
Diameter: 81 km / 50 ¼ miles
Depth: 2.1 km / 1 ¼ miles
Photograph: LRO (USA, 2014)

THE THEORY OF LUNAR ECLIPSES

↓
During the 3rd century
BC Aristarchus of Samos
estimated the sizes of
the Earth, Sun and Moon
(and therefore the distance
to the other bodies),
using measurements of
phenomena, including lunar
eclipses. He recorded his
calculations in *On the Sizes
and Distances*. Although
his results were wrong by
a factor of twenty, they still
revealed that the Sun was
much larger than either
the Earth or Moon.

↱
Two plates from Zakariyya
al-Qazwini's *Wonders
of Creation* (c. 1717) show
the origin of lunar eclipses
when the Moon passes
through the cone of the
Earth's shadow, and the
formation of the crescent
Moon, vital to the Islamic
religious calendar.

جاوا
Eclipse table from
the *Kalendarium of
Regiomontanus* (1474).
German astronomer
Johannes Müller of
Königsberg (known
by his Latinized name
Regiomontanus) published
several influential
astronomical works,
including this almanac.
These pages show the timing
and extent of lunar eclipses
visible from Germany
between 1483 and 1530.

clipses of the Moon and Sun have played an important
role in human experience of the cosmos since prehistoric
times. The sight of the full Moon becoming engulfed in
blackness or unexpectedly turning blood red must have
been a memorable one for people who had no idea of the
mechanism behind it – and an unexpected eclipse of all or
part of the Sun would have been even more awe-inspiring.
Some cultures attributed such events to mythological causes
(the Inca of South America, for example, believed that lunar
eclipses were due to a cosmic jaguar eating the Moon),
and there is a well-known story that Genoese explorer
Christopher Columbus used his foreknowledge of one lunar
eclipse to intimidate hostile Jamaican islanders during his
fourth and final voyage to the Americas in 1504.[1]

Columbus' trick was only possible thanks to a practical
understanding of eclipses that had developed in Europe

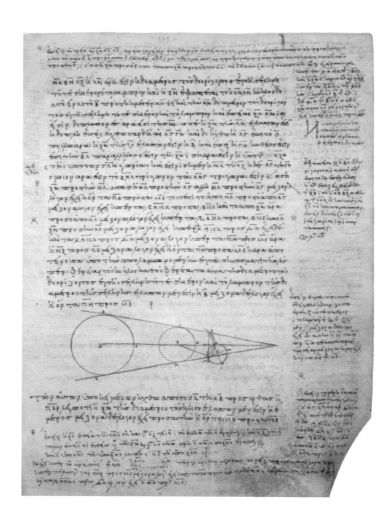

and elsewhere over the preceding 2,000 years, enabling surprisingly accurate eclipse predictions long before the true relationship between Sun, Moon and Earth in space was known. Yet the development of eclipse prediction was itself driven in part by superstition, since eclipses of both kinds were seen as events of great astrological significance.

The first systematic calendars of both solar and lunar eclipses began to be kept by the stargazers of Mesopotamia (the region between the Tigris and Euphrates rivers, mostly in present-day Iraq and Turkey) from around the mid-8th century BC. These and other systematic observations soon revealed patterns that allowed the true nature of eclipses to be recognized, such as the fact that lunar eclipses occur precisely at full Moon and solar ones at new Moon. Working using arithmetic alone (with no geometric models), the Mesopotamians also noted that eclipses of either type were only possible when the Moon's latitude relative to the ecliptic (the Sun's annual track around the sky) is close to zero.[2]

There is little evidence for Mesopotamian cosmology – a physical model of what was actually going on in the heavens – but Greek philosophers of classical times (and independently Chinese and Indian stargazers) certainly realized that both types of eclipse were caused by the shadow of one body blocking sunlight from reaching another. Eclipse calendars and calculations of the precise motion of Sun and Moon against the more distant stars also revealed a period known in modern times as the saros, which marks the time taken for Earth, Sun and Moon to return to the same relative points in space and thus predicts when eclipses of the same kind may repeat. This period of 223 lunar months is equivalent to 6,585.32 days (a little over 18 years), so eclipses in one saros happen 8 hours later than in the preceding one. Thus a 'triple saros' period of just over 54 years and 1 month is also significant, since it marks the cycle after which eclipses repeat at the same time of day.[3]

Because (by sheer coincidence) the Moon and Sun appear almost exactly the same size in Earth's sky, solar eclipses require a very precise alignment. The Moon casts its shadow only on a small band of Earth's surface, and even eclipses linked by the saros cycle are unlikely to be visible from the same location. Instead the timing of lunar eclipses (which, thanks to the geometry involved, are often visible across an entire hemisphere) would have played a key role in the discovery of eclipse patterns.

The earliest report of an actual eclipse prediction comes from the ancient Greek historian Herodotus (d. c. 420s BC), who (writing more than a century after the events in question) claimed that the philosopher Thales of Miletus had successfully foretold a solar eclipse that halted a battle between the armies of the Medes and Lydians in 585 BC.[4] Thales' prediction became well known throughout the ancient world but there is no evidence for precisely how he achieved it – knowledge of the various eclipse cycles could not have pinpointed the visibility of a solar eclipse at a particular location, and so if the story is true, luck may well have played a considerable part.

Throughout the classical period, philosophers put lunar eclipses to good use. The curve of Earth's shadow encroaching on the full Moon was seen as evidence for Earth's spherical shape. The 3rd-century BC astronomer Aristarchus of Samos used measurements of the shadow's relative size and speed in his calculations of the sizes of the Earth, Moon and Sun and their distances from each other.[5] Later, in the 2nd century AD, Ptolemy of Alexandria used the different times at which lunar eclipses were seen to begin at different locations as evidence for a spherical Earth with what we might today call 'time zones'.[6]

Ptolemy's influential *Almagest* secured the Earth-centred (geocentric) cosmology for more than a millennium, in large part because of the accuracy his complex system seemed to offer for constructing tables (or 'ephemerides') for the future motion of celestial bodies. Lunar eclipses thus remained one of the key types of astronomical event that could be predicted with at least some accuracy – as the tale of Columbus in Jamaica demonstrates.

However, a lingering lack of precision eventually played a significant role in undermining the Ptolemaic system. Polish canon Nicolaus Copernicus (1473–1543), whose Sun-centred cosmology ultimately displaced the older model, is said to have taken inspiration from his observations of a lunar eclipse at Rome in the year 1500, and his 1543 book *On the Revolutions of the Heavenly Spheres* devotes significant space to an explanation of lunar motions and the nature, duration and prediction of both solar and lunar eclipses.[7]

While Copernicus' model was widely adopted as a tool for making more accurate predictions, it was still burdened with complexities thanks to its adherence to the idea that orbits must be strictly circular in nature. Not until the early 17th century did Johannes Kepler's description of elliptical orbits allow the true nature of eclipses to be properly modelled. And even then, it was only in the 18th century that improved 'lunar theories' were able properly to take account of minor influences on lunar motion due to the varying gravitational pulls of Earth and Sun, allowing truly accurate predictions of both lunar and solar eclipses.

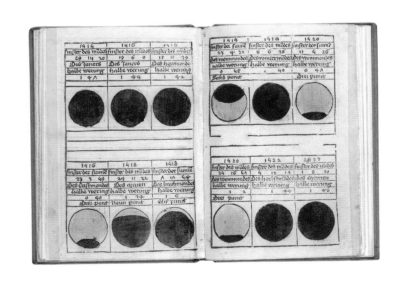

Folios from Cyprian von Leowitz's *Eclipses luminarium* (Book of Eclipses) (1555). Born in modern-day Czechia as Cyprián Lvovický, von Leowitz (c. 1514–1574) was an influential astrologer who spent much of his career working in Germany, attempting to construct a sound mathematical and astronomical basis for astrological prognostications at a time when the boundaries between mysticism and science remained fluid. His spectacular *Book of Eclipses*, listing the timing and visibility of eclipses from various locations up until 1605, contains plates illustrating the relative positions of Sun and Moon during key eclipses, alongside colourful landscape scenes showing the predicted lighting conditions during each event. Von Leowitz's work drew him to the attention of the Holy Roman Emperor Maximilian II, who sponsored his 1564 publication of a detailed series of ephemeris tables for the motions of the Sun, Moon, planets and stars.

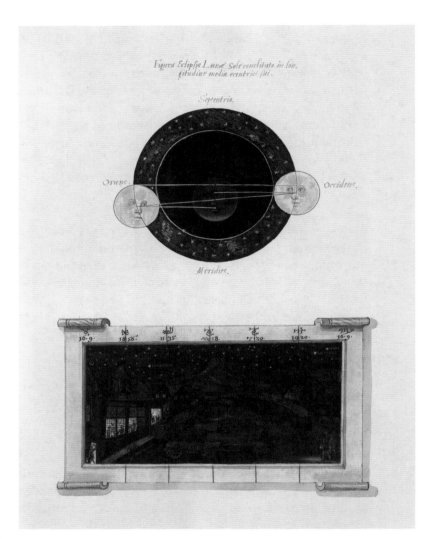

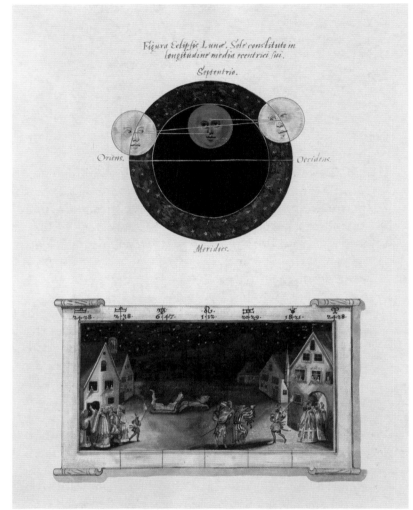

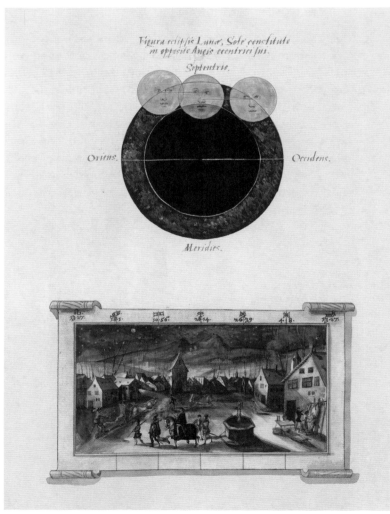

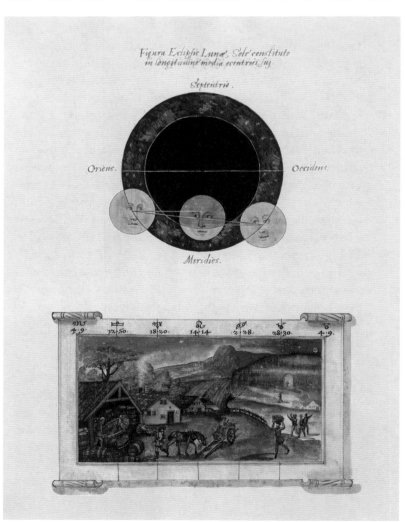

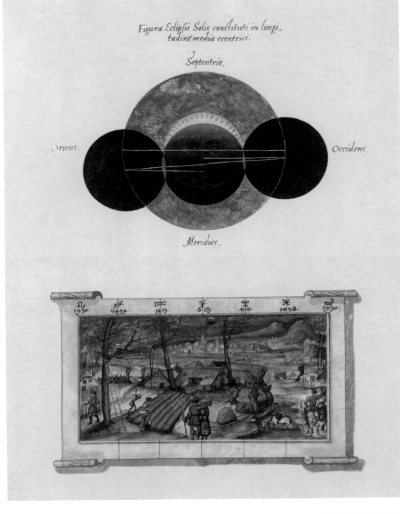

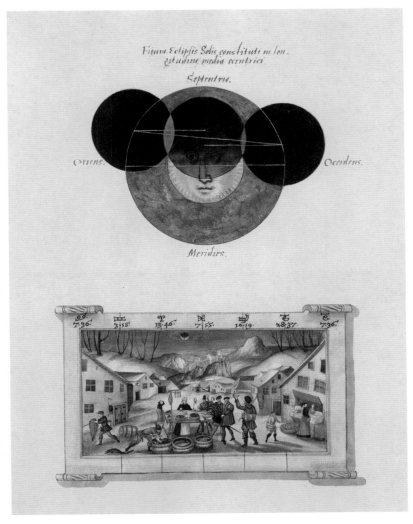

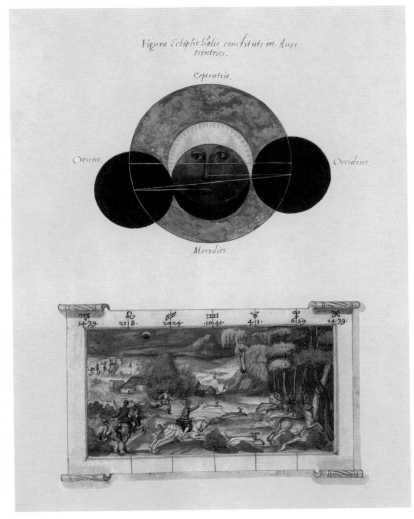

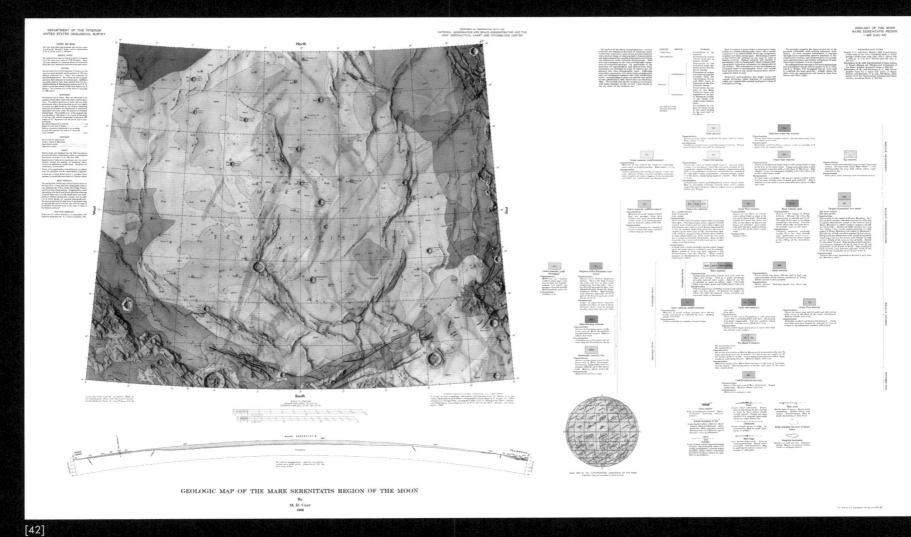

GEOLOGIC MAP OF THE MARE SERENITATIS REGION OF THE MOON
By
M. H. Carr
1966

[42]

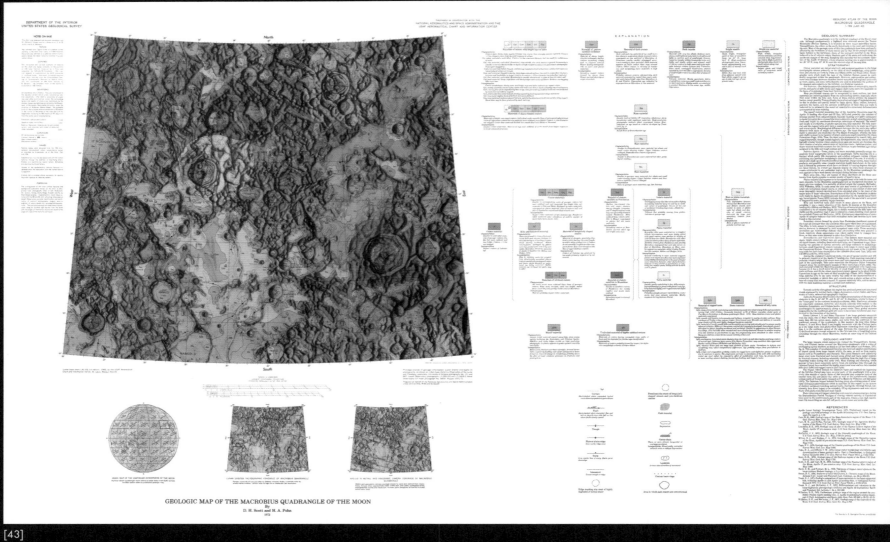

GEOLOGIC MAP OF THE MACROBIUS QUADRANGLE OF THE MOON
By
D. H. Scott and H. A. Pohn
1972

[43]

42
Facsimile:
Geologic Map of the Mare Serenitatis Quadrangle of the Moon, 1966, by M. H. Carr

43
Facsimile:
Geologic Map of the Macrobius Quadrangle of the Moon, 1972, by D. H. Scott and H. A. Pohn

44
Facsimile:
Geologic Map of the Cleomedes Quadrangle of the Moon, 1972, by Clarence J. Casella and Alan B. Binder

56
Facsimile:
Geologic Map of the Hevelius Quadrangle of the Moon, 1967, by John F. McCauley

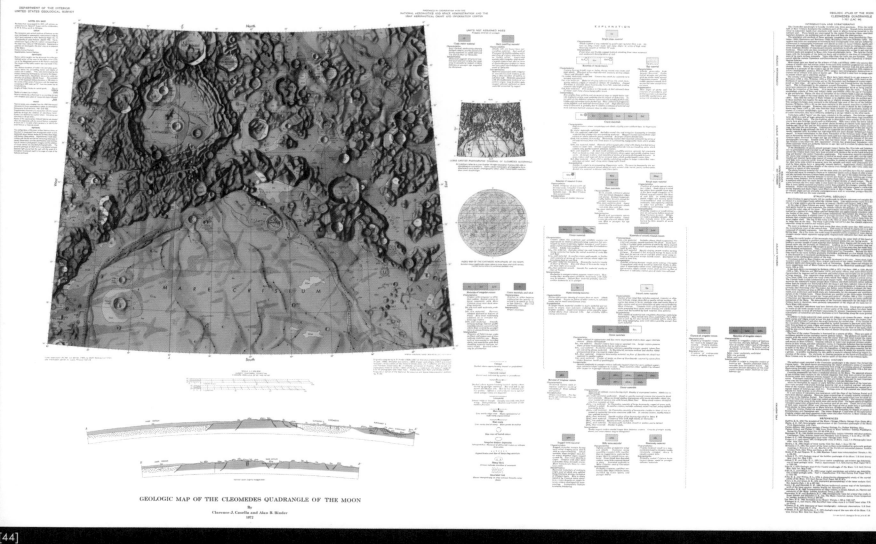

GEOLOGIC MAP OF THE CLEOMEDES QUADRANGLE OF THE MOON
By
Clarence J. Casella and Alan B. Binder
1972

[44]

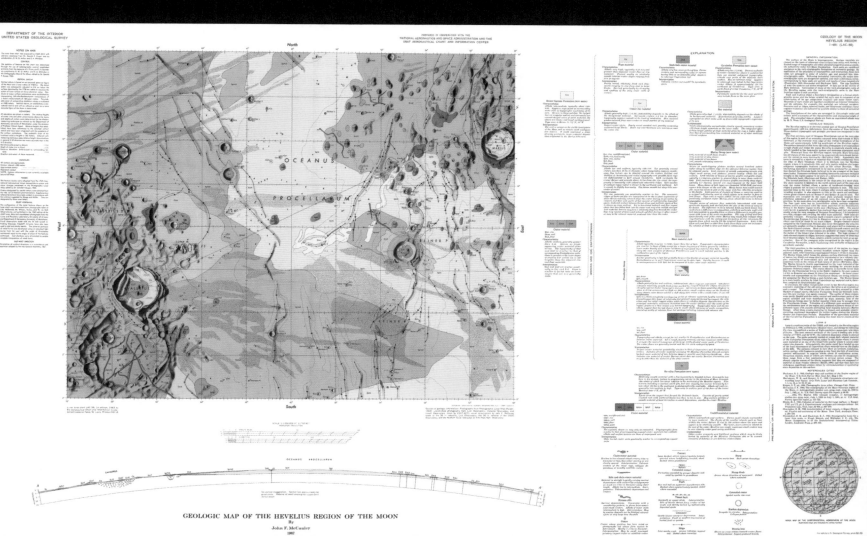

GEOLOGIC MAP OF THE HEVELIUS REGION OF THE MOON
By
John F. McCauley
1967

[56]

THE MOON AND THE HELIOCENTRIC WORLDVIEW

↱
In 1543, Nicolaus Copernicus published *De revolutionibus orbium coalestium* (On the Revolutions of the Heavenly Spheres). In it, Copernicus presented a mathematical argument for a Sun-centred universe. His heavenly spheres are perfectly circular, as were Ptolemy's, but transformed Earth into a planet with motion.

↳
This 17th-century depiction by Andreas Cellarius of the heliocentric universe reproduced in a celestial atlas incorporates new knowledge about the Earth and its continents, as well as old ideas about the astrological significance of the constellations.

↳↳
Tycho Brahe was impressed by Copernicus' mathematics, but he refused to believe that the Earth – a large and heavy sphere – moved. Brahe produced his own system, in which the planets revolved around the Sun while the Sun and Moon revolved around a stationary Earth.

entral to popular understanding of the Scientific Revolution is the overthrow of the ancient geocentric (Earth-centred) view by the new heliocentric (Sun-centred) model by a small cohort of upstart Renaissance European astronomers and mathematicians. The Polish astronomer Nicolaus Copernicus (1473–1543) first proposed moving the Sun to the centre of the cosmos in his 1543 book, *De revolutionibus orbium coelestium* (On the Revolutions of the Heavenly Spheres). He suggested that this rearrangement would simplify the cumbersome mathematics of the system associated with the 2nd-century AD Roman astronomer Ptolemy of Alexandria. Next, the Italian astronomer Galileo Galilei pointed his telescope at the night sky and collected the first observational evidence that the Moon and planets were other worlds, publishing his discoveries in his 1610 *Sidereus nuncius* (Starry Messenger). Meanwhile, Galileo's German contemporary Johannes Kepler transformed astronomy with his 1609 book, *Astronomia nova* (New Astronomy), which introduced mathematical laws describing elliptical planetary orbits.

There is more to the story: more people, more competition, and more time. Copernicus employed mathematical tools developed by 13th-century Persian and Syrian astronomers Nasir al-Din al-Tusi (1201–1274) and Mu'ayyad al-Din al-Urdi (d. 1266).[1] By the time Galileo and Kepler made their contributions, the geocentric system they faced was not Ptolemy's but the hybrid system developed by Tycho Brahe (1546–1601).[2] Brahe had followed Copernicus' lead in placing the other planets in orbit around the Sun, but kept the Earth at the centre of the orbits of the Moon, Sun

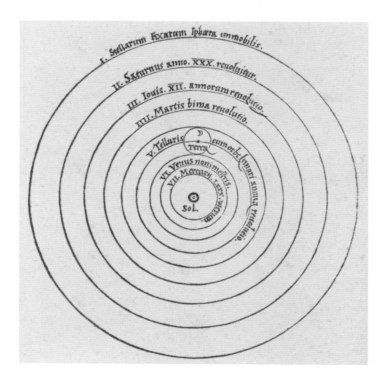

and stars. Copernicus, Galileo and Kepler did not move the Earth overnight. Convincing observational evidence remained to suggest an unmoving Earth – evidence Giovanni Battista Riccioli (1598–1671) considered when he wrote his 1651 *Almagestum novum* (New Almagest). Riccioli's book included the most detailed maps of the Moon yet produced, along with a measured scientific argument *against* the motion of the Earth.

Galileo and Kepler's contributions to the heliocentric worldview were eventually celebrated as models of revolutionary science, even if they at first appealed primarily to a small circle of Copernican astronomers. More immediately felt was their enthusiasm for the Moon and the new knowledge it might yield – knowledge they were sure would bolster the Copernican cause.

When Galileo (1564–1642) wrote his *Sidereus nuncius*, he began with a discussion of the Moon. What he had seen through the telescope was revelatory: the Moon's surface was 'full of inequalities, uneven, full of hollows and protuberances, just like the surface of the Earth itself'.[3] As he looked at the Moon from one night to the next, Galileo had witnessed changes in the terminator (the line separating lunar day and night). He had seen shadows lengthen and shorten along the terminator as it passed over those areas of the Moon that appear brightest when illuminated. He argued that these bright areas were covered in tall mountains and deep valleys, making the surface of these regions of the Moon even more rugged than Earth. 'It is clear', he wrote, 'that the prominences of the Moon are loftier than those of the Earth.'[4] The darker areas of the Moon were smooth by comparison. Galileo proposed that these smooth, dark regions were possibly seas surrounded by mountains. We know today that he overestimated the size of the lunar mountains and mistook the nature of the maria.

Later, in Galileo's *Dialogo sopra i due massimi sistemi del mondo* (Dialogue Concerning the Two Chief World Systems), the fictional character Salviati revisited the astronomer's lunar observations as he made his case for the supremacy of the Copernican system. Galileo's discovery of mountains on the Moon was a reminder that there were many things unknown to the old astronomy – literally new territory to be mapped. The character Simplicio, who argued on behalf of the geocentric system, refused even to believe that the mountains on the Moon cast shadows. Meanwhile, Salviati suggested that some of the dark regions of the Moon could be forests, that the mountains could be covered in snow, and went so far as to admit that he sometimes imagined beings that might live on the Moon.

Johannes Kepler (1571–1630) had poor eyesight due to a childhood bout with smallpox. He found his calling in the world of mathematics and geometry, rather than nightly observations of the Moon and stars. He shared Galileo's passion for making astronomy more empirical. While a student, Kepler became convinced that an observer on the Moon would be able to witness the Earth's motion and could confirm the validity of the Copernican system. He struggled to write this into a proper thesis, and eventually produced a heavily footnoted novel he titled *Somnium* (The Dream), published after his death.[5]

In his *Somnium*, Kepler dreams that he is reading a true account of the adventures of an Icelandic boy named Duracotus. Like Kepler, Duracotus studies astronomy under the tutelage of the Danish astronomer Tycho Brahe. When he returns to Iceland, Duracotus reunites with his mother, a witch, who introduces him to a demon who is able to move between the Earth and Moon. Tellingly, the demon knows Earth as 'Volva', so named because the Moon's inhabitants see it steadily turning in their sky. The reader is treated to speculative descriptions of the Moon as a place of plant and animal life, and to a description of the changing face of the rotating Earth and the features visible from the Moon.

Duracotus also finds confirmation of the Copernican world system by combining what the observer on the Moon would see of the motions of the Earth and planets with what the observer on Earth would see of the motions of the Moon and planets – using two vantage points, rather than one. The demon even suggests that they can transport human devotees to the Moon to see these things for themselves – an allegory for the power promised by Galileo and Kepler's new astronomy.[6] Kepler awakens before any such trip is made.

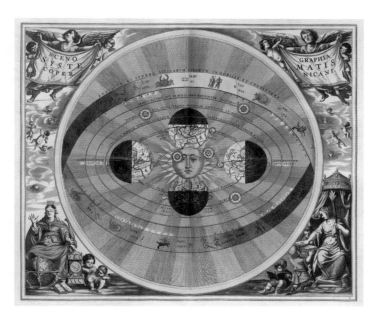

[LAC-42]

I-489

Mare Serenitatis Quadrangle

Maps

Lunar

094

GEOLOGIC ATLAS OF THE MOON

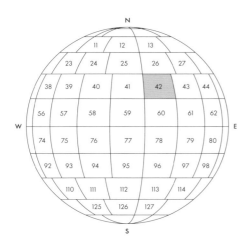

This quadrangle is characterized almost entirely by the dark lava plains of Mare Serenitatis (Sea of Serenity), a distinctly round lunar mare that, historically, formed one of the eyes of the 'Man in the Moon'. The plains receive some noticeable texture from ridges, scarps and small simple craters that dot the otherwise smooth surface. The mare is bordered in the southwest by the Montes Haemus mountain range and the nearby bowl-shaped crater Sulpicius Gallus. This crater has a relatively bright appearance, a sharp-edged rim and a small central rise in its floor. Another relatively young crater, Menelaus, appears at the eastern end of Montes Haemus. It has a sharp rim and terraced walls, and a ray system that extends across the mare. To the north of Menelaus, in Rima Menelaus, is a system of faults. On the eastern edge of this quadrangle, mountains, ridges and fault systems form a boundary between Mare Serenitatis and Mare Tranquillitatis (Sea of Tranquillity).

GEOLOGIC CROSS SECTION OF THE MARE SERENITATIS QUADRANGLE
APPROXIMATE HORIZONTAL SCALE 1:1 000 000 (NO VERTICAL EXAGGERATION)

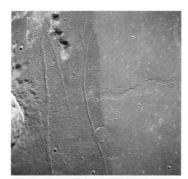

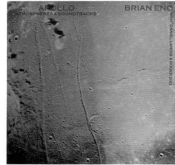

APOLLO PHOTO ART
Left: Photograph of Mare Serenitatis taken during the Apollo 17 mission on 12 December 1972 (above). The image formed the basis of the cover art for Brian Eno's 1983 album: *Apollo: Atmospheres and Soundtracks* (below).

THE MAN IN THE MOON
Above: Photograph of the full Moon taken by Luc Viatour on 7 October 2006 (left). In 2012 the photograph was edited by Peter Donohue who imagined a face in the Moon (right).

MARE SERENITATIS QUADRANGLE: KEY CHARACTERISTICS

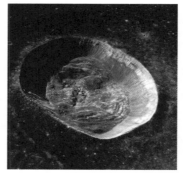

1. MARE SERENITATIS
Latitude: 28°N / Longitude: 17.5°E
Diameter: 674 km / 418 ¾ miles
Photograph: Apollo 17 (USA, 1972)

2. BESSEL CRATER
Latitude: 21.8°N / Longitude: 17.9°E
Diameter: 16 km / 10 miles
Depth: 1.7 km / 1 mile
Photograph: Apollo 15 (USA, 1971)

3. DAWES CRATER
Latitude: 17.2°N / Longitude: 26.4°E
Diameter: 18 km / 11 ¼ miles
Depth: 2.3 km / 1 ½ miles
Photograph: Apollo 17 (USA, 1972)

4. RIMA LITTROW
Latitude: 22.1°N / Longitude: 29.9°E
Length: 115 km / 71 ½ miles
Photograph: Lunar Orbiter IV (USA, 1967)

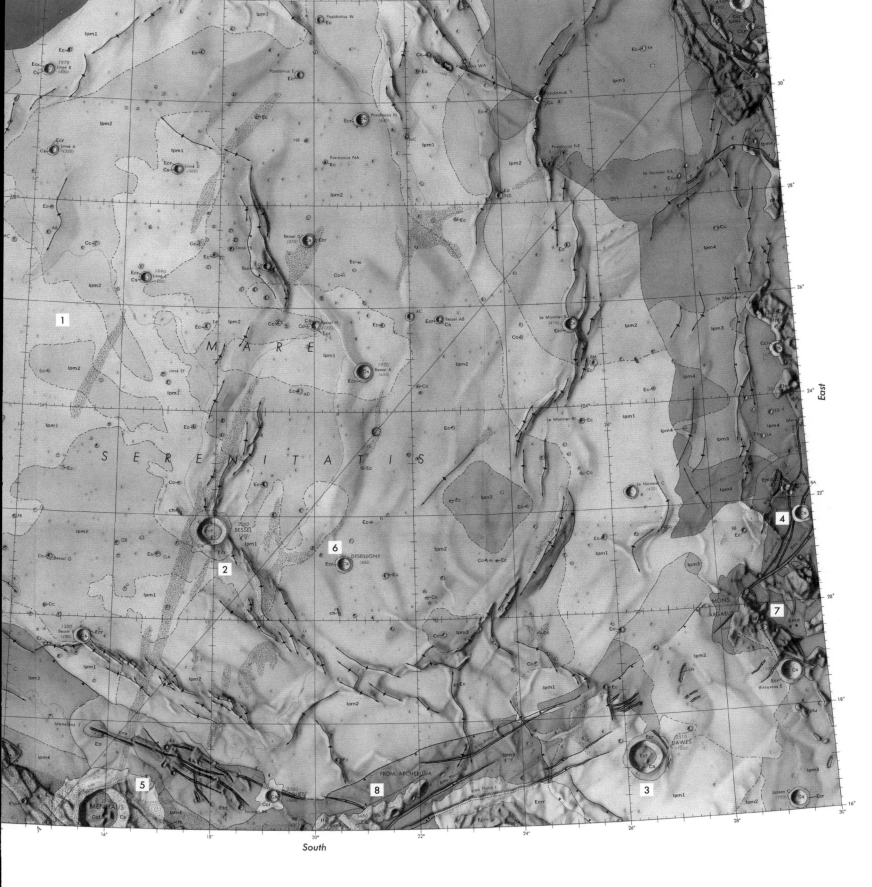

5. MENELAUS CRATER
Latitude: 16.3°N / Longitude: 16°E
Diameter: 27 km / 16 ¾ miles
Depth: 3 km / 1 ¾ miles
Photographed from Earth, 2019

6. DESEILLIGNY CRATER
Latitude: 21.1°N / Longitude: 20.6°E
Diameter: 6.6 km / 4 miles
Depth: 1,200 m / 3,937 ft
Photograph: Apollo 15 (USA, 1971)

7. MONS ARGAEUS
Latitude: 19.33°N / Longitude: 29.01°E
Diameter: 65 km / 40 ½ miles
Height: 2.56 km / 1 ½ miles
Photograph: Apollo 17 (USA, 1972)

8. PROMONTORIUM ARCHERUSIA
Latitude: 16.8°N / Longitude: 21.9°E
Diameter: 10 km / 6 ¼ miles
Height: 1,500 m / 4,921 ft 3 in
Photograph: LRO (USA, 2010)

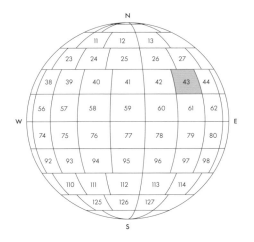

This quadrangle is made up primarily of highland terra centred around the Montes Taurus mountains. Three maria – Tranquillitatis to the south, Serenitatis to the west and Crisium on the east – mark its boundaries. The early history of the terra in this region was shaped by the formation of the basins associated with these maria. The impacts that formed the basins covered this quadrangle in a series of overlapping ejecta blankets. Large nearby craters also contributed ejecta material. Gravity over time rounded the terrain into hills. Volcanic activity filled some depressions and crater floors with smooth plains material. The prominent crater Macrobius, after which this quadrangle is named, is relatively well preserved, with some slumping along its rim, but dates from some time in the middle of the Moon's geologic history. Smaller, younger craters, such as Römer, still display visible ray systems and can be dated to the Moon's latest geologic period, the Copernican. On 10 December 1972, the Apollo 17 Lunar Module landed in the Taurus Littrow highlands, at coordinates 20.16°N, 30.77°E, south of Littrow crater. The crew collected rocks and soil from these highlands. The next year, the Soviet Union's robotic Luna 21 lander set down in Le Monnier crater, carrying the second successful lunar rover, Lunokhod 2.

APOLLO 17	ROCKET	Saturn V AS-512
	CREW	E. A. Cernan, H. H. Shmitt, R. E. Evans
	LAUNCH DATE	05:33:00, 07 December 1972 (UTC)
	LAUNCH SITE	Cape Kennedy Air Force Station
	PAYLOAD	America (CM-114), Challenger (LM-12)
	LUNAR ORBIT INSERTION	10 December 1972
	LANDING DATE	19:54:57, 11 December 1972 (UTC)
	LANDING SITE	Taurus Littrow Highlands
	LANDING COORDINATES	20.16°N, 30.77°E (A)

LUNA 21	ROCKET	Proton K/D
	CREW	N/A
	LAUNCH DATE	06:55:00, 08 January 1973 (UTC)
	LAUNCH SITE	Tyuratam (Baikonur Cosmodrome)
	PAYLOAD	Lunokhod 2 Rover, Imaging Systems
	LUNAR ORBIT INSERTION	12 January 1973
	LANDING DATE	22:35:00, 15 January 1973 (UTC)
	LANDING SITE	Le Monnier Crater
	LANDING COORDINATES	25.99°N, 30.41°E (B)

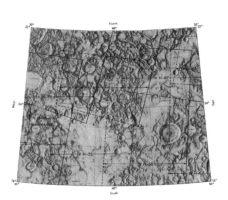

LUNAR ORBITER PHOTOGRAPHIC COVERAGE OF MACROBIUS QUADRANGLE

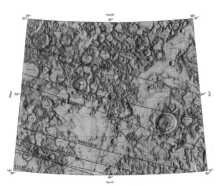

APOLLO 15 METRIC AND PANORAMIC CAMERA COVERAGE OF MACROBIUS QUADRANGLE

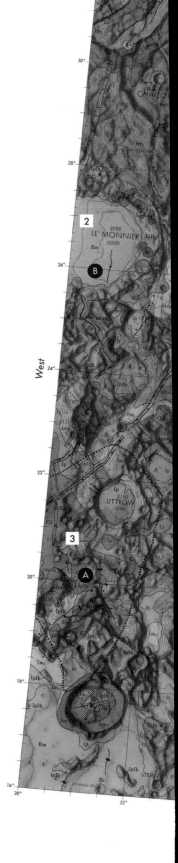

APOLLO 17 LUNAR LANDING: 11 DECEMBER 1972

OPENING THE SOLAR PANELS
Jack Schmitt reaches down to open the solar panels on an electrical transmitter during EVA 1. Photograph taken by Gene Cernan on 11 December 1972.

MAKESHIFT FENDER
A damaged fender on a rover is patched up with plasticized maps taped together during EVA 2. Photograph taken by Gene Cernan on 12 December 1972.

TRACY'S ROCK
Jack Schmitt retrieves a gnomon and heads back to the rover around the south side of Tracy's Rock during EVA 3. Photograph taken by Gene Cernan on 13 December

STATION 6
Jack Schmitt works close to a rover, parked at Station 6, during EVA 3. He holds onto a rock while he takes photographs. Taken by Gene Cernan on 13 December 1972.

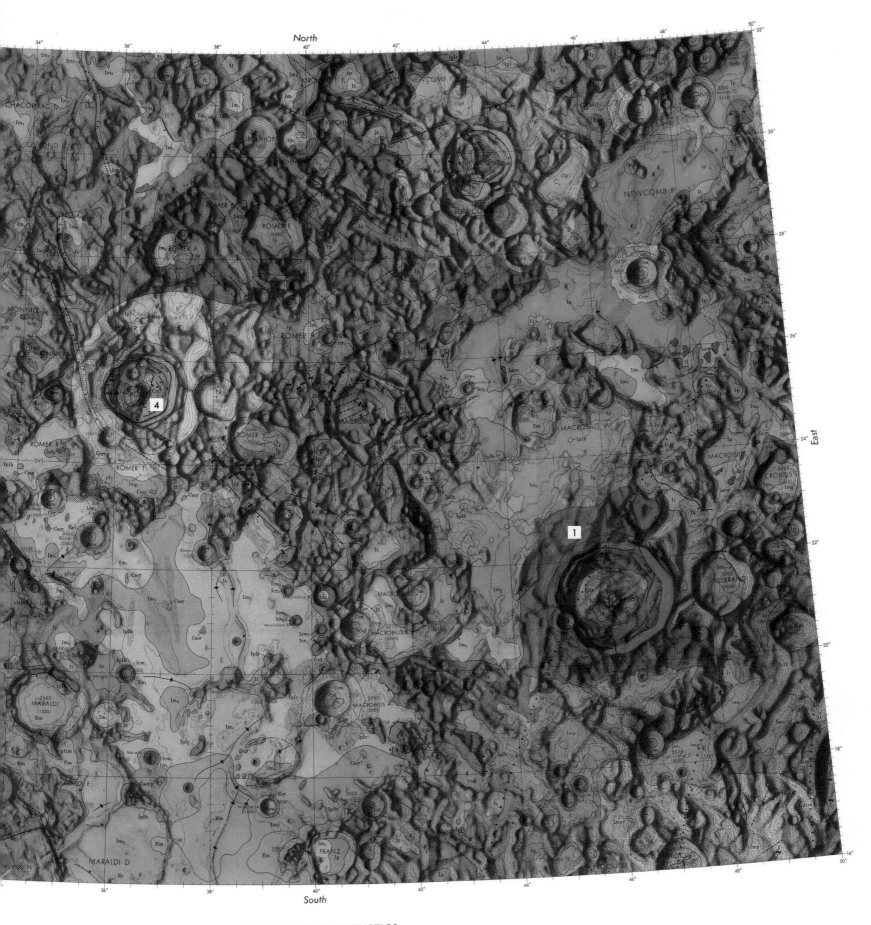

MACROBIUS QUADRANGLE: KEY CHARACTERISTICS

1. MACRBOBIUS CRATER
Latitude: 21.3°N / Longitude: 46°E
Diameter: 63 km / 39 ¼ miles
Depth: 3.9 km / 2 ½ miles
Photograph: Apollo 17 (USA, 1972)

2. LE MONNIER CRATER
Latitude: 26.6°N / Longitude: 30.6°E
Diameter: 61 km / 38 miles
Depth: 2.4 km / 1 ½ miles
Photograph: Lunar Orbiter IV (USA, 1967)

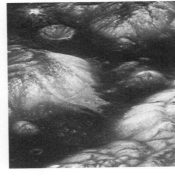

3. TAURUS LITTROW VALLEY
Latitude: 20.07°N / Longitude: 30.79°E
Diameter: 21.03 km / 31 miles
Photograph: LRO (USA, 2018)

4. RÖMER CRATER
Latitude: 25.4°N / Longitude: 36.4°E
Diameter: 43.7 km / 27 ¼ miles
Depth: 3.3 km / 2 miles
Photograph: Apollo 17 (USA, 1972)

EARLY TELESCOPIC STUDIES OF THE MOON

↓
Commissioned by the Holy Roman Emperor Rudolf II, the frontispiece of Johannes Kepler's *Rudolphine Tables* (1627) depicts the giants of pre-telescopic astronomy, including Hipparchus, Ptolemy, Nicolaus Copernicus and Kepler's own mentor, Tycho Brahe.

↱
Three depictions of the Moon by Claude Mellan (1637). After being invited by astronomers Nicholas Claude Fabri de Peiresc and Pierre Gassendi to study the Moon through a telescope in 1635, the artist and engraver produced what are widely regarded as the earliest detailed and realistic depictions of the lunar surface.

↳
Originally published in *Almagestum novum* (1651) Francesco Maria Grimaldi and Giovanni Battista Riccioli's Moon map was revised and published as a stand-alone print in 1665. Riccioli provided names for some 240 features (some borrowed from Van Langren), introducing the famous toponym Mare Imbrium (Sea of Showers).

he invention of the telescope in the early 17th century changed forever the way that people understood the Moon. The discovery that a pair of glass lenses could be arranged in such a way as to produce an enlarged image of a distant object is usually attributed to Dutch spectacle-maker Hans Lippershey (c. 1570–1619), who unsuccessfully applied to patent such a device in October 1608. The precise details of its invention, however, are lost to history.

Word of the new device spread rapidly among Europe's natural philosophers, who rushed to buy or build their own. In the early summer of 1609, the renowned Italian physicist Galileo Galilei built the first in a series of such instruments, improving them with each iteration through trial and error and boosting their power from a mere three times magnification to an impressive twenty times by November. When he turned this instrument to the heavens, the Moon was, unsurprisingly, one of his first targets.

Galileo immediately discovered previously unknown features. In addition to the dark patches that could be seen with the naked eye, he spotted smaller ones whose appearance changed with the angle of illumination from the Sun. He concluded that these features were evidence of mountains, valleys and circular pits on the lunar landscape. The Moon, in other words, was a rugged world of varying terrain – more akin to Earth itself than to the perfect pearlescent sphere described by the Aristotelian teachings that had dominated European philosophical thought for almost two thousand years.

Galileo was not the first to sketch the Moon through a telescope. The English astronomer Thomas Harriot (1560–1621) made private sketches as early as July 1609. But Galileo was the first to bring the Moon's features to public attention in his 1610 work *Sidereus nuncius* (Starry Messenger).[1] In the following years, many other astronomers took up their telescopes to study the Moon. Initially, their observations focused on the changing appearance of individual lunar regions or features, ultimately confirming the Moon's Earthlike topography.

It was not until the 1630s that attempts were made to compile detailed maps of the entire lunar surface. This effort was driven by commercial ambitions as well as the quest for scientific understanding; with European powers expanding their voyages of trade and exploration around the globe, many astronomers hoped that if the Moon's surface were mapped with sufficient accuracy, its changing appearance could be used as a sort of celestial clock. In the absence

of reliable seagoing timepieces, repeating astronomical cycles such as the phases of the Moon were seen as the only hope for finding the precise standard of time needed by navigators to establish the longitude (east–west position) of ships at sea.

The first attempt to draw up an accurate map remains one of the finest – a triptych of lunar views engraved in 1637 by the artist Claude Mellan (1598–1688) after studying the Moon through the telescope of astronomers Pierre Gassendi (1592–1655) and Nicholas Claude Fabri de Peiresc (1580–1637). The illustrations, depicting the Moon in its first-quarter, full and last-quarter phases, display a stunning level of accuracy that would not be bettered for centuries. Even more remarkably, Mellan created his features without crude outlines, using instead variations in tone created by meticulously controlling the width of horizontal lines.

Despite its beauty, Mellan's map did not become widely used. In part this was because Gassendi abandoned the vital project of naming its features following Peiresc's death later in 1637, but also the map was perhaps too accurate for its own good. The harsh shadows shown along the terminator line of the first and last quarters, and the washed-out, shadowless landscape of the full Moon made it hard to envisage the features other lunar observers might find interesting – let alone those a shipboard navigator might wish to pick out.

More practical lunar charts soon followed, however – the first from the hand of Dutch cartographer Michael van Langren (1598–1675), published in 1645.[2] Van Langren's single circular map was the first to provide a system of nomenclature for lunar features (echoing a comment of Galileo's by naming the dark grey lunar plains as maria, or 'seas') but it too failed to have much lasting influence. Van Langren named many features after Catholic monarchs and saints, and this limited his scheme's appeal in Protestant parts of Europe. The map also lacked detail, and was soon superseded.

In 1647, Polish astronomer Johannes Hevelius (1611–1687) published *Selenographia*, the first work entirely dedicated to the astronomy of the Moon.[3] Hevelius used a technique similar to Mellan's; his illustrations depicted not only the entire lunar disc, but also its appearance at different phases, and even the extreme eastern and western edges that are only occasionally visible owing to the effects of libration. He named areas of the surface after classical attributes or (somewhat confusingly) as mirrors of Earth's own geographical features.

Then in 1651 Jesuit astronomer Giovanni Battista Riccioli's influential *Almagestum novum* (New Almagest) included a map of the entire lunar disc by his colleague Francesco Maria Grimaldi (1618–1663).[4] Drawing on a mixture of direct observations and earlier sources, Grimaldi's map used both shading and outlining, sacrificing strict representation for ease of interpretation. Riccioli introduced a naming system that finally caught on, with the Moon's plains described as 'seas' with various attributes, and craters named after famous astronomers and philosophers.

Riccioli and Hevelius thus became the standard authorities on lunar astronomy throughout the 17th and well into the 18th century. The dream of using the Moon's changing appearance as a timer for longitude calculations, however, soon foundered due to the limitations of telescope technology and the availability of easier astronomical (and ultimately mechanical) clocks. Nevertheless, the Moon remained a beguiling target for astronomers, occasionally leading to the creation of stunning depictions such as those of the Nuremberg astronomer Maria Clara Eimmart (1676–1707) in the early 18th century. Even so, astronomical cartography would not take a major leap forward until new advances in telescope technology took place in the mid-18th century, and a standard coordinate system for the positions of lunar features was introduced.[5] This system was used by Tobias Mayer (1723–1762) on maps that were drawn around 1750 but not published until 1775.[6]

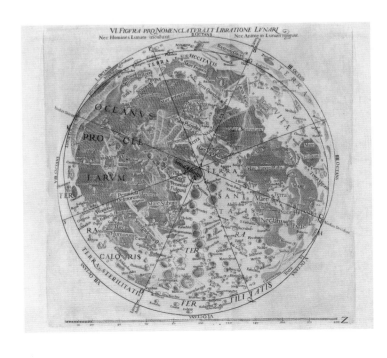

←
A fanciful frontispiece to Galileo's *Collected Works*, first published in 1656, depicts the Italian astronomer demonstrating his primitive telescope to three muses while pointing to various proofs of the heliocentric system displayed in the sky.

↵
A depiction of Johannes Hevelius' 46-metre (150-ft) telescope from his *Machina coelestis* (1673). Before 18th-century optical advances allowed the creation of lenses that bent all colours of light equally, instruments using weakly curved lenses with very long focal lengths were the only way to obtain clear, highly magnified views of the heavens.

→
Eight images reveal various ways of depicting the lunar surface throughout the first century of telescopic observations.
1: An unpublished sketch of the Moon by Thomas Harriot (1609). By around 1611 he had compiled a complete map of the Moon (2) with annotated features.
3/4: Galileo's lunar sketches in *Sidereus nuncius* (1610).
5: Claude Mellan's illustration of the full Moon (1637).
6: Michael van Langren's 1645 lunar map adds names to features for the first time.
7: One of several lunar maps in Hevelius' *Selenographia* (1647) highlights the bright 'ray' features now known to be made of material ejected during the formation of impact craters.
8: The unique Moon map found in John Seller's *Atlas terrestris* (1700) shows how long it took for a common understanding of the Moon's major features – not to mention an agreed nomenclature – to become established.

[LAC-44]

I-707

Cleomedes Quadrangle

I-707

Maps

Lunar

GEOLOGIC ATLAS OF THE MOON

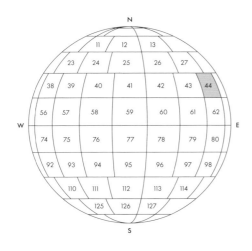

This quadrangle holds the smooth terrain of the northern half of Mare Crisium (Sea of Crises) and a large portion of Mare Anguis (Serpent Sea). Rugged terrain of cratered terra includes the crater Cleomedes, one of the largest of the Moon's surviving ancient craters. Geologists inferred that the earliest event visible in this quadrangle was the impact that formed the Crisium basin and produced a series of concentric and radial ridges and troughs. Ejecta from this impact may have buried pre-existing craters in this region. The Crisium basin was later partially filled with lava. Impact cratering then became the dominant force in shaping this region. There are some young craters visible in this quadrangle, such as two of the smaller craters inside of the Cleomedes crater, both of which are simple bowl-shaped craters with bright ejecta rays. On 21 July 1969, the Soviet Union's robotic Luna 15 crashed into this quadrangle near coordinates 17°N, 60°E.

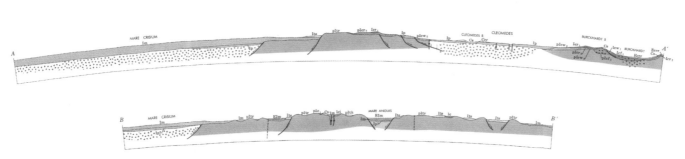

GEOLOGIC CROSS SECTIONS OF THE CLEOMEDES QUADRANGLE
APPROXIMATE HORIZONTAL SCALE 1:1 000 000 (NO VERTICAL EXAGGERATION)

LUNA 15	ROCKET	Proton K/D
	CREW	N/A
	LAUNCH DATE	02:54:42, 13 July 1969 (UTC)
	LAUNCH SITE	Tyuratam (Baikonur Cosmodrome)
	PAYLOAD	Remote Arm for Sample Collection
	LUNAR ORBIT INSERTION	17 July 1969
	LANDING DATE	15:50:40, 21 July 1969 (UTC)
	IMPACT SITE	Mare Crisium
	LANDING COORDINATES	17.00°N, 60.00°E Ⓐ

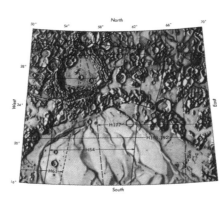

LUNAR ORBITER PHOTOGRAPHIC COVERAGE OF CLEOMEDES QUADRANGLE

CLEOMEDES QUADRANGLE: KEY CHARACTERISTICS

1. CLEOMEDES CRATER
Latitude: 27.60°N / Longitude: 55.5°E
Diameter: 130.77 km / 81 ¼ miles
Depth: 2.7 km / 1 ¾ miles
Photograph: LRO (USA, 2017)

2. DELMOTTE CRATER
Latitude: 27.16°N / Longitude: 60.2°E
Diameter: 32.16 km / 20 miles
Depth: 3 km / 1 ¾ miles
Photograph: LRO (USA, 2015)

3. MARE CRISIUM
Latitude: 16.8°N / Longitude: 59.1°E
Diameter: 556 km / 345 ½ miles
Photograph: LRO (USA, 2015)

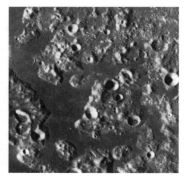

4. MARE ANGUIS
Latitude: 22.43°N / Longitude: 67.58°E
Diameter: 146 km / 90 ¾ miles
Photograph: LRO (USA, 2020)

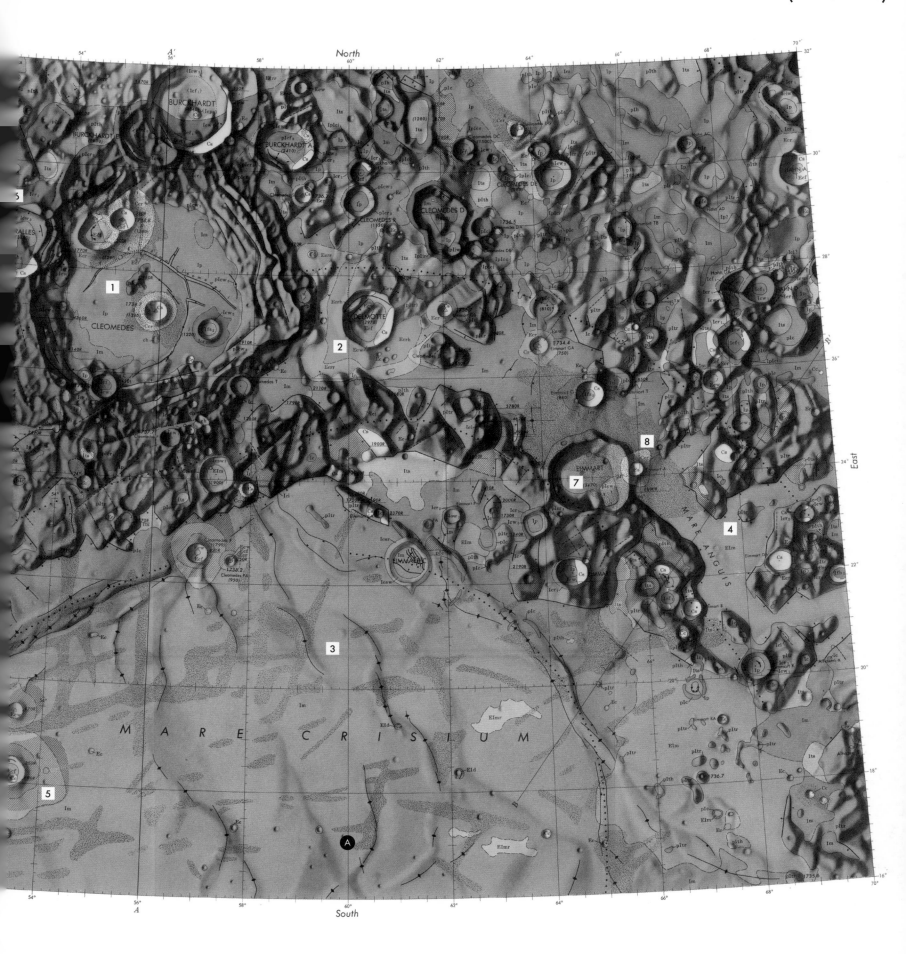

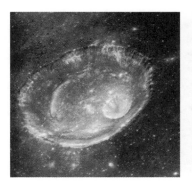

5. PEIRCE CRATER
Latitude: 18.26°N / Longitude: 53.35°E
Diameter: 18.86 km / 11 ¾ miles
Depth: 1.8 km / 1 mile
Photograph: Apollo 17 (USA, 1972)

6. TRALLES CRATER
Latitude: 28.32°N / Longitude: 52.85°E
Diameter: 44.16 km / 27 ½ miles
Depth: 3.4 km / 2 miles
Photograph: Lunar Orbiter IV (USA, 1967)

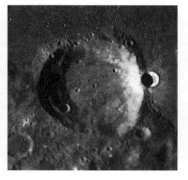

7. EIMMART CRATER
Latitude: 23.97°N / Longitude: 64.8°E
Diameter: 45 km / 28 miles
Depth: 3.2 km / 2 miles
Photograph: LRO (USA, 2017)

8. EIMMART A CRATER
Latitude: 24.11°N / Longitude: 65.66°E
Diameter: 7.34 km / 4 ½ miles
Photograph: LRO (USA, 2010)

GEOLOGIC ATLAS OF THE MOON

This quadrangle is named for the ancient crater Hevelius located in its southwest corner. The crater is filled with basaltic lava and is shown here to be crisscrossed by a system of rilles. Overlaying a portion of Hevelius' northern rim is the younger crater Cavalerius. Cavalerius has a high rim and a low central peak, and its walls are partially terraced. Just to the northeast of Cavalerius, marked with the words 'LUNA 9', is the landing site of the USSR's Luna 9 spacecraft, the first spacecraft to soft-land successfully on the Moon. The probe's landing coordinates, 8°N, 64°W, were later named Planitia Descensus (Plain of Descent) in its honour. These features form the western edge of the Oceanus Procellarum (Ocean of Storms) lava plains. To the northeast is a complex system of lava flows and domes known as the Marius hills, named after the nearby lava-flooded Marius crater. Also prominent in this quadrangle is the well-preserved impact crater Reiner, which probably dates to some time in the middle of the Moon's geologic history (evidenced by the absence of a visible ray system). This quadrangle also saw the crash of the unsuccessful Luna 8 lander.

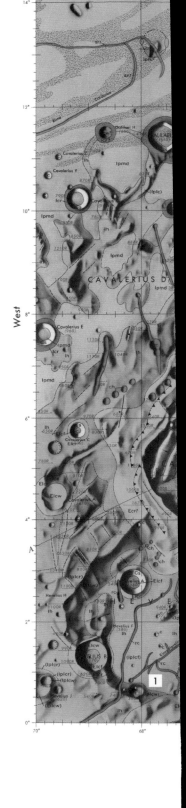

LUNA 8		
ROCKET	Molniya-M	
CREW	N/A	
LAUNCH DATE	10:46:14, 03 December 1965 (UTC)	
LAUNCH SITE	Tyuratam (Baikonur Cosmodrome)	
PAYLOAD	Imaging System, SBM-10 Radiation Detector	
LUNAR ORBIT INSERTION	N/A	
LANDING DATE	21:51:30, 06 December 1965 (UTC)	
IMPACT SITE	Oceanus Procellarum	
LANDING COORDINATES	9.8°N, 63.18°W	A

LUNA 9 PANORAMAS
The first photographs of the Moon's surface were transmitted back to Earth by Luna 9. The craft triumphed over Surveyor 1 to soft land on the Moon on 4 February 1966. Surveyor 1 reached the Moon on 2 June 1966.

LUNA 9		
ROCKET	Molniya-M	
CREW	N/A	
LAUNCH DATE	11:41:37, 31 January 1966 (UTC)	
LAUNCH SITE	Tyuratam (Baikonur Cosmodrome)	
PAYLOAD	Imaging System, Gamma-ray Spectrometer	
LUNAR ORBIT INSERTION	N/A	
LANDING DATE	18:45:30, 03 February 1966 (UTC)	
LANDING SITE	Oceanus Procellarum	
LANDING COORDINATES	8°N, 64°W	B

HEVELIUS QUADRANGLE: KEY CHARACTERISTICS

1. HEVELIUS CRATER
Latitude: 2.2°N / Longitude: 67.3°W
Diameter: 113.87 km / 70 ¾ miles
Depth: 1.8 km / 1 mile
Photograph: Lunar Orbiter IV (USA, 1967)

2. CAVALERIUS CRATER
Latitude: 5.1°N / Longitude: 66.9°W
Diameter: 59.35 km / 37 miles
Depth: 3 km / 1 ¾ miles
Photograph: Lunar Orbiter IV (USA, 1967)

3. PLANITIA DESCENSUS
Latitude: 7.2°N / Longitude: 64.15°W
Landing site of Luna 9
Photograph: Lunar Orbiter III (USA, 1967)

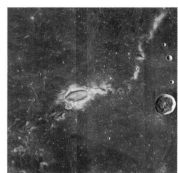

4. REINER GAMMA
Latitude: 7.39°N / Longitude: 59°W
Diameter: 73.44 km / 45 ¾ miles
Photograph: LRO (USA, 2011)

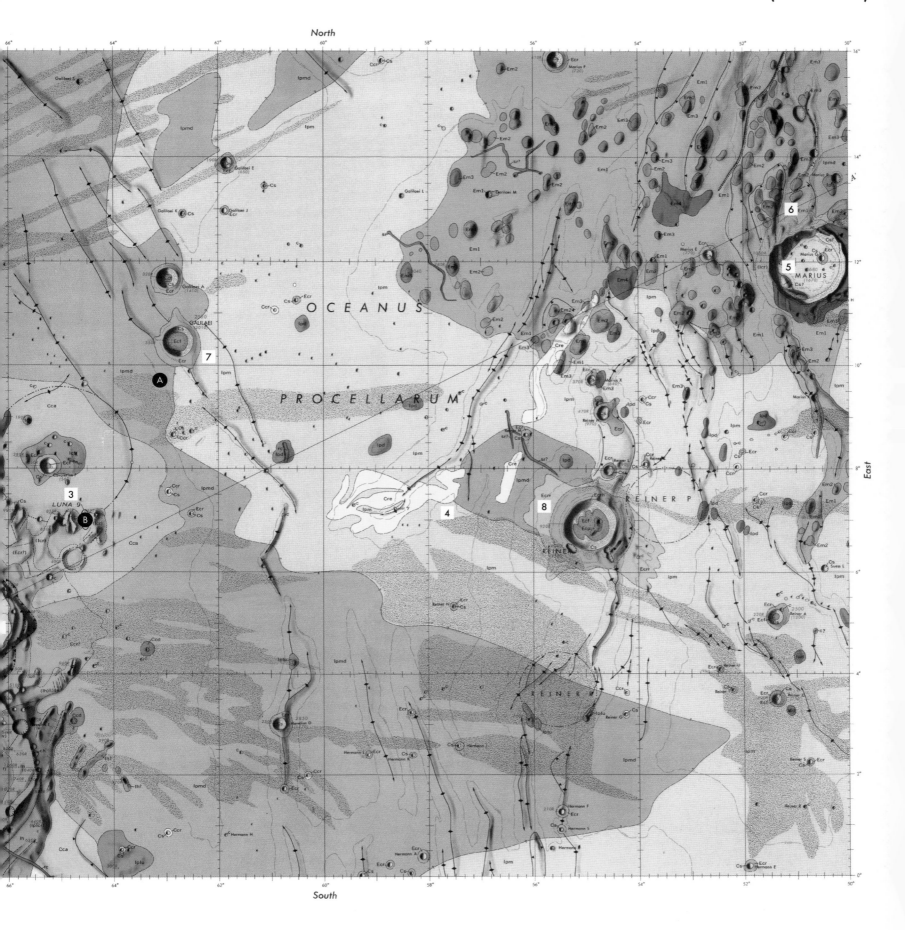

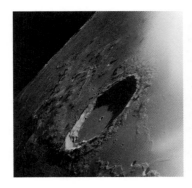

5. MARIUS CRATER
Latitude: 11.9°N / Longitude: 50.84°W
Diameter: 40.1 km / 25 miles
Depth: 1.7 km / 1 mile
Photograph: Apollo 12 (USA, 1969)

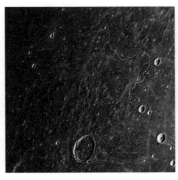

6. MARIUS HILLS
Latitude: 16.37°N / Longitude: 49.54°W
Length: 283.54 km / 176 ¼ miles
Photographed from Earth (Greece, 2011)

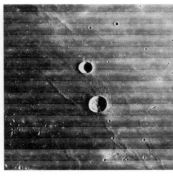

7. GALILEI CRATER
Latitude: 10.48°N / Longitude: 62.83°W
Diameter: 15.99 km / 10 miles
Depth: 1,400 m / 4,593 ft 2 in
Photograph: Lunar Orbiter IV (USA, 1967)

8. REINER CRATER
Latitude: 6.92°N / Longitude: 54.98°W
Diameter: 29.85 km / 18 ½ miles
Depth: 2.6 km / 1 ½ miles
Photograph: LRO (USA, 2011)

THE MOON IN LITERATURE AND FICTION

Myth

Lunar

'For the Moon never beams without bringing me dreams', Edgar Allan Poe, 'Annabel Lee', 1849

←
An illustration by Gustav Doré from an 1862 French edition of Rudolf Erich Raspe's *Aventures du Baron Münchausen*. Doré's darkly gothic images are among the best-known illustrations of the 18th-century fantastical tale of two journeys to the Moon. Here, the Baron is seen lowering himself down on his return to Earth, using a rope and a hatchet.

↱
The frontispiece and title page of the second edition of Francis Godwin's *The Man in the Moone* (1657), which was first published in 1638. In the engraving we see the protagonist Domingo Gonsales being transported up into the air – and eventually to the Moon – by a flimsy-looking contraption with a tiny sail, carried by large birds. The initials on the title page, *F.G. B. of H.*, stand for Francis Godwin, Bishop of Hereford.

↱↱
Frontispiece and title page of *Selenarhia, or The government of the world in the moon: a comical history written by that famous wit and caveleer of France, Monsieur Cyrano Bergerac* (1649). In this first English translation of de Bergerac's tales, we see him reaching for the Moon, which he hopes to reach by strapping phials of bottled dew to his body.

→
Frontispiece from a 1687 English edition of Cyrano de Bergerac's tales, *The Comical History of the States and Empires of the Worlds of the Moon and Sun*. Here we see de Bergerac's second attempt to reach the Moon, this time with a construction that heralds some later ideas about spaceships and rockets. The Moon is imagined as an inhabited, busy world similar to Earth.

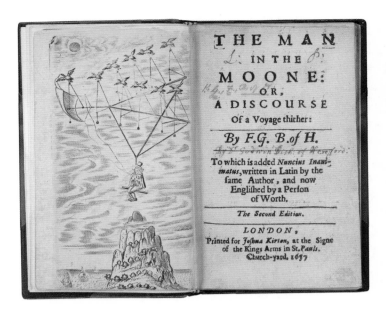

he history of imagined journeys to the Moon is almost as long as the history of creative writing. Humanity's desire to tell stories about the Moon is linked to its being an object of ancient fascination and inspiration. That glowing disc in the sky with visible contours, shadows and patterns made the ancestors of modern humans wonder whether it might be a place resembling Earth, or whether it was home to different life forms, landscapes and rules – a world beyond our terrestrial boundaries, yet visually tantalizingly close. The Moon holds many mysteries, but for writers, poets and storytellers in particular it has also been a rich source of material for narratives and poetic images.

While readers may associate lunar voyages predominantly with 20th-century science fiction, there has in fact been literature about the Moon for thousands of years. The ancient Greek poet Sappho (c. 630–c. 570 BC) observed the Moon and stars and used them as metaphors in her poetry. Some of the earliest descriptions of the Moon, its landscape and possible civilizations are found in Orphic poetry, attributed to the legendary and possibly mythical poet and musician Orpheus, from at least the 5th century BC. In the 1st century AD Plutarch discussed the lunar surface and possible lifeforms in Concerning the Face Which Appears in the Orb of the Moon,[1] introducing the common analogy of the Moon as a mirror. Classical literature also offers the first fantastical journey to the Moon in the style of satire: Lucian of Samosata's (c. AD 125–180) lunar stories in Vera Historia (True Stories) are, in fact, quite the opposite of true and serve to shine an unsparing light on our own world and its inhabitants.

With the invention of the telescope in the early 17th century stories about the Moon multiplied, although drawing the Moon closer to us with the aid of a scientific instrument did not lead to less fanciful descriptions of lunar civilizations. Even the scientist Johannes Kepler, who had worked with the astronomer Tycho Brahe and was the first to understand the Moon's influence on the tides (see p. 122 [Tides]), dreamt up a story of a lunar voyage, Somnium (The Dream) (see p.93 [Heliocentric Woldview]).[2]

In lunar literature the real focus is not always the Moon itself. Stories about journeys to the Moon are closely related to humankind's dreams of flying, and before the advent of air and space travel the means of getting to the Moon are particularly endearing, though they hardly bear scientific scrutiny.[3] Many fictional pre-aviation space travellers reach the Moon with the help of whirlwinds, hitch a ride with daemons or flocks of birds (especially those with large wings, like swans or geese), or use kites or winged flying machines. In his first-person narrative L'Autre Monde ou les états et empires de la Lune (The Other World: Comical History of the States and Empires of the Moon),[4] Cyrano de Bergerac (1619–1655) first attempts to reach space by strapping

phials of bottled dew to his body. He quickly realizes that something more powerful is needed and constructs a device that vaguely resembles a jet engine. In Rudolf Erich Raspe's (1736–1794) Baron Münchhausen (1785),[5] the pompous Baron pontificates about his alleged great adventures, including two journeys to the Moon. On the first occasion, he climbs to the Moon on a stringy bean plant, neatly hooked to the sickle Moon, but upgrades to a boat carried by a powerful storm on his second visit. Münchhausen's Moon is inhabited by humanoid creatures with detachable eyes and heads. Intended as a social satire, this hilarious book of lies quickly became an international success.

The Moon also features prominently in drama. Shakespeare (1564–1616) frequently uses lunar metaphors, and sets many an important scene at night, lit by the Moon. Aphra Behn's (1640–1689) play The Emperor of the Moon (1687)[6] is a farce about human obsession with the Moon and involves a telescope: the main protagonist, Dr Baliardo, is convinced that the Moon is inhabited and spends much time spying on the Moon's emperor through the device, neglecting his family on Earth.

The Moon is a common trope and symbol in the Gothic and Romantic imagination, as well as in fairy tales of modern times, many of which have much older origins in folklore. In many cases, the Moon acts as a mood-setter in Romantic fiction and poetry, or as a symbolic illuminator of good or evil, reflecting human emotions, such as love and longing, or fear and foreboding. Mary Shelley (1797–1851) was reportedly inspired to write *Frankenstein* (1818)[7] when she woke up one night and contemplated the moonlight flooding into her room.

In August 1835, the 'Great Moon Hoax' tapped into the public appetite for lunar science and presented satirical fiction as fact. The New York *Sun* newspaper published a series of articles purporting to report the discovery of life on the Moon by the astronomer John Herschel, illustrated with fantastical engravings.[8] The descriptions of flying humanoids and other bizarre creatures and features had their roots in Lucian of Samosata. A few months earlier, Edgar Allan Poe (1809–1849) had published his short story 'Hans Pfaall: A Tale'[9] about a balloon journey to the Moon. Poe was possibly deterred from writing further instalments after the Herschel hoax story, but he was later credited with having influenced the first author who injected a significant dose of science into fiction, Jules Verne (1828–1905). The French author described a lunar voyage with remarkable prescience in *De la Terre à la Lune* (From the Earth to the Moon) in 1865 and its 1870 sequel *Autour de la Lune* (Around the Moon),[10] sending his protagonists into space with a powerful projectile-shaped spaceship, foreshadowing the shape of rockets that would in reality take humans to the Moon 100 years later. Verne's popular novels have never been out of print; he was famously referenced by Neil Armstrong on the Apollo 11 mission, and inspired Georges

Méliès' (1861–1938) film *Le Voyage dans la Lune* (A Trip to the Moon, 1902).

In English-language fiction, Verne found a rival and successor in H. G. Wells (1866–1946), who wrote his science-fiction novels at the beginning of the age of motorized cars, air travel and telecommunication. *In The First Men in the Moon* (1901),[11] a home-made spaceship features, and his spacemen discuss zero gravity, weightlessness and thin atmosphere, making their endeavours sound scientific and plausible. Once on the Moon, though, the imagined Selenites and plants that inhabit it are just as bizarre as his predecessors' and the Moon becomes a battleground, as in many other lunar fantasies. Colonial desires, territorialism and exploitation are never far away in fictional stories that are set on the Moon.

Science fiction became an established and lucrative genre in the decades after Wells, with such pulp writers as Edgar Rice Burroughs (1875–1950) and Raymond Z. Gallun (1911–1994) freely picking and mixing from a pseudo-realistic box of space toys and tropes. A slightly different take on this is the novel *Frau im Mond* (1928) by Thea von Harbou.[12] It formed the basis of her partner Fritz Lang's movie *Woman in the Moon* (1929). While the film's technical accuracy is impressive, the story is essentially a melodrama set on the Moon, with the spacecraft named after the love interest *Friede* (incidentally also meaning 'peace'), a woman who while on Earth resembles the Moon in appearance (pale skin and diaphanous white clothes) but reveals herself as a strong, decisive character who eventually claims the

narrative and the Moon, now dressed in sturdy shoes and trousers.

By the mid-20th century, rocket science and space exploration had changed the style of lunar fiction significantly. Gone were the whimsical flying machines and visions of aggressive or winged Selenites. Enough scientific knowledge was now available to make such things unbelievable, yet there was still uncertainty about the exact structure of the lunar surface. Certain themes remained popular, such as the idea of colonizing and mining the Moon, while the romantic associations of the Moon faded away, except for lunar metaphors in poetry. The post-war and space-age generation of science fiction writers had seen some of Verne's and Wells' ideas come true, and others overridden, and they began to focus on the psychological and greater philosophical aspects of space travel and Moon exploration. In some works, the sublime and subliminal dread of humans being close to reaching the Moon is palpable, fuelled by Cold War-era existential threats. At the height of the Space Race, Robert A. Heinlein (1907–1988) published *The Moon is a Harsh Mistress* (1966),[13] a dystopian futuristic story of the Moon as a colony for criminals from Earth, while Arthur C. Clarke (1917–2008) co-wrote the screenplay for Stanley Kubrick's (1928–1999) sophisticated and deeply disturbing film *2001: A Space Odyssey* (1968). Clarke had a serious interest in the possibilities of space exploration, but – like most lunar writers before him – he was also using the Moon as a reflection of our own world and society, a mirror of ourselves, and perhaps of our darkest fears.

Myth

Lunar

↖↖
One of the illustrations relating to *The Great Moon Hoax* (1835), published by *The New York Sun*. Bat-like flying creatures with human features share a lush lunar valley with stork-like birds, unicorns and horned animals that seem to have been inspired by Earthly giraffes.

↙
The cover of the 1898 edition of Jules Verne's *De la terre à la lune* and *Autour de la lune* (From the Earth to the Moon and Around the Moon), featuring Émile-Antoine Bayard's and Alphonse de Neuville's beautiful illustrations.

↙↙
The frontispiece from an English edition of Jules Verne's *From the Earth to the Moon* (1874), showing the spaceship *Columbiad*'s approach to the Moon. Although described as a 'projectile train for the moon', it resembles the rockets used 100 years later by NASA.

↙↙↙
The title page of Jules Verne's first tale of a lunar voyage, *De la terre à la lune* (From the Earth to the Moon), from an 1872 edition. The reasonably accurate lunar map is a sign of Verne's more scientific approach to space travel in fiction.

↖
An illustration for Edgar Allan Poe's 'The Unparalleled Adventure of One Hans Pfaall', from Jules Verne's *Edgar Poe et Ses Oeuvres* (1864). In his tale Poe imagines a journey to the Moon via a balloon with the aid of a new device that can turn the vacuum of space into breatheable air.

↱
An illustration from H. G. Wells' *The First Men in the Moon* (1901), showing the men entering the spherical spaceship before their adventure. Wells combined strong narratives about journeys to the Moon with scientific detail, which made them sound more plausible.

↳
Plates from an early English edition of Jules Verne's *From the Earth to the Moon direct in ninety-seven hours and twenty minutes; And a Trip Round It* (1874). It features eighty darkly beautiful illustrations by Émile-Antoine Bayard and Alphonse de Neuville, which combine realism with romanticism.

'The arrival of the projectile at Stone's Hill.'

'Fire.'

'The Director at his post.'

'This plain would then be nothing but an immense cemetery.'

'What giant oxen.'

'He could distinguish nothing but desert beds.'

'He distinguished all this.'

'Can you picture to yourselves?'

'A violent contraction of the lunar crust.'

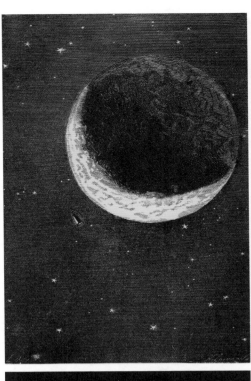

'It was an enormous disc.'

'It was the body of a satellite.'

'The telescope at Parsontown.'

'It is the fault of the moon.'

'Nothing could equal the splendour of this starry world.'

'What a sight.'

'Around the projectile were the objects which had been thrown out.'

'Arden applied the lighted match.'

'I fancy I see them.'

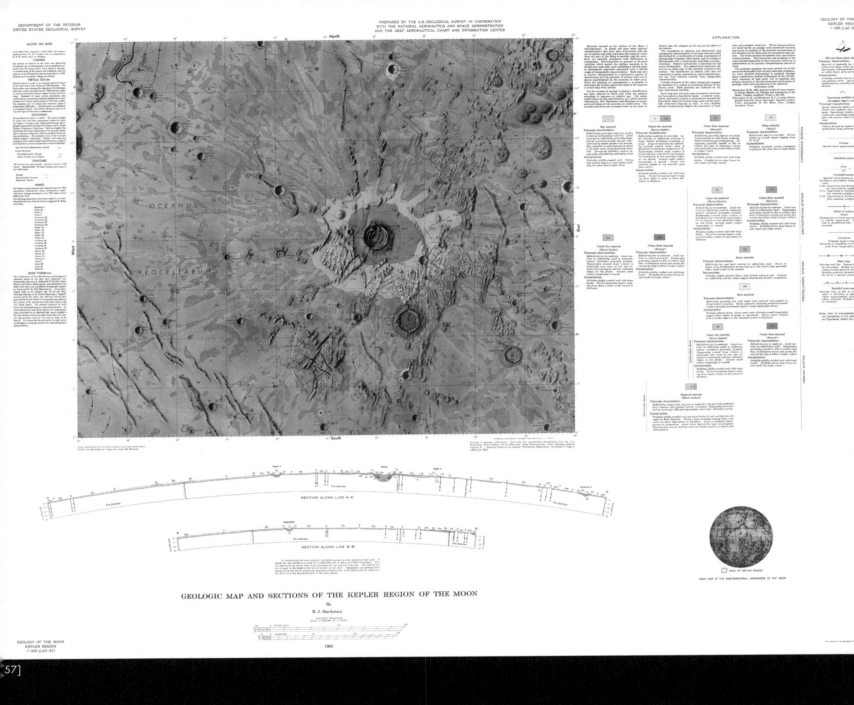

GEOLOGIC MAP AND SECTIONS OF THE KEPLER REGION OF THE MOON

By

R. J. Hackman

1962

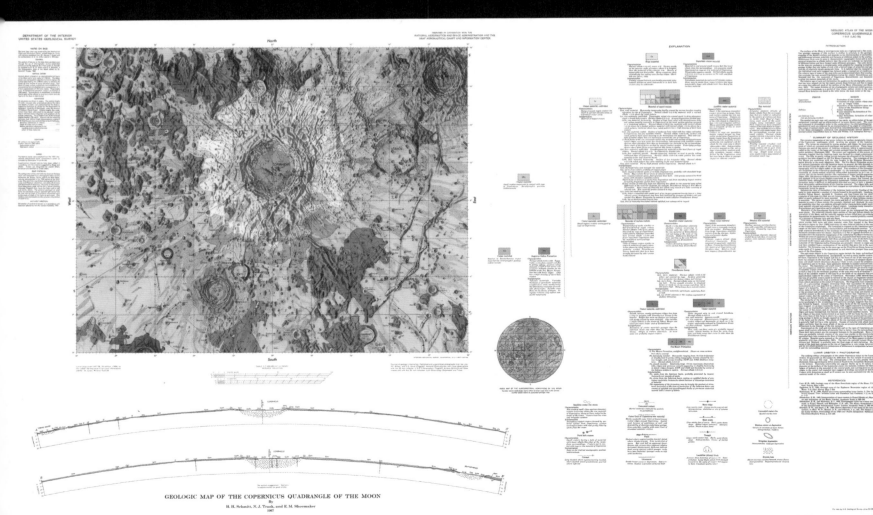

GEOLOGIC MAP OF THE COPERNICUS QUADRANGLE OF THE MOON

By

H. H. Schmitt, N. J. Trask, and E. M. Shoemaker

1967

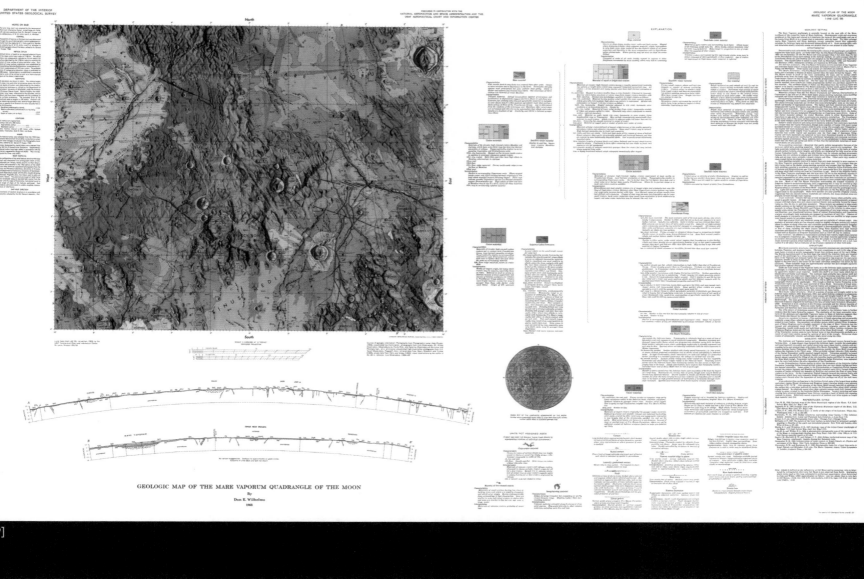

GEOLOGIC MAP OF THE MARE VAPORUM QUADRANGLE OF THE MOON

By

Don E. Wilhelms

1968

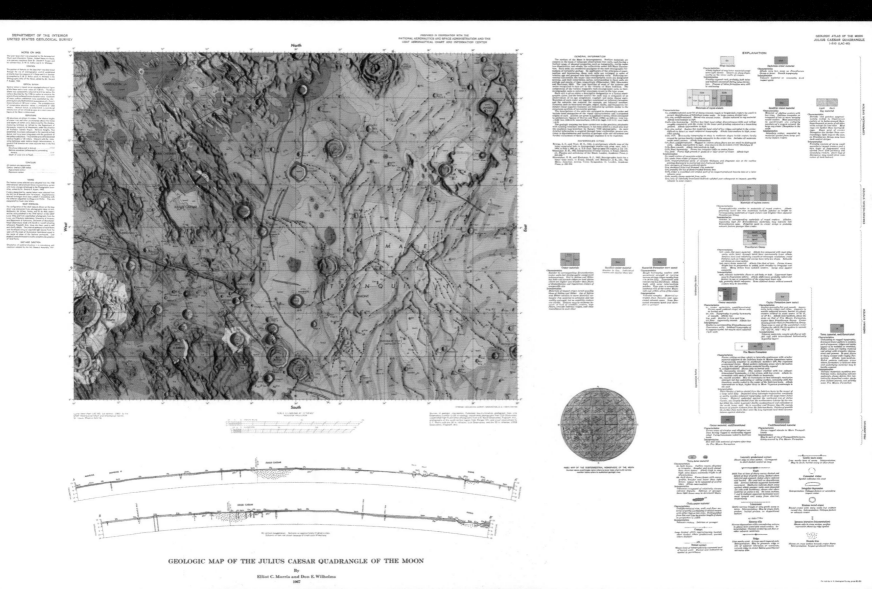

GEOLOGIC MAP OF THE JULIUS CAESAR QUADRANGLE OF THE MOON

By

Elliot C. Morris and Don E. Wilhelms

1967

MOONSCAPES— THE MOON IN 19TH-CENTURY PAINTING

↓/↰
Caspar David Friedrich, *Two Men Contemplating the Moon* (c. 1825–30). This romantic depiction of the Moon typically connects the landscape to interiority. The Moon, at the focal point of the painting, is also a mirror into the two men's feelings and thoughts.

↱
Aert van der Neer, *Moonlit Landscape with Bridge* (c. 1648–50). Van der Neer became famous for his night and crepuscular landscapes, with restricted palettes and light playing across water.

↱↱
Claude-Joseph Vernet, *Night: A Mediterranean Coast Scene with Fishermen and Boats* (1753). While his predecessor Claude Lorraine shaped a style of peaceful, harmonious marine landscapes, Vernet's works added drama to the genre by showing landscapes in darkness or stormy weather.

↳
James Abbott McNeill Whistler, *Nocturne in Black and Gold, the Falling Rocket,* (1875). Whistler's *Nocturnes* were abstract and far from accepted conventions of landscape painting in the late 19th century, and they created a sensation when they were first exhibited in the 1870s.

↳↳
Etienne Léopold Trouvelot, *Total Eclipse of the Sun. Observed July 29, 1878, at Creston, Wyoming Territory.* Touvelot's pictures bridged art and science, combining the precision of observation with the pure aesthetic enjoyment of contemplation.

'far from modern life, far from everything, at the far reaches of painting'),[2] the lure of night also had very deep roots in the more concrete aspects of *fin-de-siècle* daily life. The massive development of artificial lighting in late 19th-century cities was beginning to challenge the very idea of nocturnal darkness, one that had been universally taken for granted for millennia.[3] Suddenly, with gas and electricity, the Moon had competition. It was not just technologies of light that were redefining night and obscurity. Other technical and scientific developments increasingly questioned the once clear separation between light and darkness, and between what those two terms metaphorically stood for in the modern era: knowledge and ignorance, the rational and the irrational. Take, for instance, the invention (or rather the accidental discovery) of X-rays: while they promised to open an unprecedented access to vision and therefore knowledge, the new 'mysterious' rays also brought along an increased sense of mystery. Invisibility seemed at once conquered, and, paradoxically, deepening. Scientific progress brought along wonderful promises, but also disorienting new truths. In this context of mixed enthusiasm and awe, French painter Etienne Léopold Trouvelot (1827–1895) built a career on the borders of art and science. His amateur research interests included electricity and photography, as well as in entomology – he is famous for inadvertently introducing the destructive spongy moth to North America. In the 1870s, the artist was given access to the prestigious Harvard college observatory, where he was

p on a slope at the edge of a wood, two silhouettes, framed by twisted tree branches, observe a thin Moon crescent, situated at the very centre of the canvas. The whole scene is enveloped in its soft, rosy glow. The painting, Caspar David Friedrich's *Two Men Contemplating the Moon* (c. 1825–30), is an image of both closeness and distance. It focuses on one of the most universally familiar phenomena, which yet retains an inextinguishable measure of mystery and awe. Friedrich's romantic depiction of the Moon brings the observer closer to this mystery, and even closer to others observing it, as the two companionable figures suggest. At the same time, it is a reminder of the place humans hold in nature and in the universe: for all its beauty, the Moon remains beyond our reach, and in many ways beyond our understanding.

Following the examples of Aert van der Neer's (1603–1677) signature night scenes in the 17th century, and Joseph Vernet's (b. 1770) nocturnal seascapes in the 18th, moonlit landscapes rose with the genre of landscape itself in the Romantic period, when many artists expressed their fascination for nocturnal imagery. By the end of the 19th century, night painting became so popular among artists that critics mocked it on a regular basis. In 1902, the art students of the Society of American Fakirs organized a parodic exhibition which, according to the *New York Times*, featured several spoofs of James Abbott McNeill Whistler's famous nocturne pictures: 'one of the take-offs is a mass of black ink, with a bunch of matches in place of the fireworks. Another shows a building burning dimly on a very dark night. A third consists of a couple of black cats on the back fence on a night so dark that they can't see their paws in front of them.'[1] Whistler (1834–1903), the American painter living in Europe, was indeed a great source of inspiration for artists on both sides of the Atlantic, who emulated the monochrome paintings of the River Thames at night that he had been painting since the 1870s.

Yet the attraction for night imagery reached far beyond the circles of Whistler's disciples. While many of Whistler's followers marketed their works as purely aesthetic forms (which Symbolist poet Joris-Karl Huysmans described as

↓
Ralph Albert Blakelock, *Moonlight Sonata* (c. 1889–92). To his contemporaries, Blakelock was the perfect image of the cursed genius: he never managed to make a living from his art, which was so unique that it puzzled many of its viewers.

After his mental health deteriorated to the extent that he had to be committed to a psychiatric hospital, many saw a financial opportunity in imitating the artist with 'Fakelock' paintings, now in high demand on the art market.

able to accumulate notes and create spectacular renditions of the solar system's planets and of the Moon. The resulting pastels were then widely circulated as chromolithographs in a deluxe volume entitled *Astronomical Drawings* published by Charles Scribner in 1882.[4] Trouvelot, whose achievements resulted in having a crater on the Moon named after him, is emblematic of the overlaps between science and art that shaped the contemporary perception of the night sky and the Moon. Making use of the best scientific tools of his time, he brought his viewers a great step closer to the shiny crescent once admired by Friedrich's onlookers.

The evocative powers of darkness, however, persisted despite, or even because of, scientific developments. As the new discipline of psychology delved into the 'dark continent' of the unconscious, associations between the deeper recesses of the mind and the night gained new traction. The figure of the 'lunatic' genius, for instance, shaped the reception of the painter Ralph Albert Blakelock (1847–1919), whose moonlit canvases were read as expressions of the artist's tortured mind. While a life of poverty had precipitated Blakelock into deep mental illness, ironically, his art began selling for record prices once he was, literally, institutionalized. With compositions centred on the bright vortex of the Moon, Blakelock's pictures are emblematic of a shift in landscape painting in general: rather than detailed views of specific locales, night paintings gradually gave more and more importance to the sky, and to effects of light and atmosphere. In doing so, they contributed to the period's evolution towards abstraction, visible for instance in Félix Valotton's *Moonlight* (c. 1894). Between the precision of Trouvelot's renditions and the abstraction of Valotton's elusive orb, the Moon image, although transformed, remains true to its enduring ambivalence – forever half-light, half-darkness.

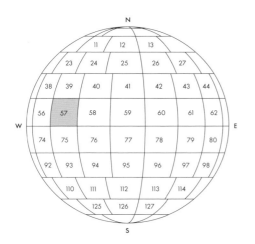

The crater Kepler, after which this quadrangle is named, stands out distinctly amid the surrounding basaltic plains of Oceanus Procellarum (Ocean of Storms) and Mare Insularum (Sea of Islands). The crater is surrounded by a very prominent ray system overlaying the maria. The visibility of these rays, along with the integrity of the crater's high walls, indicate that the impact that excavated the crater occurred relatively recently, during the Moon's latest geologic period (the Copernican). Several satellite craters created from the impact of ejecta material are also found in this quadrangle, as well as the older craters Encke and Kunowsky, which are covered by Kepler's ray material. Also visible are rills and chain crater systems that geologists interpreted as evidence of tectonic faults in this region. In the northwest corner of this quadrangle are volcanic domes, similar to shield volcanoes, created when lava extruded through the surface of the mare and solidified. One unsuccessful robotic mission, the Soviet Union's Luna 7, crashed in this quadrangle.

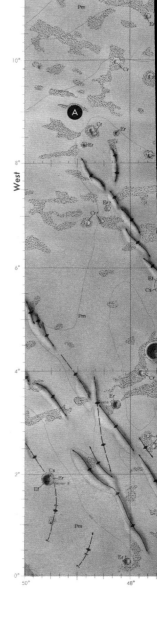

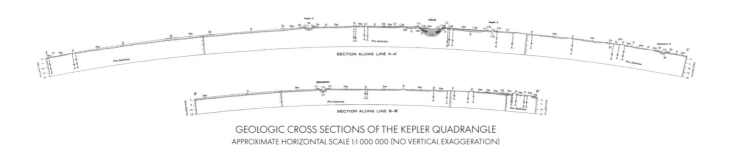

GEOLOGIC CROSS SECTIONS OF THE KEPLER QUADRANGLE
APPROXIMATE HORIZONTAL SCALE 1:1 000 000 (NO VERTICAL EXAGGERATION)

LUNA 7	ROCKET	Molniya-M 8KK78M
	CREW	N/A
	LAUNCH DATE	07:56:40, 04 October 1965 (UTC)
	LAUNCH SITE	Tyuratam (Baikonur Cosmodrome)
	PAYLOAD	Imaging System, SM-10 Radiation Detector
	LUNAR ORBIT INSERTION	N/A
	LANDING DATE	22:08:24, 07 October 1965 (UTC)
	IMPACT SITE	Oceanus Procellarum
	LANDING COORDINATES	9°N, 49°W Ⓐ

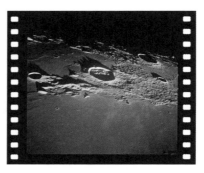

APOLLO 12 IN-ORBIT IMAGE
Oblique view of craters Kunowsky, Encke and Kepler, captured by Apollo 12, 19 November 1969.

KEPLER QUADRANGLE: KEY CHARACTERISTICS

1. KEPLER CRATER
Latitude: 8.12°N / Longitude: 38°W
Diameter: 29.5 km / 18 ½ miles
Depth: 2.6 km / 1 ½ miles
Photograph: Lunar Orbiter III (USA, 1967)

2. KEPLER RAY MATERIAL
Area south of Kunowsky crater
Latitude: 3°N / Longitude: 32.5°W
Photograph: Apollo 14 (USA, 1971)

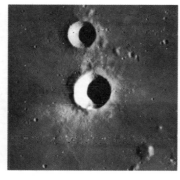

3. BESSARION CRATER
Latitude: 14.85°N / Longitude: 37.3°W
Diameter: 9.84 km / 6 miles
Depth: 2 km / 1 ¼ miles
Photograph: Lunar Orbiter IV (USA, 1967)

4. KUNOWSKY CRATER
Latitude: 3.22°N / Longitude: 32.53°W
Diameter: 18.27 km / 11 ¼ miles
Depth: 900 m / 2,952 ft 9 in
Photograph: Apollo 12 (USA, 1969)

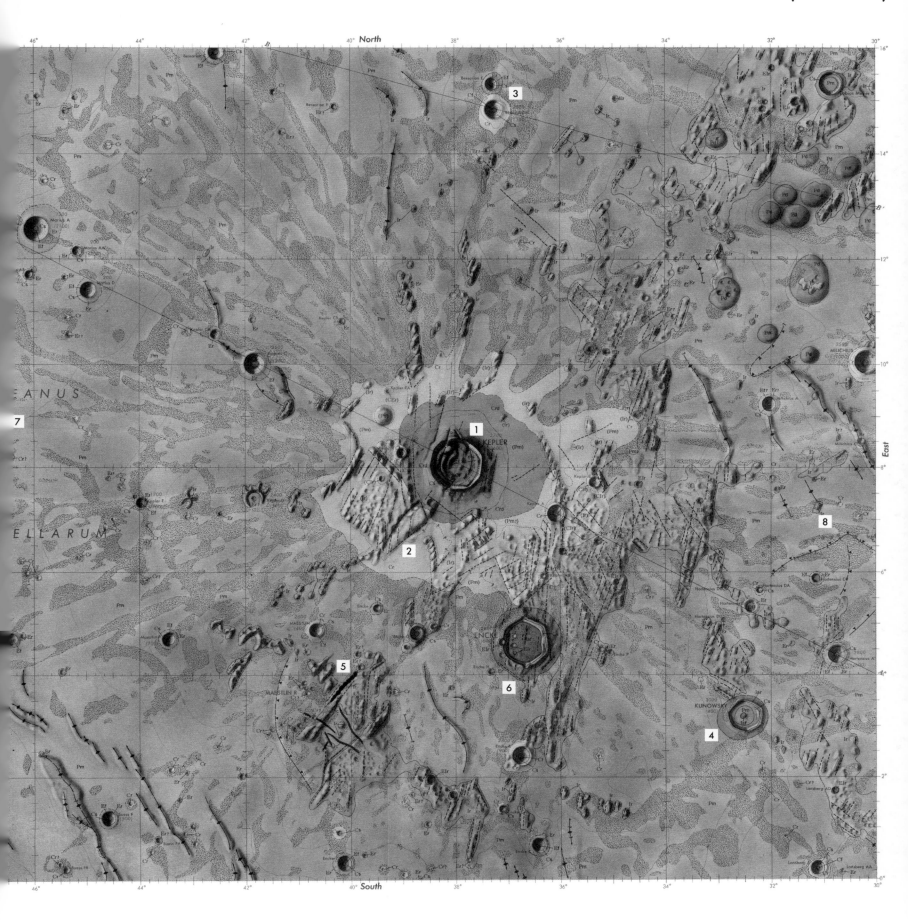

5. RIMAE MAESTLIN
Latitude: 2.88°N / Longitude: 40.48°W
Length: 71 km / 44 miles
Photograph: LRO (USA, 2015)

6. ENCKE CRATER
Latitude: 4.57°N / Longitude: 36.68°W
Diameter: 28.27 km / 17 ½ miles
Depth: 800 m / 2,624 ft 8 in
Photograph: Lunar Orbiter IV (USA, 1967)

7. OCEANUS PROCELLARUM
Latitude: 20.67°N / Longitude: 56.68°W
Diameter: 2592.24 km / 1610 ¾ miles
Photograph: LRO (USA, 2014)

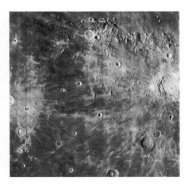

8. MARE INSULARM
Latitude: 7.8°N / Longitude: 30.64°W
Diameter: 511.93 km / 318 miles
Photograph: LRO (USA)

The large impact crater Copernicus, located in the eastern reaches of Oceanus Procellarum (Ocean of Storms), is the archetypical example of a crater formed during the Moon's latest geologic period, the Copernican. The crater is mostly pristine, with a well-defined rim, a terraced inner wall, a sloping exterior rampart and three central peaks. The crater's ray system is extensive, overlaying this and other adjoining quadrangles. It is surrounded by several secondary craters and crater chains related to its formation. North of the crater is the Montes Carpatus mountain range (named for the Carpathian Mountains), which separates Oceanus Procellarum from Mare Imbrium (Sea of Showers). To the east are the nearly erased crater remnant Stadius and, to its north, the younger crater Eratosthenes. Like Copernicus, Eratosthenes lends its name to the geologic period in which it formed, the Eratosthenian (which preceded the Copernican). Typical of craters of this period, Eratosthenes has a well-defined rim, a terraced inner wall, central peaks and an outer rampart of ejecta, but has no visible rays. Two failed robotic missions, the Soviet Union's Luna 5 and the USA's Surveyor 2, crashed in this quadrangle.

SURVEYOR 2	ROCKET	Atlas LV-3C Centaur-D
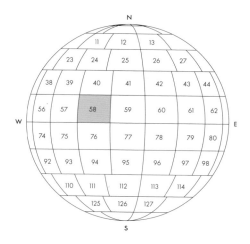	CREW	N/A
	LAUNCH DATE	12:32:00, 20 September 1966 (UTC)
	LAUNCH SITE	Cape Kennedy Air Force Station
	PAYLOAD	Imaging System
	LUNAR ORBIT INSERTION	TBC
	LANDING DATE	09:34:00, 22 September 1966 (UTC)
	IMPACT SITE	Southeast of Copernicus crater
	LANDING COORDINATES	5.30°N, 12.00°W Ⓐ

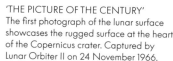

'THE PICTURE OF THE CENTURY'
The first photograph of the lunar surface showcases the rugged surface at the heart of the Copernicus crater. Captured by Lunar Orbiter II on 24 November 1966.

MOSAIC MOONSCAPE OF COPERNICUS
This image of the Copernicus crater is composed of multiple Lunar Orbiter images. It shows with visible rims, mounds and ejecta rays. The crater belongs to the youngest group of rocks on the Moon – the Copernican System.

COPERNICUS QUADRANGLE: KEY CHARACTERISTICS

1. COPERNICUS CRATER
Latitude: 9.62°N / Longitude: 20.08°W
Diameter: 96.07 km / 59 ¾ miles
Depth: 3.8 km / 2 ¼ miles
Photograph: LRO (USA, 2010)

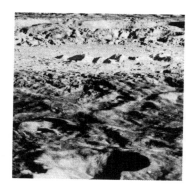

2. COPERNICUS CRATER RAYS
Latitude: 9.62°N / Longitude: 20.08°W
Diameter: 96.07 km / 59 ¾ miles
Depth: 3.8 km / 2 ¼ miles
Photograph: Lunar Orbiter II (USA, 1966)

3. ERATOSTHENES CRATER
Latitude: 14.47°N / Longitude: 11.32°W
Diameter: 58.77 km / 36 ½ miles
Depth: 3.6 km / 2 ¼ miles
Photograph: Apollo 17 (USA, 1972)

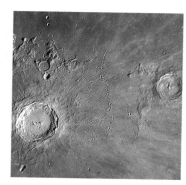

4. STADIUS CRATER
Latitude: 10.48°N / Longitude: 13.77°W
Diameter: 68.48 km / 42 ½ miles
Photographed from Earth (Italy, 2021)

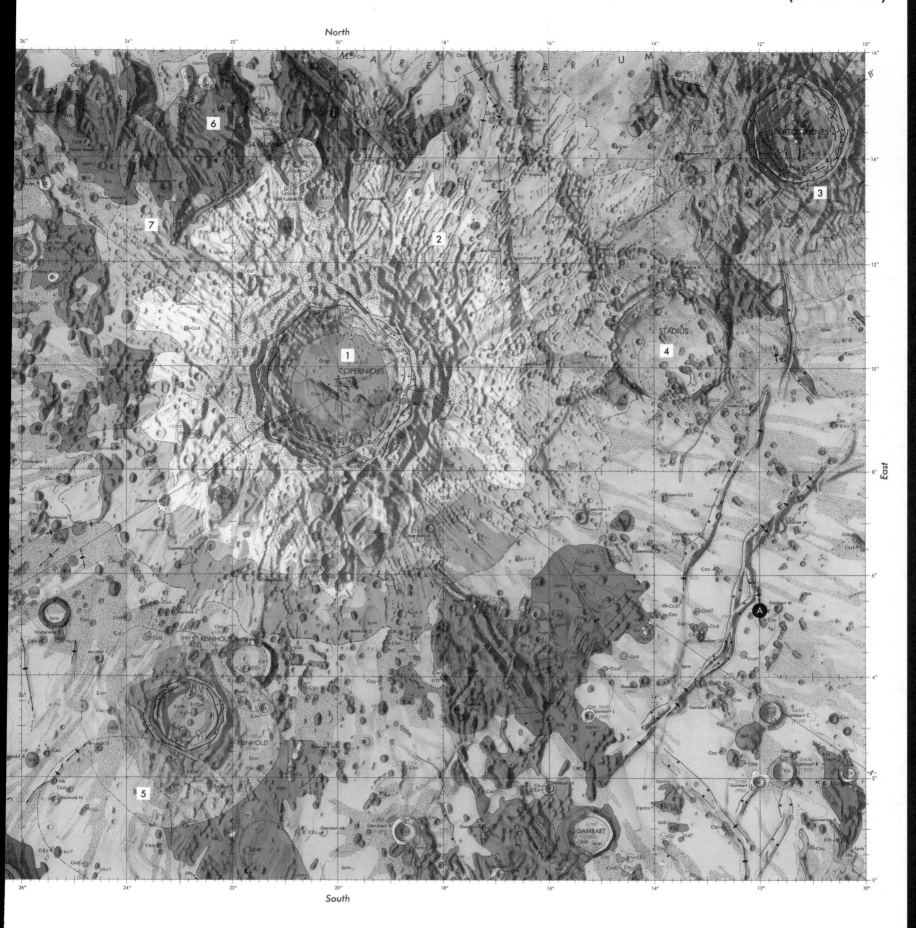

North

South

East

5. REINHOLD CRATER
Latitude: 3.28°N / Longitude: 22.86°W
Diameter: 43.28 km / 27 miles
Depth: 3.3 km / 2 miles
Photograph: Lunar Orbiter IV (USA, 1967)

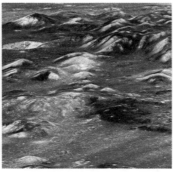

6. MONTES CARPATUS
Latitude: 14.57°N / Longitude: 23.62°W
Diameter: 333.59 km / 207 ¼ miles
Max Height: 2.4 km / 1 ½ miles
Photograph: LRO (USA, 2017)

7. RIMA GAY-LUSSAC
Latitude: 13.18°N / Longitude: 22.33°W
Length: 40.04 km / 25 miles
Photograph: Lunar Orbiter IV (USA, 1967)

8. TOBIAS MAYER CRATER
Latitude: 15.54°N / Longitude: 29.17°W
Diameter: 33.15 km / 20 ½ miles
Depth: 2.9 km / 1 ¾ miles
Photograph: Lunar Orbiter IV (USA, 1967)

HOW THE MOON GOVERNS THE TIDES

↓
Folio from René Descartes' *Principia philosophiae* (1644). In order to explain the influence of celestial bodies on one another without the need for mystical 'action at a distance', René Descartes envisaged space to be filled with aetheric fluid whose swirling vortices transmitted motion from one body to another – for instance from the motion of the Moon to the seas of Earth.

↱
Illustration from Athanasius Kircher's *Mundus subterraneus* (1665). In his complex explanation of the tides, sunlight reflecting off the Moon gives rise to 'nitrous effluvia' that pass from the Moon to Earth, pushing or pulling at ocean water, which flows from one side of the Earth to the other through the series of underground caverns and passages illustrated here.

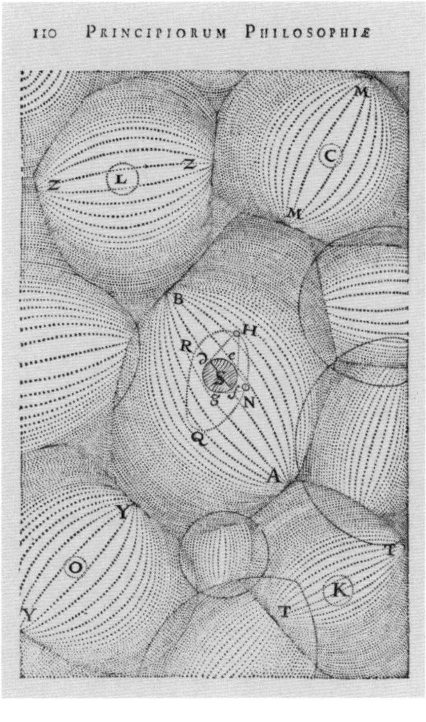

110 PRINCIPIORUM PHILOSOPHIÆ

he daily rise and fall of tides is a familiar part of the natural world that people in coastal regions have lived with – and made use of – since prehistoric times. Many ancient cultures learned to associate tides with the Moon because of clear monthly cycles in both their timing and intensity (with the strongest 'spring tides' around new and full Moons, and relatively moderate 'neap tides' around the first and last quarter phases).

Nevertheless, the complexity of tidal patterns in different locations (mostly caused by geographical factors) left plenty of room for argument on the issue. For instance, in 1616 Italian physicist and astronomer Galileo Galilei argued at length that tides were an effect of Earth's daily motion, resulting from the inertia of ocean waters as our planet spins on its axis and orbits the Sun. Other theories were somewhat more colourful, such as German scholar Athanasius Kircher's (1601–1680) proposal in 1665 that the tides were caused by 'nitrous effluvia' in moonlight driving Earth's water in and out of a network of subterranean passages.[1]

A major challenge to recognizing the Moon's true role lay in the absence of a satisfactory theory for how *any* celestial body could influence Earth. For some 2,000 years, European philosophers and scholars had accepted that there was a strict division between the changeable Earth and the unchanging cycles of the celestial realm. What was more, the adoption of the Copernican or Sun-centred model of cosmology from the 16th century onwards (see p. 92) made it clear that celestial bodies – even the Moon – were separated from the mundane world by vast distances.

In 1644, French philosopher René Descartes (1596–1650) proposed a possible way for objects to influence each other over these distances – through a swirling material called aether that supposedly filled space. The motion of celestial bodies, he theorized, created waves of aetheric pressure that could be transmitted to other worlds, and the Moon's influence on Earth's fluid oceans was a prime example of this effect at work.[2]

Descartes' idea of aether as a physical connection between distant objects became widely adopted at a time when most philosophers rejected the idea of invisible forces exerting so-called 'action at a distance'. Despite this, some scholars were inspired by new investigations of another curious natural phenomenon – magnetism – to consider the idea that massive bodies might exert an attractive force. In 1608, for example, the German scholar Johannes Kepler (now

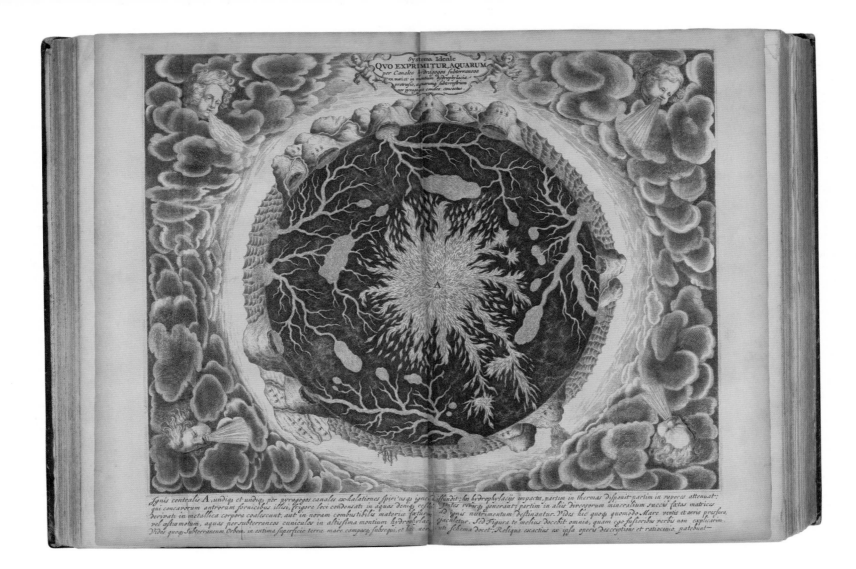

Ignis centralis A undiqs et undiqs per pyragogos canales exhalationes spiritus qs ignei [...] qui concavorum antrorum fornicibusqs illesi, frigore loci condensata in aquas deniqs res [...] derivata in metallica corpore coalescunt, aut in novam combustibili materia foetus [...] vel astrum metam, aquas per-subterraneos cuniculos in altissima montium hydrophylaci [...] Vides quoqs Subterraneum Orbem, in extima superficie terrae mare compacto, sed [...]

best known for his descriptions of planetary motion around the Sun) speculated that 'the causes of the ocean's swell seem to be the bodies of the Sun and Moon which attract the waters of the sea by a certain force similar to magnetism'.[3]

The true nature of both the tides and the more general forces between Earth, Sun, Moon and planets was finally clarified in the late 17th century by the English polymath Sir Isaac Newton (1643–1727), in his 1687 work *Mathematical Principles of Natural Philosophy*. In Newton's model of gravitation, the attractive force exerted by a massive body is proportional to its mass, but diminishes with distance in accordance with a so-called 'inverse square' relationship (in which doubling the distance to an object diminishes its gravitational influence by a factor of four). Coupled with Newton's greatly improved understanding of the physics affecting motion of bodies, this could not only describe astronomical phenomena such as the shapes and periods of orbits (including that of the Moon), but also explain the pattern of tides.[4]

Newton explained tidal motion through his 'equilibrium theory' – a model that treats Earth as a static object enveloped in a simplified, uniform layer of water. The inverse square law of gravitation means that the Moon's gravitational attraction is stronger on the side of Earth closest to it, weaker in the middle of the planet and weaker still on the opposite side. Hence, oceans on one side of Earth are drawn towards the Moon more than Earth itself, while our entire planet is attracted more than the waters of the opposite side, which form a second bulge due to their inertia.

As the Moon orbits Earth each month, the orientation of the tidal bulges gradually shifts. Meanwhile, the Sun creates a similar, but weaker, pair of tidal bulges, which boost the overall height of tides when they align to the lunar bulges (around the new and full Moon), and level out the variation at other times.

Newton's theory is sufficient to explain many different aspects of the tides. If Earth's daily spin is considered as a rotation 'beneath' the tidal bulges, it provides an intuitive explanation for the main twice-daily and monthly tidal patterns. It can also explain why successive high tides reach similar heights at certain times (when the Moon is in line with Earth's equator) and at other times may be markedly different.

However, there are many other complicating factors, and tides in the real world rarely if ever behave in line with the predictions of equilibrium theory. Throughout the early 18th century, as Newtonian physics spread across Europe, scholars worked to take some of these factors into account (in particular Earth's rotation, which meant the tidal bulges could effectively be treated as enormous moving waves). Such investigations were driven by practical as well as academic interest; accurate tables of tides could be commercially valuable in an age of burgeoning sea trade.

In 1775, this work culminated with French mathematician Pierre-Simon Laplace's (1749–1827) development of the first dynamic theory of tides.[5] Laplace provided a theoretical and mathematical framework to consider factors such as Earth's rotation, the shape of the continents constraining the seas, the friction and flow of water, and the fact that the volume of moving water in ocean basins can develop a natural harmonic resonance. The resulting model highlighted the importance of water flows across and within the oceans. This not only amplified the size of the predicted tides (resolving one major issue with Newton's theory) but also predicted that tides in different regions would form so-called 'amphidromic' systems, whose oscillations are driven by a combination of lunar gravity and their own harmonics.

Laplace's theory marked a huge step forward, but work to refine our understanding of the Moon's influence over Earth's oceans has continued to the present day. Modern models of changing tides rely on ever-improving mathematical models of both ocean circulation and the lunar orbit, detailed oceanographic maps, and precision measurements from satellites orbiting high above the Earth.

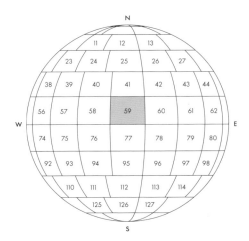

This quadrangle is named for the Mare Vaporum (Sea of Vapours) lava plains, a significant portion of which appear in the northeast of this map. But in fact this quadrangle includes two additional small maria, Sinus Aestuum (Seething Bay) and Sinus Medii (Middle Bay). Still, the majority of this quadrangle is occupied by relatively rugged terrain. This includes a portion of the Montes Apenninus mountain range. Several young and well-preserved craters are also visible in this quadrangle. These include the Triesnecker crater in Sinus Medii, which has a well-formed rim, a terraced wall, a central peak and a visible ray system. The somewhat older crater Manilius in Mare Vaporum has many of these features, and is bright enough to stand out from its older surroundings, but has no visible rays. The craters Pallas and Murchison, by contrast, are heavily eroded and have been flooded by lava. In Sinus Aestuum, geologists used dotted lines to indicate the presence of scarps suggestive of the structure of the mare's underlying basin. On 10 January 1967, the USA's Surveyor 6 mission successfully soft-landed at the coordinates 0.47°N, 1.43°W (landing site not marked).

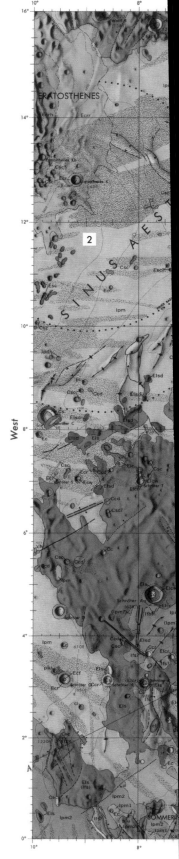

SURVEYOR 6	ROCKET	Atlas SLV-3C Centaur-D
	CREW	N/A
	LAUNCH DATE	07:39:01, 07 November 1967 (UTC)
	LAUNCH SITE	Cape Kennedy Air Force Station
	PAYLOAD	TV Camera, Footpad Magnets
	LUNAR ORBIT INSERTION	N/A
	LANDING DATE	01:01:04, 10 November 1967 (UTC)
	LANDING SITE	Sinus Medii
	LANDING COORDINATES	0.47°N, 1.43°W (A)

PRELAUNCH
A Surveyor 6 footpad undergoes a magnet experiment prior to its launch in November 1967.

TOUCHDOWN
Television image mosaic of the Surveyor 6 footpad on the surface of the Moon. The setting Sun casts a shadow.

SURVEYOR 6 PANORAMA
A panorama of Sinus Medii composed of images taken by Surveyor 6's TV cameras, 10 November 1967.

MARE VAPORUM QUADRANGLE: KEY CHARACTERISTICS

1. MARE VAPORUM
Latitude: 13.2°N / Longitude: 4.09°E
Diameter: 242.46 km / 150 ¾ miles
Photograph: Apollo 17 (USA, 1972)

2. SINUS AESTUUM
Latitude: 12.1°N / Longitude: 8.34°W
Diameter: 316.5 km / 196 ¾ miles
Photograph: Apollo 12 (USA, 1969)

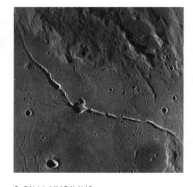

3. RIMA HYGINUS
Latitude: 7.62°N / Longitude: 6.77°E
Length: 219 km / 136 miles
Photograph: LRO (USA, 2010)

4. TRIESNECKER CRATER
Latitude: 4.18°N / Longitude: 3.60°E
Diameter: 24.97 km / 15 ½ miles
Depth: 2.8 km / 1 ¾ miles
Photograph: Lunar Orbiter IV (USA, 1967)

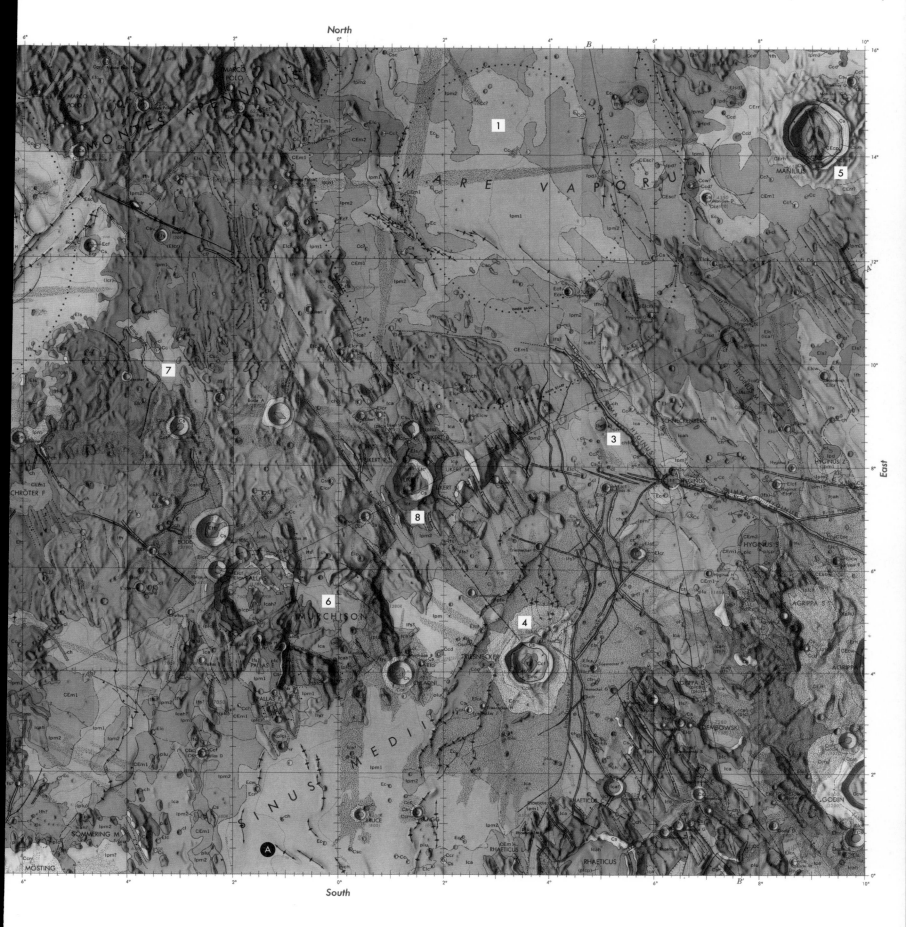

North

South

MARE VAPORUM

SINUS MEDII

5. MANILIUS CRATER
Latitude: 14.45°N / Longitude: 9.07°E
Diameter: 38.34 km / 23 ¾ miles
Depth: 3.05 km / 2 miles
Photograph: Apollo 17 (USA, 1972)

6. MURCHISON CRATER
Latitude: 5.07°N / Longitude: 0.21°W
Diameter: 57.83 km / 36 miles
Depth: 1.8 km / 1 miles
Photograph: Lunar Orbiter III (USA, 1967)

7. RIMA BODE
Latitude: 9.54°N / Longitude: 3.22°W
Diameter: 233 km / 144 ¾ miles
Photograph: LRO (USA, 2010)

8. UKERT CRATER
Latitude: 7.71°N / Longitude: 1.37°E
Diameter: 21.71 km / 13 ½ miles
Depth: 2.9 km / 1 ¾ miles
Photograph: Lunar Orbiter III (USA, 1967)

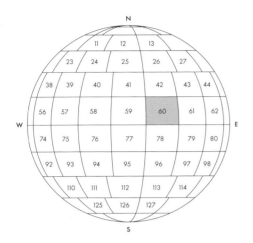

Although this quadrangle is named for the heavily worn Julius Caesar crater, its best-known feature is Mare Tranquillitatis (Sea of Tranquillity). The Apollo 11 Lunar Module landed here on 20 July 1969, making it the first location on another world to be visited by humans. Two robotic probes had explored this site previously. The 1965 Ranger 8 impact site is noted at the coordinates 2.43°N, 24.38°E. In 1967, Surveyor 5 soft-landed at the coordinates 1.42°N, 23.20°E, but neither this nor the Apollo landing site of Statio Tranquillitatis (Tranquillity Base) at 0.69°N, 23.43°E is noted as these landings occurred after this map was made. The relatively smooth terrain of the mare in this quadrangle is interrupted by ridges, impact craters, small crater chains and volcanic domes. The basin is ringed by a system of faults and scarps. Dominating the region north of the Mare is the imposing sharp-rimmed crater Plinius, enclosing a central peak and both smooth and hummocky terrain.

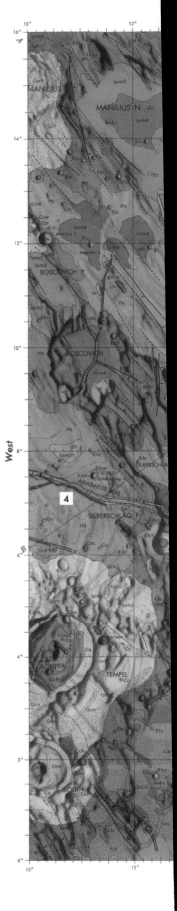

RANGER 8	ROCKET	Atlas LV-3 Agena-B 196D/AA13
	CREW	N/A
	LAUNCH DATE	17:05:00, 17 February 1965 (UTC)
	LAUNCH SITE	Cape Kennedy Air Force Station
	PAYLOAD	Imaging System (Six TV Cameras)
	LUNAR ORBIT INSERTION	N/A
	LANDING DATE	09:57:37, 20 February 1965 (UTC)
	IMPACT SITE	Mare Tranquillitatis
	LANDING COORDINATES	2.43°N, 24.38°E Ⓐ

RANGER 8 PROBE
NASA technicians complete their final checks two weeks before the launch of the Ranger 8 probe, 2 February 1965.

SURVEYOR 5	ROCKET	Atlas Centaur
	CREW	N/A
	LAUNCH DATE	07:57:01, 08 September 1967 (UTC)
	LAUNCH SITE	Cape Kennedy Air Force Station
	PAYLOAD	TV Camera, Alpha-Scattering Instrument
	LUNAR ORBIT INSERTION	N/A
	LANDING DATE	00:46:42, 11 September 1967 (UTC)
	LANDING SITE	Mare Tranquillitatis
	LANDING COORDINATES	1.42°N, 23.20°E Ⓑ

SURVEYOR 5 LANDER
The Surveyor 5 lander ahead of its journey to the Moon. It will be the first spacecraft to analyse lunar soil.

APOLLO 11	ROCKET	Saturn V AS-506
	CREW	N. Armstrong, E. Aldrin Jr, M. Collins
	LAUNCH DATE	13:32:00, 16 July 1969 (UTC)
	LAUNCH SITE	Cape Kennedy Air Force Station
	PAYLOAD	Columbia (CSM–107), Eagle (LM–5)
	ORBITAL INSERTION	19 July 1969
	LANDING DATE	20:17:00, 20 July 1969 (UTC)
	IMPACT SITE	Mare Tranquillitatis
	LANDING COORDINATES	0.69°N, 23.43°E Ⓒ

APOLLO 11 LANDER
Landing gear fixture complete, the lunar module is ready to be taken to the spacecraft-lunar module adapter.

APOLLO 11 LUNAR LANDING: 20 JULY 1969

THE EAGLE HAS LANDED
Buzz Aldrin just seconds before becoming the second man to walk on the Moon. The moment was captured by the first man, Neil Armstrong, 20 July 1969.

BOOTPRINT ON THE MOON
Photograph by Buzz Aldrin of his bootprint on the surface of the Moon an hour into humankind's first moonwalk during EVA 1, 20 July 1969.

PHOTO OF THE CENTURY
Portrait photograph of Buzz Aldrin as he poses next to the lunar module – Eagle. Neil Armstrong took this photograph with a 70-mm (2 ¾ in) lunar surface camera.

FLYING THE FLAG
Buzz Aldrin salutes the flag of the United States at Statio Tranquillitis (Tranquillity Base), where Apollo 11 landed. It was the first flag to fly on the surface of the Moon.

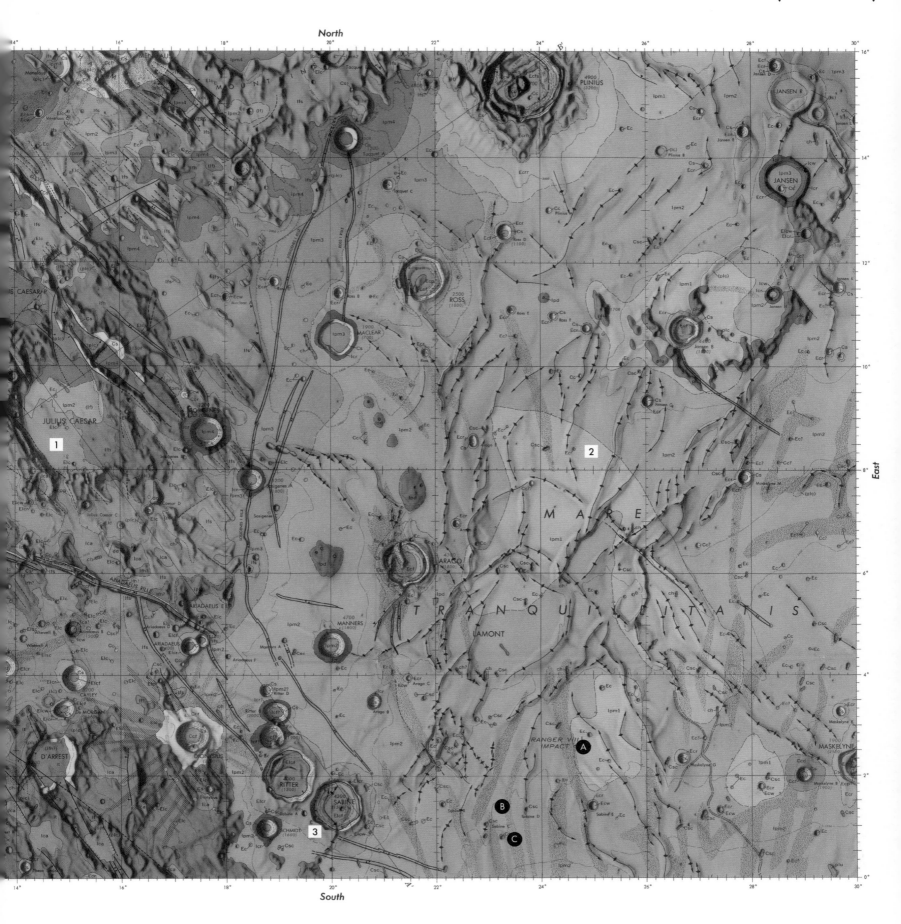

JULIUS CAESAR QUADRANGLE: KEY CHARACTERISTICS

1. JULIUS CAESAR CRATER
Latitude: 9.17°N / Longitude: 15.21°E
Diameter: 84.72 km / 52 ¾ miles
Depth: 3.4 km / 2 miles
Photograph: Lunar Orbiter IV (USA, 1967)

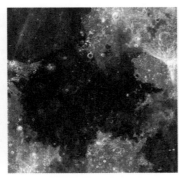

2. MARE TRANQUILLITATIS
Latitude: 8.35°N / Longitude: 30.83°E
Diameter: 875.75 km / 544 ¼ miles
Photograph: unknown

3. SABINE CRATER
Latitude: 1.38°N / Longitude: 20.07°E
Diameter: 29.75 km / 18 ½ miles
Depth: 1,300 m / 4,265 ft 1 in
Photograph: Apollo 11 (USA, 1969)

4. ARIDAEUS RILLE
Latitude: 6.48°N / Longitude: 13.44°E
Length: 247.45 km / 153 ¾ miles
Max Width: 5 km / 3 miles
Photograph: Apollo 10 (USA, 1969)

LUNAR PHOTOGRAPHY— OPPORTUNITY AND CHALLENGE

← The brothers William and Frederick Langenheim, Philadelphia photographers, took these first images of a total solar eclipse ever captured in the United States. They made eight daguerreotypes of the partial phases of the 26 May 1854 eclipse. Seven survive in the collections of the Metropolitan Museum in New York.

↱ Daguerreotype of the Moon taken by J. A. Whipple on 26 February 1852. He used the 38-centimetre (15-inch) Great Refractor at the Harvard College Observatory in Cambridge, Massachusetts. Whipple was an early manufacturer and inventor of photographic processes in the United States, and was a prominent photographer of people and buildings in Boston.

↱↱ The Great Refractor at Harvard, photographed by G. W. Pac (c. 1879). It was one of the two largest telescopes in the world around 1850, when Whipple collaborated with the director of the Harvard College Observatory, William Cinch Bond, on a series of astronomical photographs.

↳ Lewis Morris Rutherfurd's photograph (c. 1865) was the frontispiece of Richard Proctor's *The Moon, her Motions, Aspect, Scenery, and Physical Condition* (1873). It shows a waxing gibbous Moon in which south is up, an inverted image typically seen in a telescope.

he promise of a new technology can take longer to be realized than desired. Astronomers quickly reacted to Louis Daguerre's 1839 announcement of his chemical recording technique, sensitizing silver-plated copper sheets by iodine vapour. The Moon was an immediate target and early results were exciting. In 1853, the Oxford geologist and proto-planetologist John Phillips (1800–1874) hailed celestial photography before the British Association, predicting that: 'If photography can ever succeed in portraying as much of the Moon as the eye can see and discriminate, we shall be able to leave to future times monuments by which the secular changes of the Moon's physical aspects may be determined.'[1] In November 1856, however, he advised astronomer Edward Sabine to move away from photography and continue with visual techniques aided by micrometer measurements, recommending sketches first, then photography, followed by a comparison of the two, engaging both photography specialists and astronomers: 'Thus a strong interest would be maintained in the subject, and it would be really making progress.'[2]

Phillips' caution reflects the fact that, for the rest of the 19th century and well into the 20th, although photography revolutionized much of observational stellar astronomy, it never fully replaced the critical eye and hand as the means to map the lunar surface. Moon mappers continued to make hand-drawn sketches of the Moon, including some who still saw cities or plant life there. But photography did add new dimensions of permanence and objectivity.

The story of Lewis Morris Rutherfurd offers a particularly poignant example of both the opportunity posed by applying photography to lunar exploration, and the challenge of realizing it.

Rutherfurd (1816–1892), a wealthy lawyer and amateur astronomer, was excited by the earliest applications of the daguerreotype processes by astronomers in the USA, France and England; especially the spectacular successes at Harvard by W. C. Bond (1789–1859) and the daguerrotypist J. A. Whipple (1822–1891) in 1850, which, in the words of Charles Nevers Holmes in 1918, 'caused a great sensation and interest' at the 1851 London Exhibition.[3] Rutherfurd was determined to improve the new technique with photographic collodion (or 'wet process') emulsions, which allowed for

shorter exposures.[4] He felt his first results were worthy using his 11.25-inch (29-centimetre) Fitz refractor at his home in New York City. But they were not an improvement in contrast or detail over drawings. He took it as a stimulus to innovate.

In the early 1860s, Rutherfurd made a significant technical advance. He modified his telescope lens to concentrate light more effectively in the spectral region to which his emulsions were sensitive. On the night of 6 March 1865, when the air above his observatory was unusually stable, his negatives of the first quarter Moon 'were remarkably fine' with exposures from 2 to 3 seconds. Even so, he concluded that 'The great obstacle which prevents the results of photography from realizing the achievements of vision is atmospheric disturbance', which only the eye could overcome in brief moments of clarity as the image danced in the field of view.[5] Rutherfurd knew that photography promised much for astronomy, such as exposures of wide areas of the sky with large photographic plates, but the atmosphere was an insurmountable problem. Focused on instrumentation, he turned his attention to more promising areas such as spectroscopy and microscopy.

Even so, Rutherfurd's images were hailed as revolutionary, and he made every effort to get them published in the popular press.[6] The well-known popularizer Richard Proctor (1837–1888) featured a Rutherfurd photograph as the

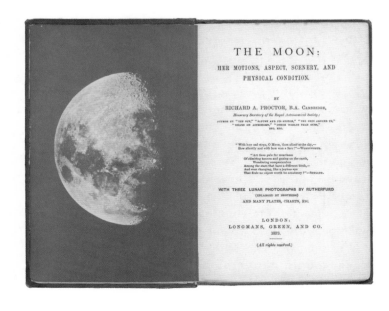

↓
James Nasmyth (1808–1890) used crayon sketches, such as this one of the crater Gassendi and its environs, as the first stage in his production of models of the lunar surface.

↓↓
James Nasmyth's 1844 plaster relief model shows the lunar craters Maurolycus and Barocius and the area around them. It is part of the collections of the Science Museum, London.

↓
Scientific American's cover for 23 December 1905 included an inset of the first quarter Moon and a close-up of Mare Crisium. These photographs were taken by Maurice Loewy and Pierre Puiseux in Paris.

↓↓
The Observatoire de Paris's large equatorial coudé telescope is shown on this postcard (c. 1900). This unusual instrument was used by Loewy and Puiseux to take lunar images.

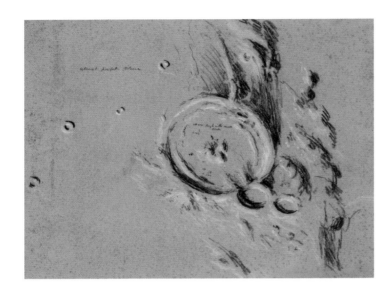

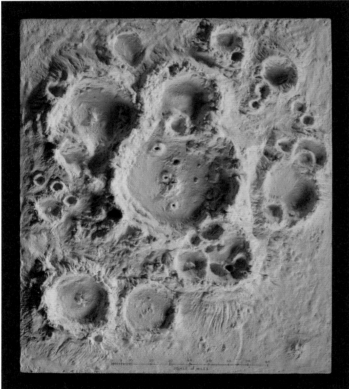

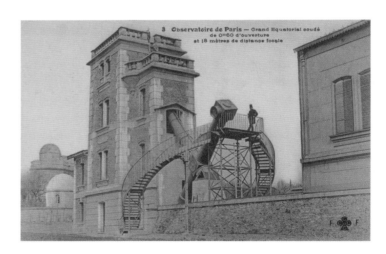

frontispiece of his book *The Moon* (1873, reprinted 1886), stating that 'it is difficult to conceive that anything superior can ever be obtained…' but still hoping that continuing efforts in America and Europe might someday do so.[7] And indeed they did, as the cover illustration for the December 1905 edition of *Scientific American* demonstrated with an image of the Moon from the Paris Observatory's Coudé refractor by Maurice Loewy (1833–1907) and Pierre Henri Puiseux (1855–1928).

Telescopes and photographic emulsions continually improved. Astronomers also took their instruments to regions of stable and clear air. Some, like the engineer James Nasmyth (1808–1890), following the ingenious creativity of polymath John F. W. Herschel (1792–1871), started with drawings, observed how shadows changed over time, created three-dimensional plaster topological models from those data, and then photographed the models. Others, seeking clearer and darker skies, put telescopes on high mountains, and devised new forms of telescopes to accommodate photography. Some of these collected more light, such as George Willis Ritchey's (1864–1945)

reflecting telescopes built for George Ellery Hale at Yerkes Observatory, Wisconsin and then (from 1904 to 1919) at Mount Wilson, so that exposures could be shortened to a fraction of a second. These efforts were aided by photographic providers, most notably the indomitable C. E. Kenneth Mees (1882–1960) at Eastman Kodak in the 20th century, who improved the sensitivity and resolution of photographic emulsions. Their collective efforts led to significant improvements in photographic imagery, but not solutions to all the difficulties.

Many of the questions posed by 19th-century astronomers about the nature and origin of lunar features remained unresolved in the 20th century. As one review observed, 'more distant objects absorbed the interest of astronomers, [and] the study of the Moon was neglected'.[8] There were still enthusiasts, to be sure. In 1925, the Carnegie Institution of Washington organized a 'Committee on Study of the Surface Features of the Moon' engaging astronomers, physicists and geophysicists. One of them, astronomer Francis G. Pease (1881–1938) from Mount Wilson, was especially passionate. At a public lecture at the Los Angeles

↓
Four plates from Gerard
P. Kuiper's *Photographic
Lunar Atlas* (1960). These
photographs of an area
near the Moon's limb under
different lighting conditions
and lunar librations show
various viewing angles
and illuminations of the
same area. The lunar atlas
contained photographs
taken at the Mount Wilson,
Lick, Pic Du Midi, McDonald
and Yerkes Observatories.

Public Library on 15 December 1927, Pease, who had been photographing the Moon with the 60-inch (1.52-metre) and 100-inch (2.54-metre) telescopes, used the Moon as one goal for building larger telescopes, made more sensitive with improved photographic films. Campaigning for larger and larger telescopes, echoing his boss George Ellery Hale's dreams, Pease referred to the hand drawings of Polish astronomer Johannes Hevelius (1611–1687) and how their fidelity was now more than matched by photography. But even beyond that, although he saw the future in larger telescopes on Earth, still he invited his audience to imagine that 'we could step into Jules Verne's projectile and pay a visit to the Moon; see it closely, with its rills and valleys, its craters and mountain peaks. Would it not be worth the while?'[9]

In Pease's day, this was a fantasy. But by the late 1950s it was becoming a possibility, so to prepare, Yerkes planetary astronomer Gerard Kuiper (1905–1973) proposed a definitive photographic lunar atlas using the 82-inch (2.08-metre) reflector at McDonald Observatory in Texas. After inspecting what Pease and others had already done, however, Kuiper realized that the atlas would not be ready for the lunar missions. So he and his staff combined the best photographs available from the major observatories, supplemented them with their own efforts, and produced a series of atlases that, following guidelines on coordinates and nomenclature set by international committees, allowed them to create a rectified map of the visible lunar surface by projecting images onto a sphere and photographing them.[10] Beyond furthering lunar science, this effort could supply precise locations and topographical characteristics to help place images from the Lunar Orbiters that would identify the best landing sites where the many questions about the nature of the Moon and its surface could be explored *in situ*.

Despite the best efforts of Kuiper and his cohorts, producing a rectified photographic map giving maximum detail, visual mapping continued to be competitive, especially the combined efforts in the 1960s of the US Air Force's Aeronautical Chart and Information Center, the US Army Map Service and the US Geological Survey. The reasoning behind this effort, no doubt, has a complex history, but the view of at least one ardent amateur group, the Lunar Recorders of the Association of Lunar and Planetary Observers, summed up popular opinion very simply in 1970:

> Even though the Orbiters have produced outstanding views, it should be remembered that high-resolution photography has left many problems still unanswered. Such photographs (though highly detailed) represent only a very limited view of an area in space and time. It is now, and will continue for some time, impossible to secure a complete high-resolution photographic record under all conditions of lighting.[11]

Passions aside, by the time the Lunar Orbiters were in operation armed with panoramic cameras searching for suitable Apollo landing sites, the Air Force team closed down their operation. As Harold 'Hal' Masursky, a leader of the Geological Survey effort, recalled, 'That made the observations through the telescope in effect obsolete'.[12]

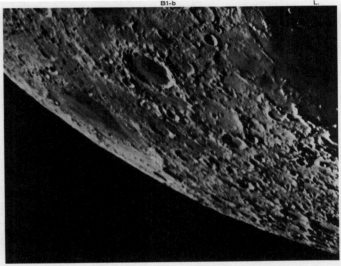
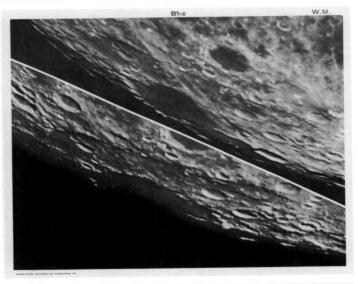
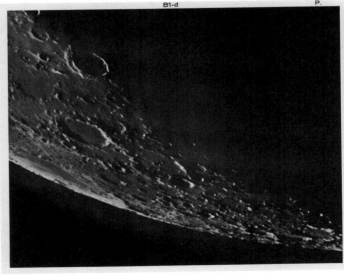

↑
James Nasmyth used photographs of his plaster models to illustrate lunar features in *The Moon: Considered as a Planet, a World, and a Satellite* (1874). Plate XXI (top) shows the exaggerated relief produced by viewing the Moon near the terminator, which produced long, dramatic shadows. Plate XIII (left) shows Nasmyth's model of Ptolemaeus, Alphonsus and Arzachel craters. Plate XI (right) shows Nasmyth's model of the Triesnecker crater.

↑
Three further plates from James Nasmyth's book, showing his plaster models. Plate XXIII (top) reveals Nasmyth's model of a group of lunar mountains, making an ideal lunar landscape. Again, the ruggedness of the terrain is exaggerated compared to what would be found in the space age. Plate IX (left) shows a model of the Lunar Apennines and Archimedes crater. Plate XVI (right) is a model of the crater Tycho and its surroundings.

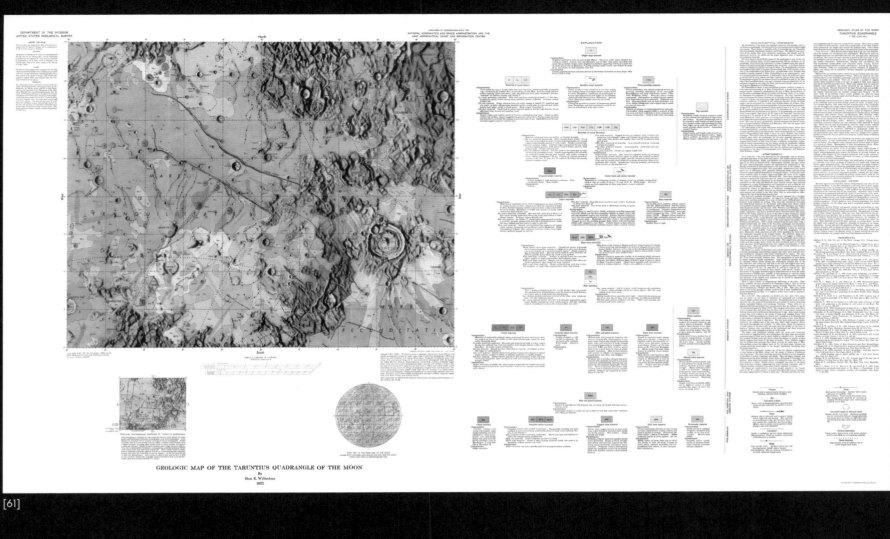

GEOLOGIC MAP OF THE TARUNTIUS QUADRANGLE OF THE MOON
By
Don E. Wilhelms
1972

[61]

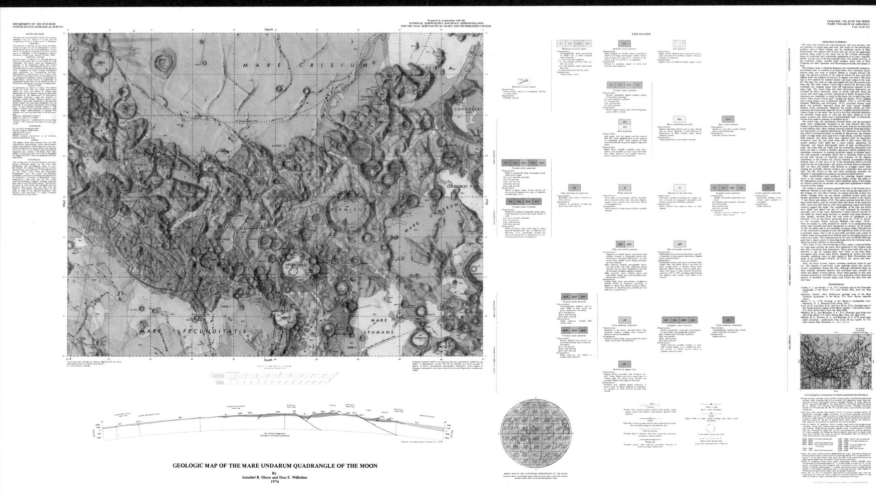

GEOLOGIC MAP OF THE MARE UNDARUM QUADRANGLE OF THE MOON
By
Annabel B. Olson and Don E. Wilhelms
1974

[62]

61
Facsimile:
*Geologic Map of the Taruntius
Quadrangle of the Moon, 1972,*
by Don E. Wilhelms

62
Facsimile:
*Geologic Map of the Mare
Undarum Quadrangle of the
Moon, 1974,* by Annabel B. Olson
and Don E. Wilhelms

74
Facsimile:
*Geologic Map of the Grimaldi
Quadrangle of the Moon, 1973,*
by John F. McCauley

75
Facsimile:
*Geologic Map of the Letronne
Quadrangle of the Moon, 1963,*
by C. H. Marshall

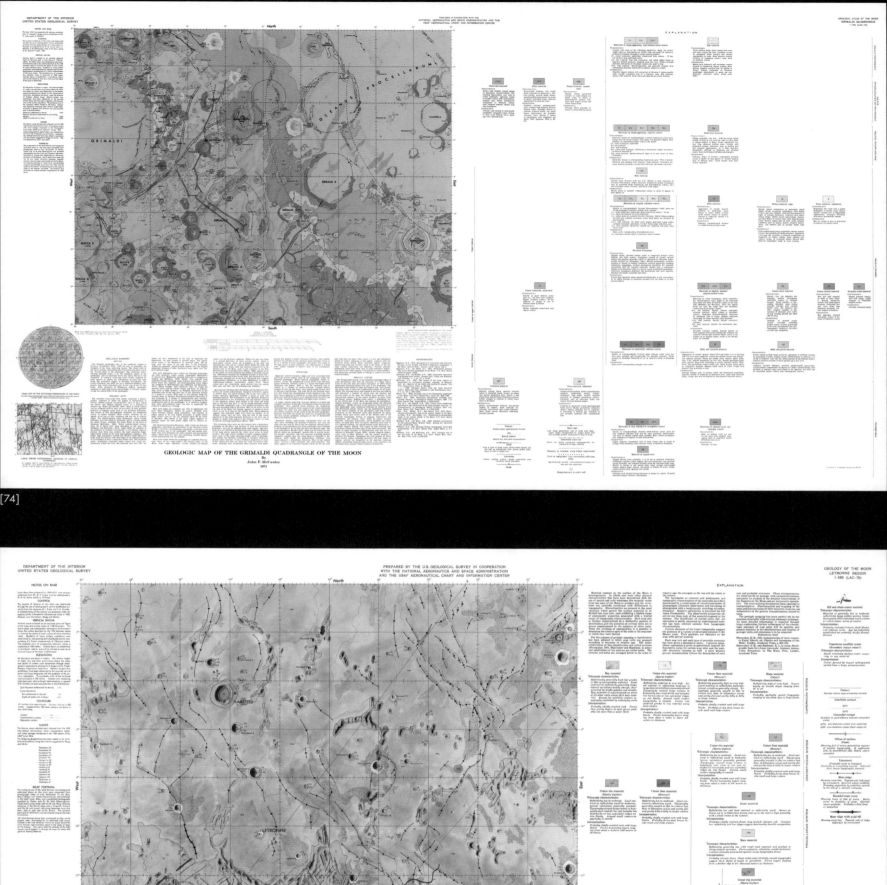

GEOLOGIC MAP OF THE GRIMALDI QUADRANGLE OF THE MOON

By

John F. McCauley

1973

[74]

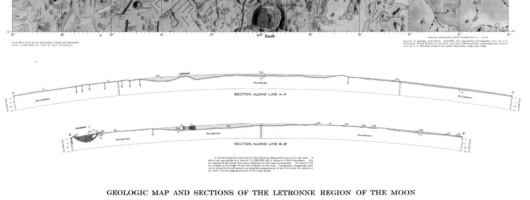

GEOLOGIC MAP AND SECTIONS OF THE LETRONNE REGION OF THE MOON

By

C. H. Marshall

1963

A PAPER MOON — CARTES DE VISITE

↱
Three 19th-century *cartes de visite* depicting a paper Moon. With the development of photographic techniques and the increasing sophistication of private studios, the public was offered fanciful ways of having their portrait taken. Reutlinger Photography Studio, founded in 1850 in Paris, produced the example, left, of a woman sitting on a crescent-shaped Moon. The Schloss Studio in New York, adopted a more fantastical approach, producing the profile of a woman's face emerging from the Moon. In 1903, a studio opted for a more humorous ensemble of a man standing with his arm around a sleeping Moon.

ollecting and exchanging sought-after miniature images known as *cartes de visite* became a popular trend in the 19th century in the wake of developments in the mass production of photographic images.

Accurate visual representation of the Moon had long been a fascination for scientists. The arrival of photography only magnified this interest. But the creation of a popular and cheap form of mass-produced photographic image not only made lunar images more widely available, but also enabled photographic studios in the late 19th century to offer their clients a playful way to engage with a fad for all things lunar.

Almost thirty years after Joseph Nicéphore Niépce (1765–1833) took what is regarded as the first photograph, 'View from the Window at Le Gras' (1826/27), the medium had become so popular that there was substantial demand for it from all sections of society. The cost of producing

an image was made significantly cheaper by André-Adolphe-Eugène Disdéri (1819–1889), who made use of the introduction of the collodion wet plate process,[1] a cheaper form of photography. He used it to create an image that could be reproduced in the format used for calling cards (these were left during social visits by the middle and upper classes). Through the use of a four-lensed camera, he was able to produce eight 8.9 × 6.35-centimetre (3.5 × 2.5-inch) negatives on one full-sized plate. The plate could be printed on relatively cheap albumen paper[2] and then cut up. The individual photographs were mounted on to affordable cards measuring 10 × 7.7 centimetres (4 × 3 inches). He patented his process in 1854 and for a time it proved extraordinarily popular. (Although Disdéri was the person to profit from them, the *cartes de visite* were actually created by fellow French photographer Louis Dodero [1824–1902] in Marseilles, in 1851.)

Though the original intention of *cartes de visite* was to produce accessible images of family members, there was soon a demand for cards featuring notable figures in society, from politicians and scientists to artists and celebrities of the day. The craze for this new form, which became known as carteomania, even extended to royalty. Britain's Queen Victoria and Prince Albert were avid collectors. They even had portraits of themselves made for their family. And then, in 1860, the royal couple gave permission to photographer John Jabez Edwin Mayall (1813–1901) to take a portrait of them that would be used on a *carte de visite* made available to the general public. The response was immense, and the format became even more enthusiastically embraced.

The American Civil War was arguably the first major conflict to be covered extensively by photographers. One of the most notable was Mathew Brady (1822–1896), who helped shape the public's perception of what life was like on the front. He and other photographers did great business in reuniting families with their loved ones through – often heroic – photographic portraits. But just as the *cartes de visite* emerged as a result of cheaper means to produce images, further changes in technology ultimately spelled the death knell of the form, whose popularity waned from the late 1860s as the larger cabinet portrait replaced it.

There were two very different uses of the *carte de visite* when it came to representations of the Moon. The first was from a purely scientific perspective. The format allowed images of the Moon to be easily transported and gave those

with an interest in the subject the opportunity to own such photographs at a fairly low cost. It also gave astronomers and scientific publishers an opportunity to create a revenue stream from their research. In many instances, the images in these *cartes de visite* were reproduced from larger, earlier photographs. This is the case with American photographer Austin Augustus Turner's (1831–1870) *Twelve Photographs of the Moon* collection, a copy of which is owned by New York's Metropolitan Museum of Art. Produced as a boxed set in 1863, the images were actually pirated from original photographs taken by Warren De La Rue (1815–1889), a chemist and inventor who was noted for his work as an astronomical photographer.

A more popular example of the Moon's presence in *cartes de visite* was in general portraiture. The photographic portrait form itself can be traced back to the late 1830s, when Robert Cornelius (1809–1893) first photographed himself. The earliest known photographic studio, for portraits of paying customers, was opened in New York, in March 1840, by Alexander S. Wolcott (1804–1844). His 'Daguerrean Parlor', named after the first commercially successful photographic process (invented by Louis Daguerre [1787–1851] in 1835 and made public in 1839), was soon followed by many others. Imaginative sets and backgrounds became popular and many sitters were depicted with a large crescent Moon. This form of portraiture was cheaper than painting, so demand was much greater; it also took less time to produce than a painting. However, until developments in technology resulted in more sensitive photographical materials, and therefore less exposure time, a sitter needed to remain completely still for a number of seconds or even minutes. This severely limited a sitter's possible poses. However, the environment surrounding the sitter could be as fanciful as the photographer and their subject desired, and the Moon – an otherworldly, fantastic and even exotic image – often featured.

Advances in science, the growing popularity of science fiction, an increased awareness of our place in the universe thanks to developments in astronomical research, and a general sense of wonderment about our celestial neighbour may have prompted a desire to include representations of the Moon in photographic portraits. But it is clear from the many *cartes de visite* that feature a (frequently anthropomorphized) crescent Moon that the feeling of wonder produced by this relatively new medium was perfectly reflected in the inclusion of an equally wondrous setting.

A selection of paper Moon portraits. The presence of the Moon in *cartes de visite* proved more than a short-term fad. Once the trend took off, it sustained a remarkable degree of popularity during the late 19th and the early 20th centuries. Unlike some of the more inventive examples of photographic magic, combining the Moon, the photographer's subject and a variety of effects, many paper Moon portraits opted for a conventional set-up. The crescent Moon, with or without a face, was a popular choice. But the longevity of this style is found less in the photographic techniques, or the changing style of the sets that subjects appear in, than in the fashions they wear.

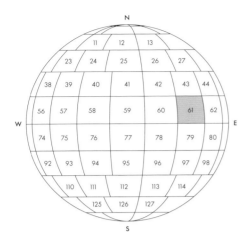

The crater Taruntius, after which this quadrangle is named, is one of the youngest and brightest prominent features on this map. Not only does it stand out from the dark surrounding lava plains of Mare Fecunditatis (Sea of Fertility) and Mare Tranquillitatis (Sea of Tranquillity), but its otherwise intact features are overlaid with a few even younger, brighter, small bowl-shaped craters and their ejecta (as are the plains). Five mare basins intersect here – Tranquillitatis, Fecunditatis, Serenitatis, Nectaris and Crisium – their structure outlined in dotted lines. Their intersections have formed much of the rugged terrain in this region, including the terra mountains and islands. The oldest of these basins is Tranquillitatis, and we can see the deformation of the basin in the ridges and scarps imposed by the formation of the newer basins. Lava domes dot the surface of Mare Tranquillitatis, as do small craters of varying ages. The oldest craters are mostly concentrated in the mountains, as this terrain has not been covered in lava plains since their formation. The older craters that can be found in the mare, such as the satellites of Maskelyne, are only partially intact, much of their rims being submerged or degraded.

PRINCIPAL PHOTOGRAPHIC COVERAGE
OF TARUNTIUS QUADRANGLE

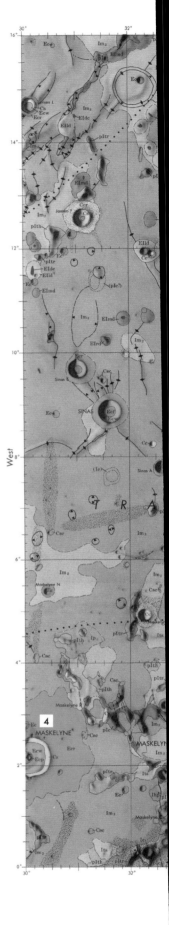

MARE CRISIUM
An Apollo 11 photograph of the surface of the Moon, taken July 1969. The large crater Taruntius is visible to the left of Mare Crisium.

TARUNTIUS QUADRANGLE: KEY CHARACTERISTICS

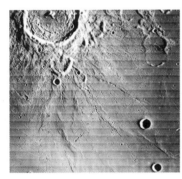

1. TARUNTIUS CRATER
Latitude: 5.50°N / Longitude: 46.54°E
Diameter: 57.32 km / 35 ½ miles
Depth: 1,000 m / 3,280 ft 10 in
Photograph: Lunar Orbiter I (USA, 1966)

2. MARE FECUNDITATIS
Latitude: 7.83°S / Longitude: 53.67°E
Diameter: 840.35 km / 522 ¼ miles
Photograph: LRO (USA)

3. MARE TRANQUILLITATIS
Latitude: 8.35°N / Longitude: 30.83°E
Diameter: 875.75 km / 544 ¼ miles
Photograph: LRO (USA)

4. MASKELYNE CRATER
Latitude: 2.16°N / Longitude: 30.04°E
Diameter: 22.42 km / 14 miles
Depth: 2.5 km / 1 ½ miles
Photograph: Apollo 10 (USA, 1969)

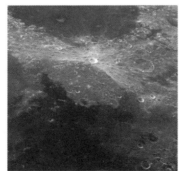

5. PALUS SOMNII
Latitude: 13.69°N / Longitude: 44.72°E
Diameter: 163.45 km / 101 ½ miles
Photographed from Earth (China, 2020)

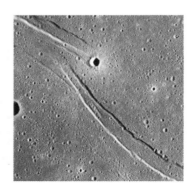

6. CAUCHY I RILLE
Latitude: 10.42°N / Longitude: 38.07°E
Diameter: 167 km / 103 ¾ miles
Depth: 2.8 km / 1 ¾ miles
Photograph: LRO (USA)

7. DA VINCI CRATER
Latitude: 9.10°N / Longitude: 44.95°E
Diameter: 37.46 km / 23 ¼ miles
Photograph: Lunar Orbiter IV (USA, 1967)

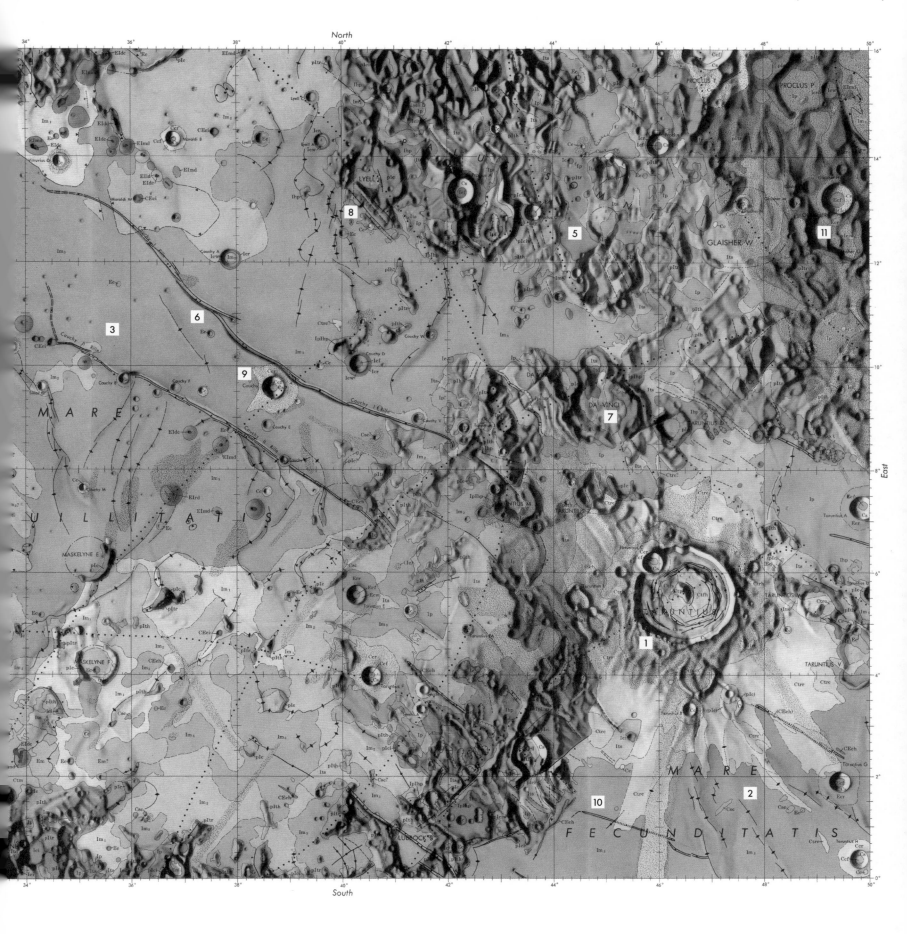

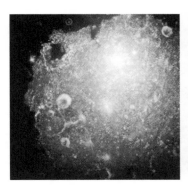

8. LYELL CRATER
Latitude: 13.63°N / Longitude: 40.56°E
Diameter: 31.17 km / 19 ¼ miles
Depth: 1,000 m / 3,280 ft 10 in
Photograph: Apollo 17 (USA, 1972)

9. CAUCHY CRATER
Latitude: 9.56°N / Longitude: 38.63°E
Diameter: 11.80 km / 7 ¼ miles
Depth: 2.6 km / 1 ½ miles
Photograph: Apollo 17 (USA, 1967)

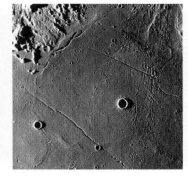

10. RIMMA SECHII
Latitude: 1°N / Longitude: 44.08°E
Length: 35 km / 21 ¾ miles
Photograph: LRO (USA, 2018)

11. GLAISHER CRATER
Latitude: 13.19°N / Longitude: 49.34°E
Diameter: 15.92 km / 10 miles
Depth: 2 km / 1 ¼ miles
Photograph: Apollo 15 (USA, 1971)

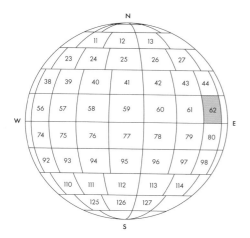

Cratered terra runs through the basaltic plains of several maria, including Mare Undarum (Sea of Waves). These highlands are the southern rim of the multi-ringed basin that contains Mare Crisium (Sea of Crises). Dotted lines mark the positions of the overlapping basin rings in this region, following the contours of mountainous rings and rugged terrae. The terrain has been heavily modified over time. Gravity has pulled materials downslope, rounding sharp edges and filling depressions. Impact cratering further modified this region, and the maria basins were filled with smooth, dark, basaltic lava plains. Some younger, sharp-rimmed craters can be found in this quadrangle, but cratering appears to have slowed considerably in this region in the Moon's later geologic history. The largest prominent craters, Firmicus and Apollonius, are considerably worn and subdued. On 11 September 1971, the Soviet Union's unsuccessful Luna 18 spacecraft crashed at coordinates 3.34°N, 56.30°E. On 21 February 1972, Luna 20 successfully landed at nearby coordinates 3.79°N, 56.62°E (marked with the words 'LUNA 20'). The probe returned soil from the Apollonius highlands region to Earth. The next successful landing was the Soviet Union's Luna 24 mission in 1976, which also returned samples.

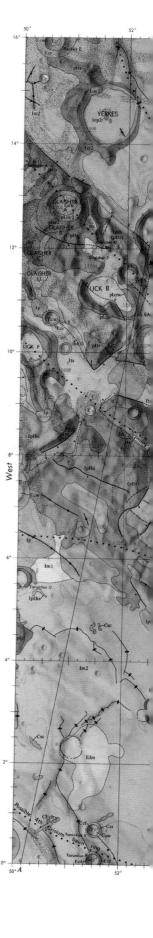

LUNA 18	ROCKET	Proton K + Blok D
	CREW	N/A
	LAUNCH DATE	13:40:40, 02 September 1971 (UTC)
	LAUNCH SITE	Tyuratam (Baikonur Cosmodrome)
	PAYLOAD	Remote Arm for Sample Sollection
	LUNAR ORBIT INSERTION	07 September 1971
	LANDING DATE	07:47:16, 11 September 1971
	IMPACT SITE	Near the edge of Mare Fecunditatis
	LANDING COORDINATES	3.34°N, 56.30°E Ⓐ

LUNA 20	ROCKET	Proton K + Blok D
	CREW	N/A
	LAUNCH DATE	03:27:58, 14 February 1972 (UTC)
	LAUNCH SITE	Tyuratam (Baikonur Cosmodrome)
	PAYLOAD	Stereo Imaging System
	LUNAR ORBIT INSERTION	18 February 1972
	LANDING DATE	19:19:00, 21 February 1972 (UTC)
	LANDING SITE	Apollonius Highlands
	LANDING COORDINATES	3.79°N, 56.62°E Ⓑ

LUNA 23	ROCKET	Proton K + Blok D
	CREW	N/A
	LAUNCH DATE	14:30:32, 28 October 1974 (UTC)
	LAUNCH SITE	Tyuratam (Baikonur Cosmodrome)
	PAYLOAD	LB09 Drill for Sample Sollection
	LUNAR ORBIT INSERTION	02 November 1974
	LANDING DATE	N/A, 06 November 1974 (UTC)
	LANDING SITE	Mare Crisium
	LANDING COORDINATES	12.41°N, 62.17°E Ⓒ

LUNA 24	ROCKET	Proton K + Blok D
	CREW	N/A
	LAUNCH DATE	15:04:12, 09 August 1976 (UTC)
	LAUNCH SITE	Tyuratam (Baikonur Cosmodrome)
	PAYLOAD	Radiation Detector, Radio Alitmeter
	LUNAR ORBIT INSERTION	14 August 1976
	LANDING DATE	06:36:00, 18 August 1976 (UTC)
	LANDING SITE	Mare Crisium
	LANDING COORDINATES	12.71°N, 62.21°E Ⓓ

MARE UNDARUM QUADRANGLE: KEY CHARACTERISTICS

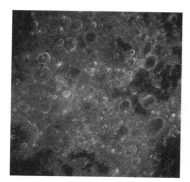

1. MARE UNDARUM
Latitude: 7.49°N / Longitude: 68.66°E
Diameter: 244.84 km / 152 ¼ miles
Photograph: Apollo 8 (USA, 1968)

2. MARE CRISIUM
Latitude: 16.18°N / Longitude: 59.10°E
Diameter: 555.92 km / 345 ½ miles
Photograph: LRO (USA, 2012)

3. MARE SPUMANS
Latitude: 1.3°N / Longitude: 65.3°E
Diameter: 143.13 km / 89 miles
Photograph: LRO (USA, 2015)

4. RIMA APOLLONIUS
Latitude: 4.39°N / Longitude: 54.33°E
Diameter: 89.64 km / 55 ¾ miles
Photograph: Lunar Orbiter I (USA, 1966)

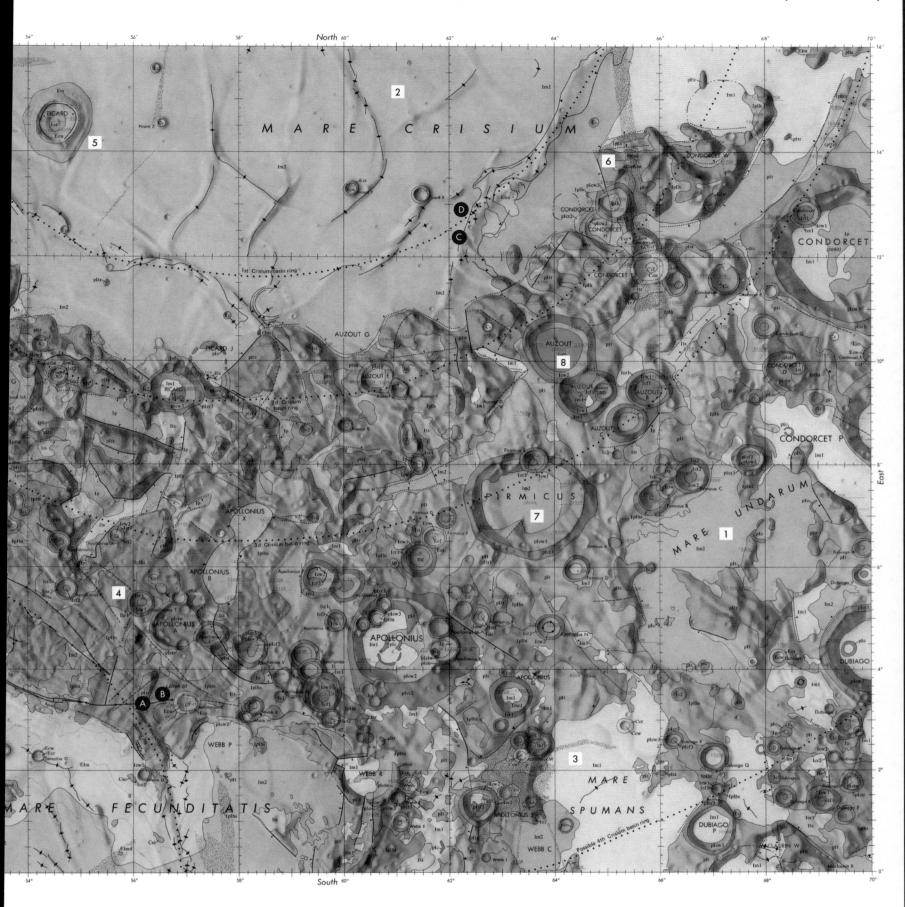

5. PICARD CRATER
Latitude: 14.57°N / Longitude: 54.72°E
Diameter: 22.35 km / 14 miles
Depth: 2.4 km / 1 ½ miles
Photograph: Apollo 17 (USA, 1972)

6. PROMONTORIUM AGARUM
Latitude: 13.87°N / Longitude: 65.73°E
Diameter: 62.46 km / 38 ¾ miles
Height: 3.65 km / 2 ¼ miles
Photograph: Lunar Orbiter IV (USA, 1967)

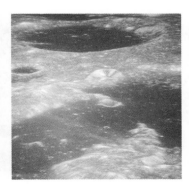

7. FIRMICUS CRATER
Latitude: 7.25°N / Longitude: 63.43°E
Diameter: 56.81 km / 35 ¼ miles
Depth: 1.5 km / 1 mile
Photograph: Apollo 10 (USA, 1969)

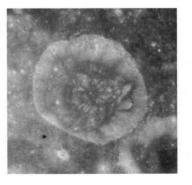

8. AUZOUT CRATER
Latitude: 10.21°N / Longitude: 64°E
Diameter: 32.92 km / 20 ½ miles
Photograph: Apollo 15 (USA, 1971)

THE FEMININE SYMBOLISM OF THE MOON

→
Aleksandra Mir, *First Woman on the Moon* (1999), video still. On the thirtieth anniversary of the 1969 Moon landing, Mir created a low-budget and humorous video to comment on the epochal event from a feminist perspective. In the work the artist recuperates the mythology of the Moon as a symbol of femininity and fertility to provide a powerful take on gender inequality.

↱
Yayoi Kusama, *Soul Under the Moon* (2002). This immersive installation consists of a purpose-built room lined with infinity mirrors containing multiple neon-coloured spheres, hanging at various heights. Kusama's Moon-inspired distinctive polka dots and orbs fuse with mirrors to create the illusion of a never-ending space, which is also a nurturing and mesmerizing world.

↳
Judy Chicago, *The Dinner Party* (1974–79). This seminal feminist artwork challenges notions of patriarchy by celebrating under-represented female historical figures. Among them is a tribute to Ashtoreth, the supreme female deity of the Phoenicians associated with the crescent Moon, an inclusion which references the enduring importance of the feminine symbolism of the Moon.

↲
Aura Satz, *The Leavitt Crater* (2014). Satz pays tribute to the lesser-known individuals who made the Moon's exploration possible. Using a Problicom projector, she showed images of the craters named after women astronomers, including that of the Leavitt crater, dedicated to the pioneering deaf astronomer Henrietta Swan Leavitt (1868–1921).

n 1999, coinciding with the 30th anniversary of the first Moon landing, artist Aleksandra Mir (b. 1967) transformed a sandy landscape near the Dutch town of Beverwijk into a low-budget lunar film set. At the end of the one-day performance, accompanied by children, Mir planted an American flag on the highest dune, becoming the *First Woman on the Moon* (1999). Staging her 'feminine attempts'[1] to problematize the male 'first step' on the lunar soil, in the twelve-minute video Mir reappropriates the feminine symbolism of the Moon to spotlight gender inequalities. She does so by bringing together two universally recognized cultural references: the first Moon landing and the enduring symbolism of the Moon as a female entity associated with traditional feminine character traits such as emotion, intuition and instinct, and life stages such as conception and birth.

By bringing young children along on the conquest of her fictional Moon, and creating a deliberately low-tech event more akin to craft than science, the artist reappropriates the mythology of the Moon as a female, natural realm. In doing so, she connects to a previous generation of feminist artists who challenged notions of patriarchy by repudiating science and technology and reclaiming a connection to nature. This is the case for Judy Chicago's (b. 1939) seminal work *The Dinner Party* (1974–79) which, among other under-represented female historical figures, makes a direct reference to the feminine mythology of the Moon by including a tribute to Ashtoreth,[2] the supreme female deity of the Phoenicians associated with the crescent Moon. In her early Mesopotamian incarnation, she was known as

Inana,[3] the goddess of love and war, daughter of Nanna, the Sumerian male god of the Moon. Throughout time and geographies, Inana was assimilated into many other deities including Artemis, the Greek goddess of hunting, who like her Roman equivalent Diana (whose name in Latin means 'the bright one'), was worshipped as the goddess of the half-Moon on Earth, representing the Maiden, youth, innocence and new beginnings. Together, the Maiden, the Mother and the Crone embodied the Triple Goddess archetype, found across cultures, reflecting the cyclical nature of the lunar phases.[4] The Crone archetype, referencing wisdom, transformation and the waning phase of life, was represented in ancient Greece by Hecate, the goddess of the dark Moon in the underworld. The Mother, represented by the Greek Moon-goddess Selene and associated with fertility, abundance and nurturing, is a clear reference point for Chicago's *The Birth Project* (1980–85).[5]

The opposition of scientific patriarchy versus natural feminine force found in the work of some exponents of feminist art can be traced back to the centuries-old Moon and Sun dualism representing that of woman and man (as well as the feminine/masculine, mother/father, night/day, soul/body). If in Greek and Roman mythology the feminine energy within the Moon was read as connecting to our intuition and emotions, the Sun was seen as the governor of justice, a power linking us to our inner strength and a source of rational wisdom. Similarly, Chang'e, the Chinese Moon-goddess, was associated with Yin, the feminine principle complementary to the masculine Yang, while Mama Quilla, the Moon-goddess in Inca mythology, was married to Inti, the god of the Sun. The Roman goddess of the Moon, Luna, was also seen as the female counterpart of the Sun; considered the celestial embodiment of feminine energy, Luna, like Selene, also represented change, growth, renewal, beauty and cosmic harmony. These enduring associations are reinterpreted by Yayoi Kusama (b. 1929) in installations such as *Soul Under the Moon* (2002) in which infinity mirrors and her Moon-inspired distinctive polka dots and orbs fuse to create an intrinsically immersive world.[6]

The duality of science and nature, machine and craft as reflecting that of man and woman, and Sun and Moon, is also at the base of Vija Celmins' (b. 1938) pencil drawings of the Moon's surface (1969–72).[7] *Moon Surface (Luna 9) #1* (1969) exemplifies the painstaking accuracy and skill with which the artist brings together the scientific precision of

her source (images of the lunar surface provided by Soviet and US robotic spacecraft) and her patient, labouring craft. Seen from a distance the images appear as impersonal reproductions of mechanical work; from up close the artist's delicate, feminine touch makes them feel tactile and emotional. Like Mir, Celmins responds to the Moon landing event by reassociating the surface of the Moon with femininity; like Chicago, she highlights the creative labour of craft, traditionally associated with female labour, which she uses to re-naturalize the geography of the Moon and provide an intimately emotional experience of it. Similarly, in the installation *Her Luminous Distance* (2014), Aura Satz (b. 1974) references the Moon's landscape but pays tribute to the lesser-known individuals who made its exploration possible by showing slides of the craters named after women astronomers, including pioneering deaf astronomer Henrietta Swan Leavitt (1868–1921). The images are projected using a Problicom (Projector Blink Comparator). Created by Ben Mayer (1925–1999), this allowed amateur astronomers to contribute to academic research; like the professional Blink Comparator (1904), it assisted observations of the changed position and brightness of objects in the night sky through a very rapid switching back and forth between two photographs. The meaningful choice of this projector allows the artist to use the lens of gender to look back to scientific discoveries while simultaneously merging the lines dividing ideas of femininity from those of masculinity and science from nature. In doing so, *Her Luminous Distance* contributes to the construction of a new feminine mythology of the Moon able to reflect the true complexity of what constitutes female identity in the present.

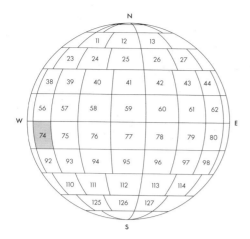

In the northwest corner of the map is the two-ring Grimaldi basin from which this quadrangle takes its name. To its east is a portion of Oceanus Procellarum (Ocean of Storms). The terra and mare south of Oceanus Procellarum seem to follow the contours of a vestigial multi-ring basin that has been flooded by lava. Much of the western portion of the quadrangle is moderately cratered highland terrain with impact craters of varying sizes and ages giving way to the plains of Lacus Aestatis (Summer Lake, here identified as Mare Aestatis). Much of this quadrangle has been covered in ejecta material thrown out by impacts in other nearby quadrangles. Volcanism followed, erasing earlier craters along with the original textures of the basin ejecta blankets. The western third of this quadrangle is covered by the more recent ejecta blanket of the nearby Orientale basin, one of the youngest such multi-ringed basins on the Moon's near side. The mare then flooded with lava, forming the dark plains. Impacts continued to crater the surface, and several young simple craters and their rays can be seen to overlay older craters, terrae and maria.

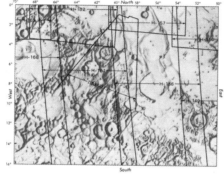

LUNAR ORBITER PHOTOGRAPHIC COVERAGE
OF GRIMALDI QUADRANGLE

DAMOISEAU CRATER, OCEANUS PROCELLARUM
The Damoiseau crater (centre right) and Oceanus Procellarum (foreground) photographed by Lunar Orbiter III on 22 February 1967.

GRIMALDI QUADRANGLE: KEY CHARACTERISTICS

1. GRIMALDI BASIN
Latitude: 5.38°S / Longitude: 68.36°W
Diameter: 235 km / 146 miles
Depth: 2.7 km / 1 ¾ miles
Photograph: Lunar Orbiter IV (USA, 1967)

2. LOHRMANN CRATER
Latitude: 0.44°S / Longitude: 67.38°W
Diameter: 31.25 km / 19 ½ miles
Depth: 1.7 km / 1 mile
Photograph: Lunar Orbiter IV (USA, 1967)

3. MARE AESTATIS
Latitude: 14.83°S / Longitude: 68.57°W
Diameter: 86.39 km / 53 ¾ miles
Photograph: Lunar Orbiter IV (USA, 1967)

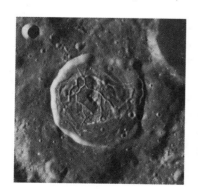

4. DAMOISEAU CRATER
Latitude: 4.85°S / Longitude: 61.25°W
Diameter: 36.66 km / 22 ¾ miles
Depth: 1,200 m / 3,937 ft
Photograph: Lunar Orbiter IV (USA, 1967)

5. SIRSALIS CRATER
Latitude: 12.5°S / Longitude: 60.5°W
Diameter: 44.17 km / 27 ½ miles
Depth: 3 km / 1 ¾ miles
Photograph: Lunar Orbiter IV (USA, 1967)

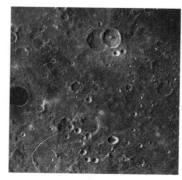

6. RIMA SIRSALIS
Latitude: 15°S / Longitude: 61.36°W
Length: 405 km / 251 ¾ miles
Photograph: LRO (USA, 2014)

7. HANSTEEN CRATER
Latitude: 11.53°S / Longitude: 52.06°W
Diameter: 45 km / 28 miles
Depth: 1,300 m / 4,265 ft 1 in
Photograph: Lunar Orbiter IV (USA, 1967)

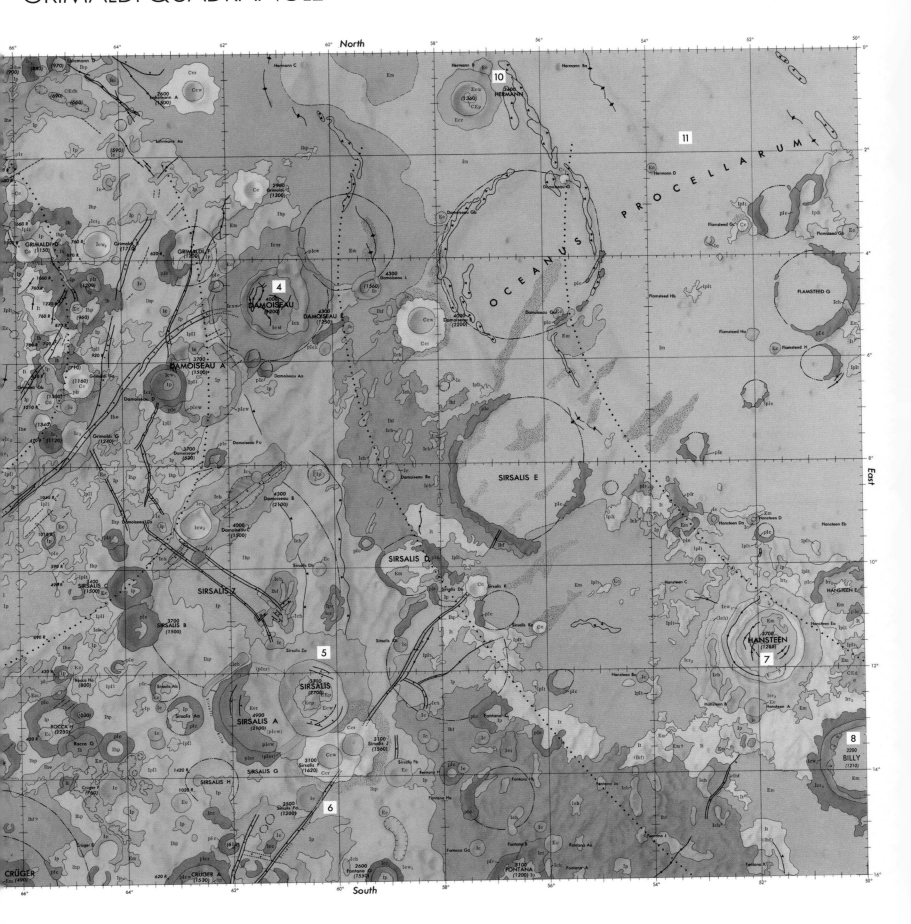

8. BILLY CRATER
Latitude: 13.83°S / Longitude: 50.24°W
Diameter: 45.57 km / 28 ¼ miles
Depth: 1,200 m / 3,937 ft
Photograph: Smart-1 mission (ESA, 2006)

9. CRÜGER CRATER
Latitude: 16.68°S / Longitude: 66.96°W
Diameter: 45.94 km / 28 ½ miles
Depth: 500 m / 1,640 ft 5 in
Photograph: Lunar Orbiter IV (USA, 1967)

10. HERMANN CRATER
Latitude: 0.33°S / Longitude: 57.25°W
Diameter: 5.84 km / 3 ¾ miles
Photograph: LRO (USA, 2012)

11. OCEANUS PROCELLARUM
Latitude: 20.67°N / Longitude: 56.68°W
Diameter: 2,592.24 km / 1,610 ¾ miles
Photograph: LRO (USA, 2014)

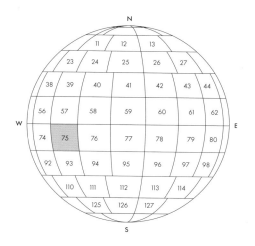

Although the name Letronne appears next to a small bowl-shaped crater, Letronne crater is in fact the larger, ridged half-circle that forms the outlines of a bay on the southwestern edge of Oceanus Procellarum (Ocean of Storms). This crater was long ago flooded with lava, leaving only the southern portion of the crater's rim visible. The smaller craters within Letronne, which also carry its name, are younger. These craters have more well-defined walls and rims, and visible ray systems overlaying the surrounding basalt plains. Most of these are simple craters without a central peak. The large Gassendi crater that lies below this quadrangle is not visible, but some of its satellite craters are, including Gassendi A, which interrupts the norther rim of Gassendi (giving the two craters the combined appearance of a gemstone in a ring). Gassendi A is much younger than the lava-filled Gassendi, and probably dates to the Moon's latest geologic period. The mare in this quadrangle is crisscrossed by ridges and rounded scarps that may have formed through the extrusion and flow of lava when the plains formed. The USA's robotic Surveyor 1 mission successfully soft-landed near Flamsteed crater on 2 June 1966.

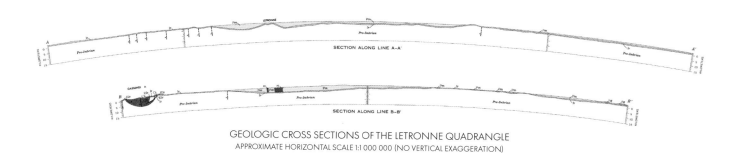

GEOLOGIC CROSS SECTIONS OF THE LETRONNE QUADRANGLE
APPROXIMATE HORIZONTAL SCALE 1:1 000 000 (NO VERTICAL EXAGGERATION)

SURVEYOR 1	ROCKET	Atlas LV-3C Centaur-D
	CREW	N/A
	LAUNCH DATE	14:41:01, 30 May 1966 (UTC)
	LAUNCH SITE	Cape Kennedy Air Force Station
	PAYLOAD	TV Camera
	LUNAR ORBIT INSERTION	N/A
	LANDING DATE	06:17:36, 02 June 1966 (UTC)
	LANDING SITE	Oceanus Procellarum
	LANDING COORDINATES	2.46°S, 43.32°W (A)

SURVEYOR 1 SELF-PORTRAIT
Narrow-angle photographs taken by the Surveyor I's TV camera on 13 June 1966 form this mosaic of the lunar terrain. The elongated shadow of the spacecraft was formed by the sun setting on the horizon behind.

SURVEYOR 1 PANORAMA
The Flamsteed region in Oceanus Procellarum is captured in this reconstructed panorama taken by Surveyor I, June 1966.

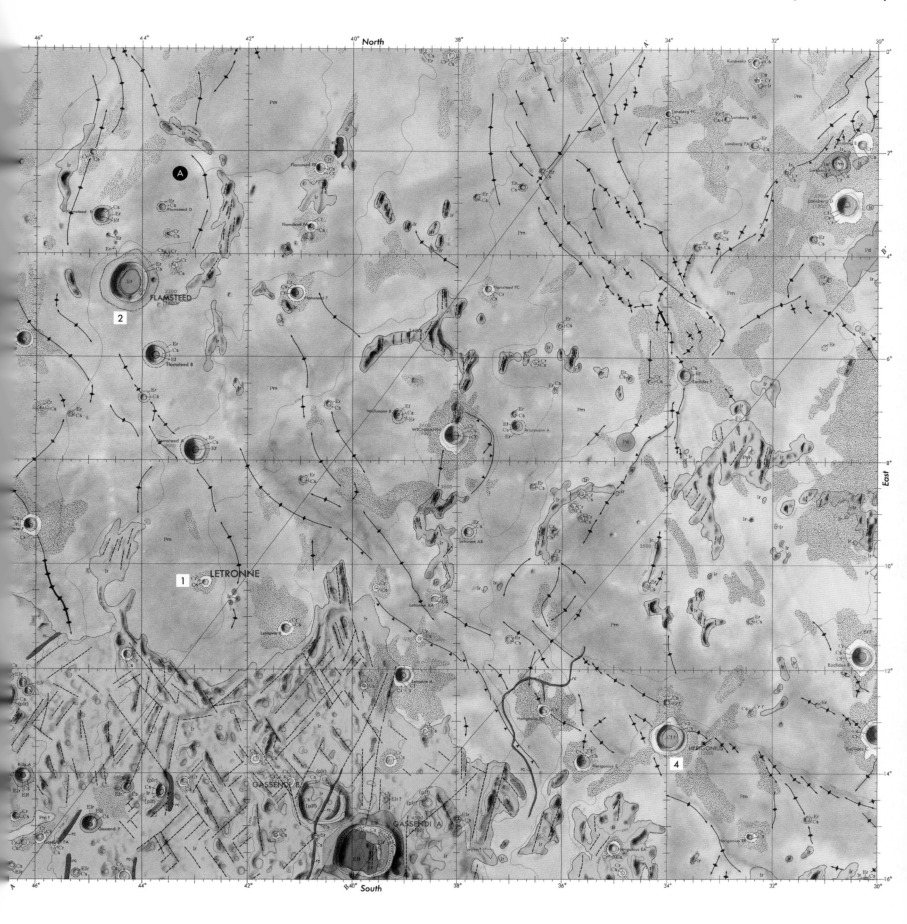

LETRONNE QUADRANGLE: KEY CHARACTERISTICS

1. LETRONNE CRATER
Latitude: 10.5°S / Longitude: 42.5°W
Diameter: 117.6 km / 73 miles
Depth: 1,000 m / 3,280 ft 10 in
Photograph: Apollo 16 (USA, 1972)

2. FLAMSTEED CRATER
Latitude: 4.5°S / Longitude: 44.34°W
Diameter: 19.34 km / 5 ¾ miles
Depth: 2.2 km / 1 ¼ miles
Photograph: Apollo 12 (USA, 1969)

3. BILLY CRATER
Latitude: 13.83°S / Longitude: 50.24°W
Diameter: 45.57 km / 28 ¼ miles
Depth: 1,200 m / 3,937 ft
Photograph: Lunar Orbiter IV (USA, 1967)

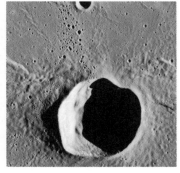

4. HERIGONIUS CRATER
Latitude: 13.36°S / Longitude: 33.97°W
Diameter: 14.86 km / 9 ¼ miles
Depth: 2.1 km / 1 ¼ miles
Photograph: Apollo 16 (USA, 1972)

Matter

Lunar

THE MOON IN POPULAR SCIENCE BOOKS

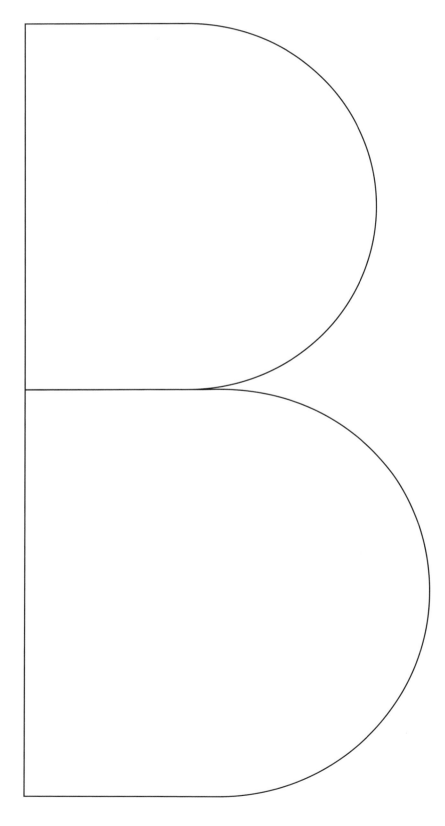

'She is very beautiful. That soft, silvery light, so unlike sunlight or gaslight, or any other light seen upon earth ...'

Agnes Giberne, *Sun, Moon and Stars*, 1884

efore the mid-19th century, publications about lunar studies and astronomy in general were mostly academic, and appeared in the form of treatises, essays published by societies, or as high-priced volumes, many illustrated with detailed engravings. Selenographic atlases, often in folio formats, were key reference works and financially out of reach for many.[1] With advances in print culture and new methods of image production, such as chromolithography, photography and photomechanical printing, illustrated books changed in style and format, and became more affordable. In the field of science, and astronomy especially, accessibly written books became a popular genre, but until recently they have been largely excluded from historical overviews of what is considered the canon of astronomical literature. Yet these books are not just aesthetically fascinating, but they also provide an insight into the huge general interest in lunar studies. Often, but not always, aimed at young readers, they developed parallel to the genre of science fiction.

The boundaries between popular science, didactic writing, specialist academic literature and indeed fiction are of course fluid, and all the more interesting for it. A large number of authors of popular science books were professional astronomers with a knack for writing in an entertaining, narrative style. It is hard to identify a first publication about the Moon that was an attempt to appeal to a wide audience, but James Nasmyth's (1808–1890) and James Carpenter's (1840–1899) 1874 work *The Moon: Considered as a Planet, a World, and a Satellite*[2] is a splendid and highly creative example of the emerging genre. The astronomer Carpenter and engineer Nasmyth teamed up to write this book about the latest lunar knowledge for a general readership, and used creative methods to explain complex facts, including ingenious visuals. While a couple of telescopic photographs of the Moon feature in the book, other photographs show plaster models of the lunar surface, made by Nasmyth and carefully lit in a studio to simulate light conditions on the Moon. The authors considered these models more realistic than even detailed drawings from telescopic observations. These photographs were still being used in other popular astronomy books until the late 1940s. The authors also used physical metaphors in their images, including a cracked glass sphere that illustrates a possible cause of the radial streaks around the crater Tycho, or a human hand and a shrivelled apple to represent lunar rilles and mountain ranges. The book went into several editions and has become a classic in lunar literature.

Many later 19th-century and early 20th-century publications about the Moon are linked to telescopic

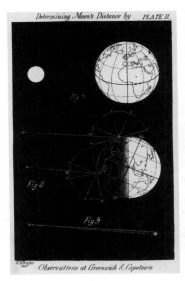

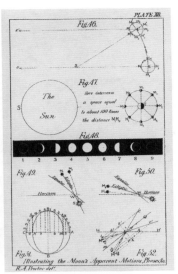

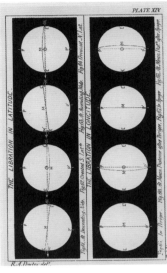

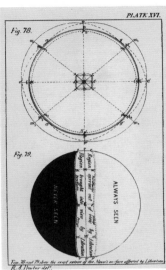

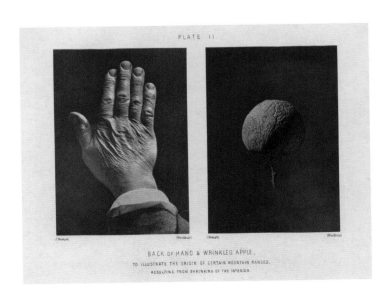

BACK OF HAND & WRINKLED APPLE.
TO ILLUSTRATE THE ORIGIN OF CERTAIN MOUNTAIN RANGES.
RESULTING FROM SHRINKING OF THE INTERIOR.

FULL MOON.
EXHIBITING THE BRIGHT STREAKS
RADIATING FROM TYCHO &c.

GLASS GLOBE.
CRACKED BY INTERNAL PRESSURE
ILLUSTRATING THE CAUSE OF THE BRIGHT STREAKS
RADIATING FROM TYCHO &c.

← Plates from Richard Procter's *The Moon: Her Motions, Aspect, Scenery, and Physical Condition* (1873) show the distance between the Earth and the Moon (plate II), the Moon's apparent motions (plate XII) and its libration (plates XIV and XVI).

↓ Plates from Nasmyth's and Carpenter's *The Moon: Considered as a Planet, a World, and a Satellite* (1885) using photographs of the Moon as well as familiar objects that resemble some of the Moon's features.

1887 to young people at the Royal Institution, including one on just the Moon. *Star-land* was illustrated with numerous small in-text vignettes, wood engravings and etchings, many of them pictorial and figurative, to appeal to those younger readers. Ball's writing and speaking style was immensely engaging, and he gave hundreds of popular astronomy lectures in Britain and Ireland. It is noticeable how the term 'popular' itself became popular in publishing, indicating that the interest in accessible literature about scientific topics was burgeoning and that the amateur and juvenile market was considered as valuable as, if not more than, a specialist readership. The large print runs of many of these books are testament to this. Of Ball's *Star-land*, for example, 17,000 copies were produced of the first edition.

One of the most successful and influential astronomy authors was Camille Flammarion (1842–1925), who seamlessly blended science with fiction and facts with fantasy, while injecting a considerable dose of spiritualism into his hugely successful books. His best-known work was the extensive tome *Astronomie populaire – description générale du ciel* (Popular Astronomy: A General Description of the Heavens), first published in 1880.[9] It includes a long section on the Moon, featuring many images of the lunar surface, including some of Nasmyth's plaster models. Many authors were inspired by Flammarion's success, including the lesser-known Théophile Moreux (1867–1954), an astronomer and meteorologist who in around 1910 published the charming *Un Jour dans la Lune* (A Day in the Moon).[10] He weaves the scientific content into an engaging narrative about a journey to the Moon, but – unlike contemporary science fiction writers – he acknowledges the technical limitations of his time. The book is illustrated with photographs and some of Moreux's own paintings of the lunar surface.

The genre of popular science books provided a significant number of female authors, some of them scientists, opportunities to publish. Mary Ward has already been mentioned, but there are many more, some not yet sufficiently researched and acknowledged. Among them is Agnes Giberne (1845–1939), an evangelical author and amateur astronomer, who wrote dozens of wildly successful books, largely for children, including several on scientific subjects. *Sun, Moon and Stars: Astronomy for Beginners* (1879)[11] was the first of these, and is remarkable for its clear and engaging style, as well as some particularly lovely

observation, a pastime that was becoming increasingly popular among amateurs. As the Moon is such an easy celestial object to observe with amateur equipment, it features prominently in books such as Mary Ward's *The Telescope* (1870)[3] or James Baikie's *Through the Telescope* (1906).[4] Telescopic observations and images (including photographs) also greatly informed popular atlases, which were quickly becoming collectible items and were marketed by publishers as gift books, many with fold-out chromolithographic plates and lavishly decorated covers. Among them are Irish Royal Astronomer Robert Stawell Ball's *Atlas of Astronomy* (1892),[5] which includes sixteen plates of lunar charts and other Moon images, and classicist Thomas Heath's *Twentieth Century Atlas of Popular Astronomy* (1903).[6]

Many of the smaller-format astronomy books had a high educational value and were aimed at children, younger readers, or adults who were not professional astronomers. Apart from his *Atlas of Astronomy*, R. S. Ball (1840–1913) also authored *The Story of the Heavens* (1885),[7] for which he used Nasmyth's Moon drawings, and *Star-land* (1893),[8] comprising the 'Christmastide lectures' he gave in 1881 and

Front cover of *The Moon: Considered as a Planet, a World, and a Satellite* by James Nasmyth and James Carpenter (John Murray, London, 1874).

Front cover of *The Story of the Sun, Moon and Stars* by Agnes Giberne (National Book Company, Cincinnati, 1879).

Front cover of *Star-Land* by Robert Stawell Bal (Cassell & Company, London, 1891).

Front cover of *An Atlas of Astronomy* by Robert Stawell Ball (George Philip & Son, London, 1892).

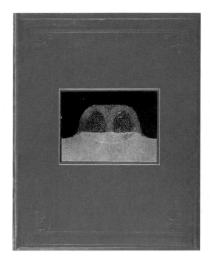 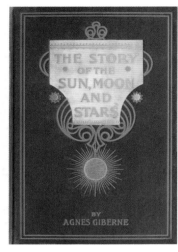 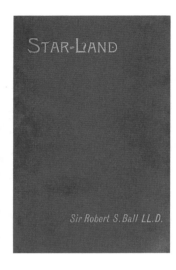 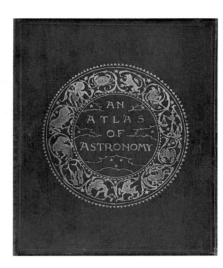

images of the imagined lunar landscape, possibly painted by Giberne herself. The book had the approval of Charles Pritchard (1808–1893), the Savilian Professor of Astronomy at Oxford University, who promptly offered to write a foreword for it.

A more high-profile and better recorded case is that of Mary Proctor (1862–1957), who, like Giberne, wrote predominantly for children. She came from an astronomical background: her father was Richard Anthony Proctor (1837–1888), a professional astronomer who had published *The Moon* in 1873,[12] a tome similar in scope and ambition to Nasmyth's and Carpenter's Moon book. One wonders whether this inspired young Mary to publish her many books on astronomy, including the attractive *Romance of the Moon* (1928)[13] and *The Children's Book of the Heavens* (1924),[14] all richly illustrated with images of lunar landscapes, charts,

and photographs taken at observatories. Mary Proctor, an outstanding communicator and lecturer, effortlessly combined science, observation, folklore and legend in her work, which is yet to be fully appreciated.

Book publishing of the late 19th and early 20th century was often aesthetically appealing, with authors and publishers benefiting from mechanical colour reproductions. In respect of popular astronomy books, it was also a fascinating time because there was still a lot left to the creative imagination, especially concerning the structure, colour and patterns of the lunar landscape. Books on the Moon from this period are peppered with dreamy, fantastical paintings of how authors imagined that other world to look, combining scientific knowledge with awe, wonder and artistic imagination. There was still mystery surrounding the Moon, something (arguably) lost in the later 20th century.

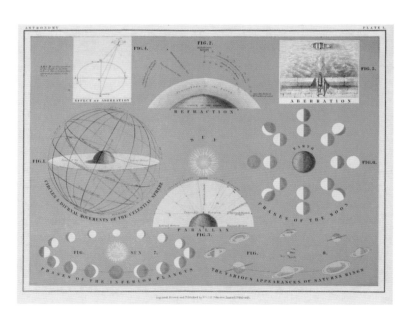 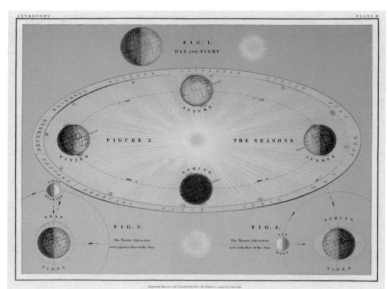

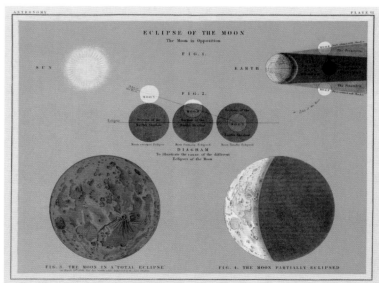 

Front cover of *In Starry Realms* by Robert Stawell Ball (Isbister and Company Limited, London, 1893).

Front cover of *The Twentieth Century Atlas of Popular Astronomy* by Thomas Heath (W. & A. K. Johnston, Edinburgh, 1903).

Front cover of *Through the Telescope* by James Baikie (Adams and Charles Black, London, 1906).

Front cover of *The Children's Book of the Heavens* by Mary Proctor (George G. Harrap & Company Ltd, London, 1924).

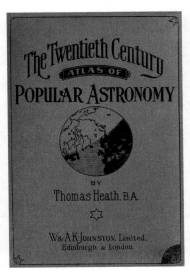
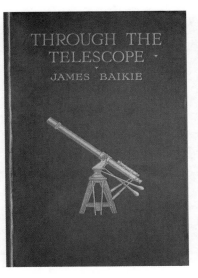
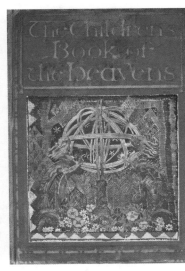

↑
In this image from *Star-Land* (1893) Ball explains the effect of the Earth's gravitational pull on the movements of the Moon.

→
Wood engravings from *Through the Telescope* (1906) by Baikie. The first is described by the author as a 'rough sketch' of the crater Clavius, while the second shows clefts and rills in the region of the Aristarchus and Herodotus craters.

⇉
Illustrations from Théophile Moreux's *Un jour dans la lune* (1913). They show a gap in the mountains of Copernicus (above), a perspective view of a mountain-ring (centre), and the linear feature Rupes Recta (below), also known as the 'Straight Wall' or 'Straight Range' of the Moon.

↓
Illustration of the Moon and Earth's movement, from Camille Flammarion's *Popular Astronomy* (1880).

←
Plates from *The Twentieth Century Atlas of Popular Astronomy* (1903), showing the Moon and celestial bodies.

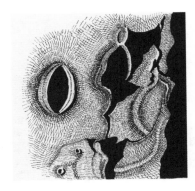

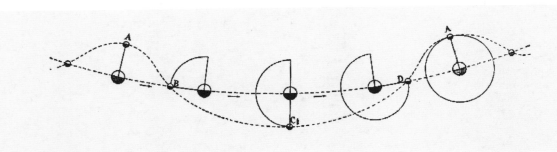
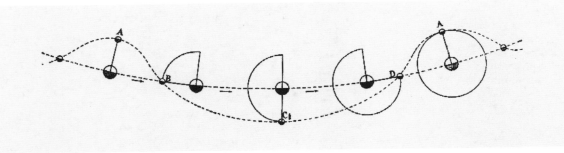

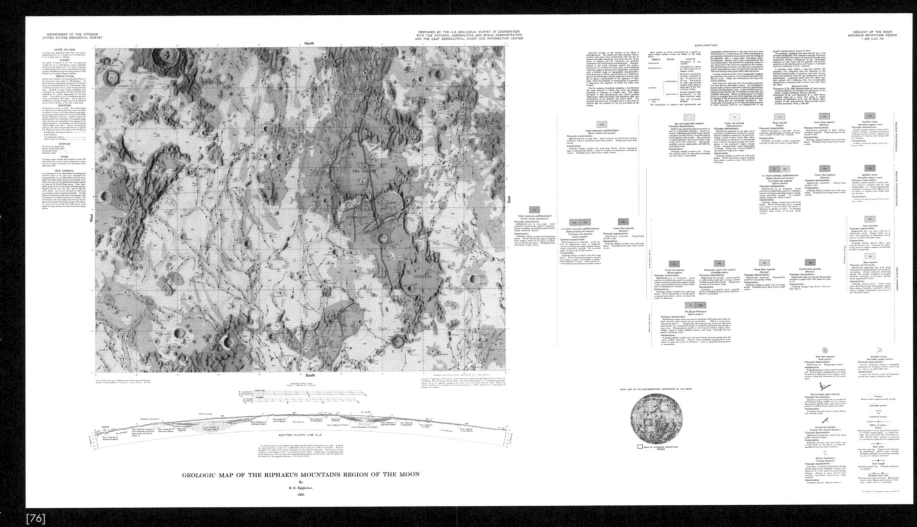

[76]

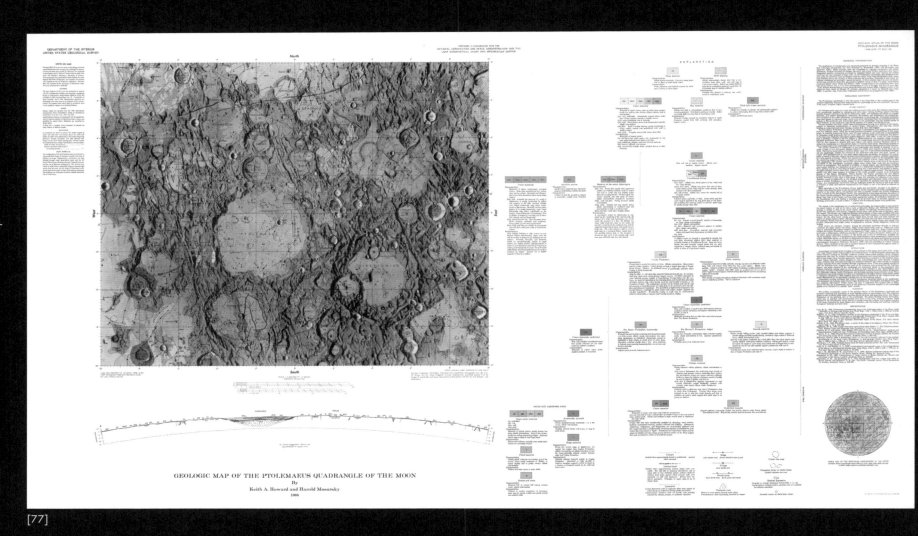

[77]

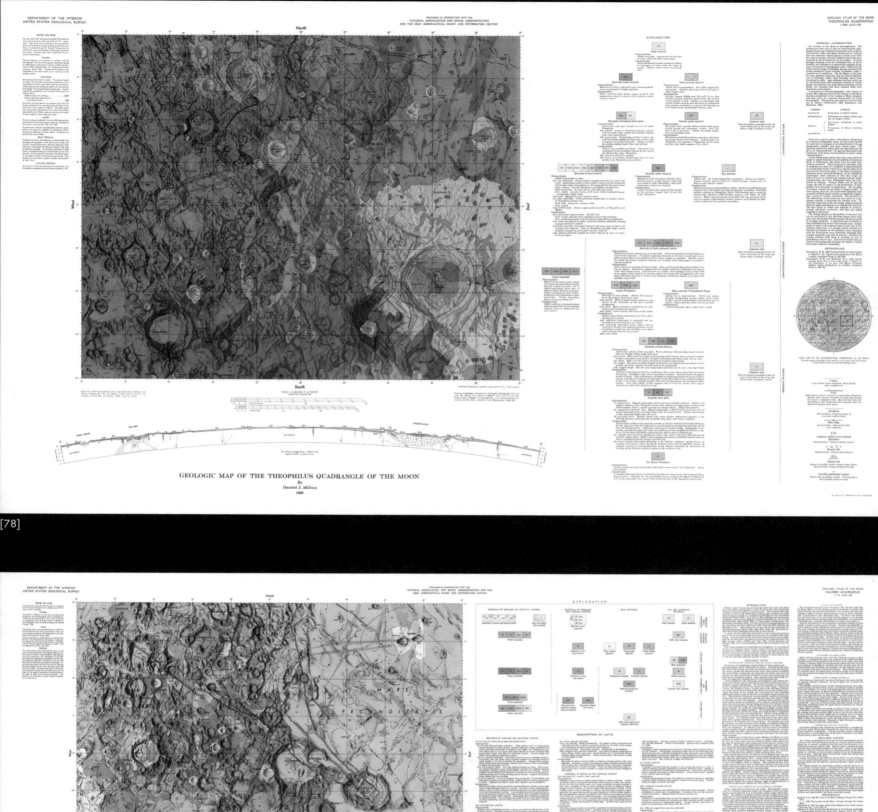

GEOLOGIC MAP OF THE THEOPHILUS QUADRANGLE OF THE MOON
By
Daniel J. Milton
1968

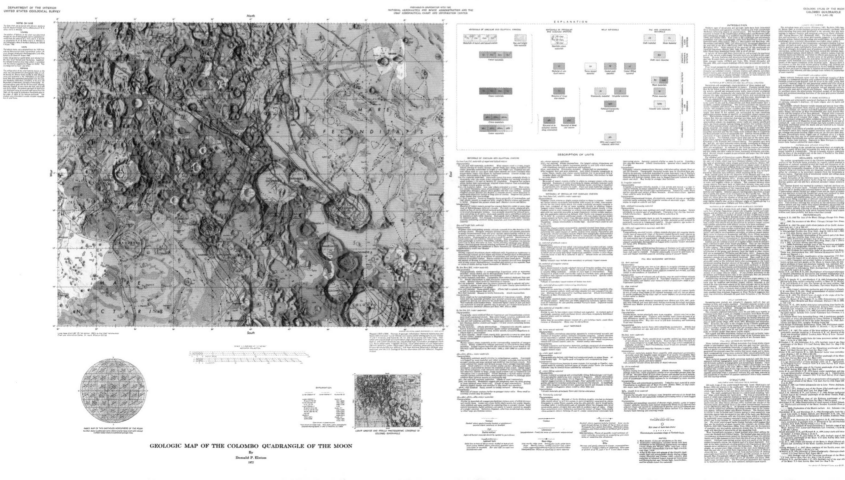

GEOLOGIC MAP OF THE COLOMBO QUADRANGLE OF THE MOON
By
Donald P. Elston
1972

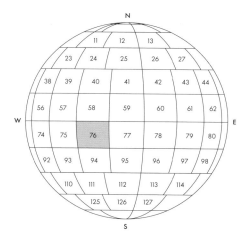

Just below the centre of this quadrangle, at the coordinates 10.38°S, 20.36°W, are the words, 'RANGER VII IMPACT'. This marks the spot where, on 31 July 1964, the Ranger 7 spacecraft deliberately crashed into the lunar surface after returning thousands of high-resolution lunar images. This was the first successful US lunar mission after six years of failure. This landing site, nestled between Mare Nubium (Sea of Clouds) and Oceanus Procellarum (Ocean of Storms), was subsequently named Mare Cognitum (Sea that has Become Known). The quadrangle's eponymous Montes Riphaeus lie inconspicuously to the northwest of the Ranger 7 impact site. The three-pointed mountain range to its east is created by the intersection of the Fra Mauro, Bonpland and Parry craters, all of which have been degraded by erosion and volcanism. The robotic soft-lander Surveyor 3 and the crewed Apollo 12 Lunar Module also landed in this quadrangle in 1967 and 1969. Apollo 14 landed in the Fra Mauro region in 1971 and sampled the breccia thrown here by the impact that formed the Mare Imbrium basin. But in 1965 when this quadrangle was mapped, geologists were still unpacking the Ranger 7 results and relied primarily on telescopic images.

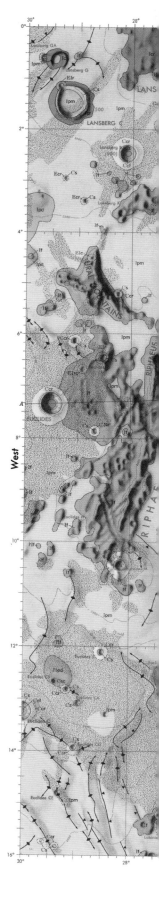

RANGER 7	ROCKET	Atlas LV-3 Agena-B 250D/AA9
	CREW	N/A
	LAUNCH DATE	16:50:07, 28 July 1964 (UTC)
	LAUNCH SITE	Cape Kennedy Air Force Station
	PAYLOAD	Imaging System (Six TV Cameras)
	LUNAR ORBIT INSERTION	N/A
	LANDING DATE	13:25:49, 31 July 1964 (UTC)
	IMPACT SITE	Mare Cognitum
	LANDING COORDINATES	10.38°S, 20.36°W (A)

SURVEYOR 3	ROCKET	Atlas Centaur
	CREW	N/A
	LAUNCH DATE	07:05:01, 17 April 1967 (UTC)
	LAUNCH SITE	Cape Kennedy Air Force Station
	PAYLOAD	TV Camera, Surface Sampler
	LUNAR ORBIT INSERTION	N/A
	LANDING DATE	00:04:17, 20 April 1967 (UTC)
	LANDING SITE	Oceanus Procellarum
	LANDING COORDINATES	03.00°S, 23.41°W (B)

APOLLO 12	ROCKET	Saturn V AS-507
	CREW	C. Conrad (Jr), R. F. Gordon (Jr), A. L. Bean
	LAUNCH DATE	16:22:00, 14 November 1969 (UTC)
	LAUNCH SITE	Cape Kennedy Air Force Station
	PAYLOAD	Lunar Surface Experiments Package
	LUNAR ORBIT INSERTION	18 November 1969
	LANDING DATE	06:54:35, 19 November 1969 (UTC)
	LANDING SITE	Oceanus Procellarum
	LANDING COORDINATES	-3.19°S, 23.38°E (C)

APOLLO 14	ROCKET	Saturn V SA-509
	CREW	A. B. Shepard (Jr), E. D. Mitchell, S. A. Roosa
	LAUNCH DATE	21:03:00, 31 January 1971 (UTC)
	LAUNCH SITE	Cape Kennedy Air Force Station
	PAYLOAD	Kitty Hawk (CM-110), Antares (LM-8)
	LUNAR ORBIT INSERTION	04 February 1971
	LANDING DATE	09:18:11, 05 February 1971 (UTC)
	LANDING SITE	Fra Mauro Formation
	LANDING COORDINATES	3.67°S, 17.46°E (D)

APOLLO 12 LUNAR LANDING: 19 NOVEMBER 1969

PREPARING TO WALK ON THE MOON
Alan Bean descends the ladder of the lunarmodule *Intrepid* on 19 November 1969, to step onto the surface of the Moon to begin the first Apollo 12 EVA.

LUNAR SURFACE EXPERIMENTS
A Lunar Surface Magnetometer (LSM) isset up during EVA 1. In the distance, other instruments from the Lunar Surface Experiments Package (ALSEP) are visible.

COLLECTING ROCK SAMPLES
Alan Bean holds up a test tube filled with samples of lunar soil during EVA 2 on 20 November 1969. Charles 'Pete' Conrad is reflected in his visor.

VISITING SURVEYOR 3
Charles 'Pete' Conrad examines the TV camera on the abandoned Surveyor III spacecraft before collecting parts during EVA 2. Apollo 12 landed just metres away.

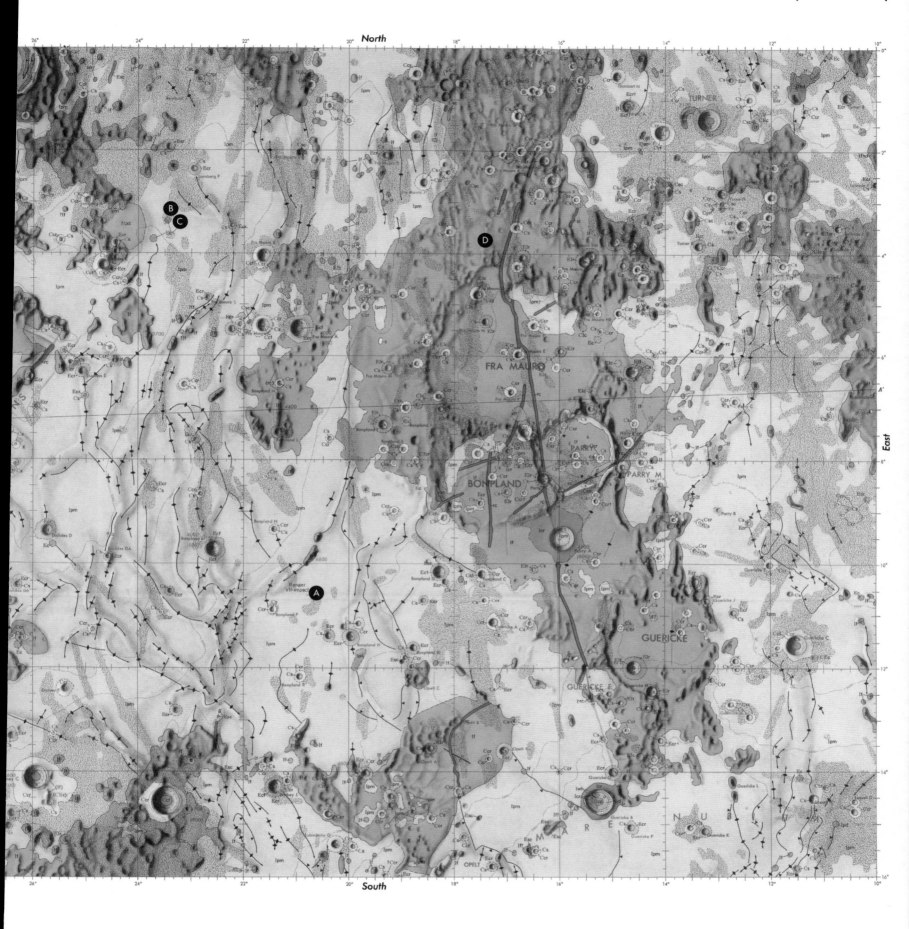

APOLLO 14 LUNAR LANDING: 5 FEBRUARY 1971

ASTRONAUT BENEATH ANTARES
Alan Shepard Jr with his right arm raised to his visor, photographed on 5 February 1971 by Edgar Mitchell from the window of the lunar module *Antares* during EVA 1.

TRACKS ON THE MOON
Photograph of *Antares* against a brilliant sun, taken by one of the astronauts from the ALSEP site during EVA 1. Vehicle tracks are visible on the Moon's surface.

CONE CRATER
A geology hammer and bag sit on top of large boulders near the rim of Cone crater on 6 February 1971 during EVA 2. Samples were excavated and returned to Earth.

CENTRAL STATION OF THE ALSEP
Alan Shepard Jr assembles a double core tube beside a makeshift workbench during EVA 2. He is photographed by Edgar Mitchell.

The ancient, lava-filled impact crater Ptolemaeus gives its name to this quadrangle. It is surrounded by other similarly ancient craters, including Alphonsus, which overlaps the southern wall of Ptolemaeus. Inside Alphonsus, at coordinates 12.83°S, 02.37°W, are the words 'RANGER IX'. On 24 March 1965, the US Ranger 9 spacecraft intentionally crashed into the lunar surface after transmitting nearly 6,000 images of the Moon taken during its descent. These images revealed rilles on the floor of Alphonsus, shown in the map. Smaller, younger craters overlie the ancient craters of this quadrangle. Mosting crater, in the north, is one of the youngest craters shown. Other notable young or well-preserved craters are Lalande, Herschel and Horrocks.

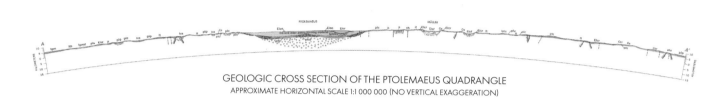

GEOLOGIC CROSS SECTION OF THE PTOLEMAEUS QUADRANGLE
APPROXIMATE HORIZONTAL SCALE 1:1 000 000 (NO VERTICAL EXAGGERATION)

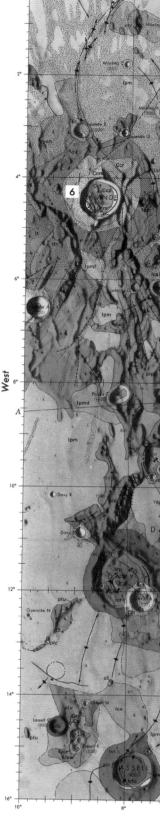

RANGER 9	ROCKET	Atlas Agena
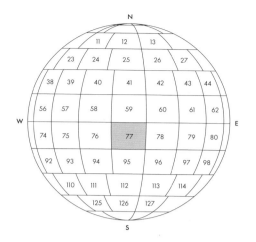	CREW	N/A
	LAUNCH DATE	21:37:02, 21 March 1965 (UTC)
	LAUNCH SITE	Cape Kennedy Air Force Station
	PAYLOAD	Imaging System (Six TV Cameras)
	LUNAR ORBIT INSERTION	N/A
	LANDING DATE	14:08:19:994, 24 March 1965 (UTC)
	IMPACT SITE	Alphonsus Crater
	LANDING COORDINATES	12.83°S, 357.63°E (A)

RANGER 9 LUNAR SURFACE SEQUENCE
A sequence of images captured by Ranger 9's Camera A as the spacecraft approaches the lunar surface on 24 March 1965.

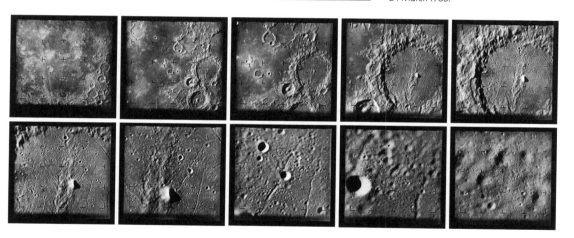

PTOLEMAEUS QUADRANGLE: KEY CHARACTERISTICS

1. PTOLEMAEUS CRATER
Latitude: 9.16°S / Longitude: 1.84°W
Diameter: 153.67 km / 95 ½ miles
Depth: 2.4 km / 1 ½ miles
Photograph: Apollo 16 (USA, 1972)

2. ALPHONSUS CRATER
Latitude: 13.39°S / Longitude: 2.85°W
Diameter: 110.5 km / 68 ¾ miles
Depth: 2.7 km / 1 ¾ miles
Photograph: LRO (USA, 2015)

3. FLAMMARION CRATER
Latitude: 3.33°S / Longitude: 3.73°W
Diameter: 76.18 km / 47 ½ miles
Depth: 1.5 km / 1 mile
Photograph: Apollo 16 (USA, 1972)

4. MÖSTING CRATER
Latitude: 0.7°S / Longitude: 5.88°W
Diameter: 24.38 km / 15 ¼ miles
Depth: 2.8 km / 1 ¾ miles
Photograph: Lunar Orbiter III (USA, 1967)

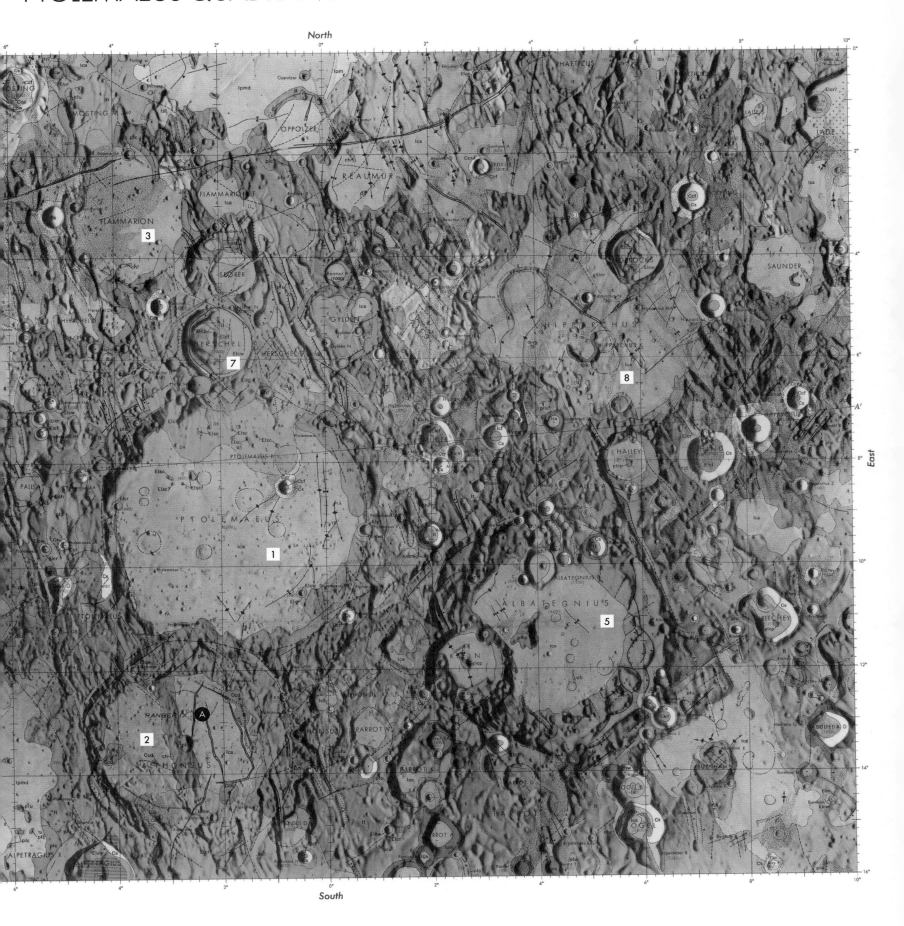

5. ALBATEGNIUS CRATER
Latitude: 11.24°S / Longitude: 4°E
Diameter: 130.84 km / 81 ¼ miles
Depth: 4.4 km / 2 ¾ miles
Photograph: Apollo 10 (USA, 1969)

6. LALANDE CRATER
Latitude: 4.46°S / Longitude: 8.65°W
Diameter: 23.54 km / 14 ¾ miles
Depth: 2.6 km / 1 ½ miles
Photograph: Apollo 16 (USA, 1972)

7. HERSCHEL CRATER
Latitude: 5.69°S / Longitude: 2.09°W
Diameter: 39.09 km / 24 ¼ miles
Depth: 3.8 km / 2 ¼ miles
Photograph: Apollo 12 (USA, 1969)

8. HIPPARCHUS CRATER
Latitude: 5.36°S / Longitude: 4.9°E
Diameter: 144 km / 89 ½ miles
Depth: 3.3 km / 2 miles
Photograph: LRO (USA, 2015)

Ian Haydn Smith

THE MOON IN SILENT CINEMA

Myth

Lunar

↓
A scene from *Frau im Mond* (Woman in the Moon) (1929), directed by Fritz Lang. The film was adapted by Lang's wife and regular screenwriting collaborator Thea von Harbou, from her 1928 novel *The Rocket to the Moon*. It followed the success of Lang's 1927 groundbreaking sci-fi epic *Metropolis*, and is one of the first feature-length films to grapple with travel beyond the Earth's atmosphere.

↱
Still from *Le Voyage dans la Lune* (A Trip to the Moon) (1902), directed by pioneering director and visual effects innovator Georges Méliès. This, his most famous film, was inspired by Jules Verne's 1865 novel *From the Earth to the Moon*. It was his most complex and longest film to date, and beneath the effects lies a stinging satire of imperialism.

↳
Le cauchemar (A Nightmare) (1896) is one of around eighty short films that Georges Méliès directed in his first year as a filmmaker. In this scene the director stars as the sleeping man, whose wildly imaginative dreams culminate in his arm being consumed by a ravenous Moon. The film was shot in the garden of Méliès's property in Montreuil, Seine-Saint-Denis.

In this scene from Méliès's *A Trip to the Moon*, the director's vivid imagination is evident. Not content with in-camera effects, Méliès was a master of the *mise-en-scène*, building impressive sets that conjured up life on the lunar surface.

A scene from Bruce Gordon and J. L. V. Leigh's *The First Men in the Moon* (1919). It was the first feature-length sci-fi film ever made. An adaptation of H. G. Wells' eponymous 1901 novel, the film is listed on the British Film Institute's list of the '75 Most Wanted' lost films.

ince the inception of cinema, filmmakers have considered the possibility of travel to the Moon and pondered over the lifeforms that may exist both on and beneath its surface.

Developments in technology since the mid-19th century enabled photographers to take increasingly detailed images of the Moon. But as impressive as these photographs were, they could not capture the sense of wonder the Moon evoked in people's imaginations. The arrival of the moving image at the end of the 19th century changed everything, making the impossible real and bringing dreams to life.

The first notable cinema screening took place in Paris on 28 December 1895, when Auguste and Louis Lumière screened their ten 'actuality' films[1] – moments captured from everyday life. The films may have been prosaic, but they represented a revolution in art and science. Theatre illusionist Georges Méliès (1861–1938)[2] was an early admirer of this new form, and the hundreds of films he subsequently wrote, directed and appeared in embraced the fantastic: none more so than the films that indulged audiences' fascination with the Earth's only natural satellite.

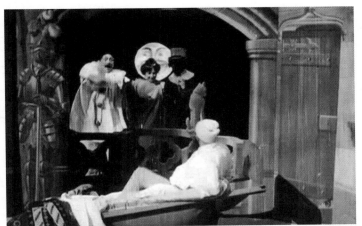

Arguably the earliest appearance by the Moon in a film was Méliès' one-minute 1896 short *Le Cauchemar* (A Nightmare). In it, a man (Méliès) has a nightmare comprising various visions, one of which is the appearance of the Man in the Moon, who proceeds to chew on the dreamer's arm. But it is Méliès' 1902 fantasy *Le Voyage dans la Lune* (A Trip to the Moon) – seen as the first science fiction film – for which the filmmaker is best remembered. He plays Professor Barbenfouillis, who leads a small team of scientists to the Moon's surface, where they encounter an insectoid alien species, the Selenites. The scientists are captured, but eventually escape, making their way back to Earth where they are celebrated as pioneering heroes. The Moon makes its first appearance some six minutes into the film, first as a drawn image and then as an actor's face, into whose left eye the scientists' rocket famously crashes.

↓
Hand-coloured prints from
A Trip to the Moon (1902).
The first shows the lunar
surface after the rocket lands
in the eye of the Man in the
Moon. Méliès used layered
sets to give depth of field to
the surface, which resembled
a forest of angular stalgmites.

↓↓
In this print the scientists
have boarded the spaceship,
which is to be fired directly
at the 'face' of the Moon.
They are given a send-off
by local dignitaries and
a group of young women.
The cannon appears to rest
on the rooves of houses.

↓↓↓
Here the scientists have
disembarked from the
spacecraft and begin
exploring the lunar surface.
Their imperialist tendencies
take over and they claim
the Earth's satellite as their
own before they realize
that it is already occupied.

↓↓↓↓
In the fourth print, the
scientists decide to leave
the Moon and its spear-
wielding, wasp-like
inhabitants. They drop off
a cliff edge in their capsule,
unaware that an alien
has joined them on their
journey back to Earth.

Méliès credited Jules Verne's novels *From the Earth to the Moon* (1865) and *Around the Moon* (1870) as inspirations. H. G. Wells' *The First Men in the Moon* (1901), published just a few months before Méliès made his film, was also a likely point of reference. Verne and Wells would prove influential over later science fiction cinema, but Méliès' use of these texts, alongside his passion for advances in technology, symbolizes a major shift in the way societies regarded both the world around them and what was once seen as the impossibly distant Moon. The Industrial Revolution, which encompassed the late 18th and early-to-mid-19th century, not only altered everyday life, but also radically changed perceptions of what could be achieved with science and engineering. Advances in mass transportation, from rail to shipping, along with the development of the automobile towards the end of the 19th century and the increasing viability of manned flight at the turn of the 20th, made the world a smaller place and spurred artists such as Verne, Wells and Méliès to think of the fantastical as possible.

While Méliès considered a journey to the Sun with his 1904 film *An Impossible Voyage* and a romance between two celestial bodies with *The Eclipse: Courtship of the Sun and Moon* (1907), other filmmakers took inspiration from his earlier lunar adventure. Spanish filmmaker Segundo de Chomón (1871–1929) offered up a remake with *An Excursion to the Moon* (1908), while Enrico Novelli (1874–1943) directed the oldest surviving Italian science fiction film, *Un Matrimonio Interplanetario* (A Marriage in the Moon, 1910). An adaptation of his 1908 novel *La Colonia Lunare* (The Moon Colony), Novelli's film tells the story of two scientists, one from Earth and the other Mars, who decide to meet on the Moon. Although the lunar landscape is wildly imagined, it is possible to see some similarity to the images captured in photographs, perhaps suggesting a broader general awareness at the time of what the lunar surface may have looked like. Almost a decade later, Bruce Gordon and J. L. V. Leigh directed a faithful feature-length adaptation of Wells' *The First Men in the Moon* (1919). It is believed to be one of the earliest science fiction features, but has long been lost.

The scientists and explorers of these early films could walk on the Moon without the aid of breathing apparatus, but German filmmakers Hanns Walter Kornblum (1878–1970), Johannes Meyer (1888–1976) and Rudolf Biebrach (1866–1938) attempted to bring science fact into the world of film with their 1925 documentary *Wunder der Schöpfung* (Our Heavenly Bodies). Seven individual chapters, combining animation, dramatic recreation and text, attempt to explain the movement of celestial bodies, how our notions regarding Earth's place in the solar system have changed through the ages and – hilariously – what our world would be like without gravity. One chapter deals specifically with possible travel to the Moon, engaging more thoroughly with the real challenges such a journey would involve. But perhaps the most compelling pre-sound film to grapple with the Moon is Fritz Lang's (1890–1976) *Frau im Mond* (Woman in the Moon, 1929). Made two years after the filmmaker's ground-breaking and hugely influential futuristic epic *Metropolis*, *Woman in the Moon* was adapted by Lang's wife and regular screenwriting collaborator Thea von

↓

Stills from *Excursion dans la lune* (Excursion to the Moon) (1908). Directed by Segundo de Chomón, it is a near-identical remake of Méliès's *A Trip to the Moon* (1902), showing his influence at the time and how popular his style of filmmaking was.

↓↓

Scenes from *Un Matrimonio Interplanetario* (A Marriage in the Moon) (1910), directed by Enrico Novelli. Based on the filmmaker's 1908 novel *The Moon Colony*, the film's visuals were inspired by Novelli's illustrations for his own novel.

↓↓↓

Three frames that show the Moon and a lunar landscape from the 1925 German production *Wunder der Scöpfung* (Our Heavenly Bodies). The film was an inventive attempt to bridge the gulf between science and speculative fiction.

↓↓↓↓

Woman in the Moon. A still (left) and poster artwork (right) for Fritz Lang's sci-fi spectacular. The artwork was created by Alfred Herrmann, who created the artwork for many German films from this period. Lang's film became a hit with rocket scientists.

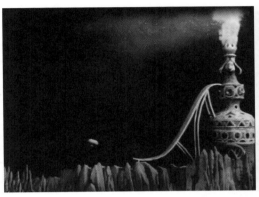

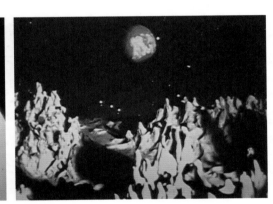

Harbou (1888–1954) from her 1928 novel *The Rocket to the Moon*. Regarded as one of the first 'serious' science fiction films, the movie features elements that would become components in real space missions carried out by the USA and USSR. But it still retains the sense of wonder that made the earliest films so compelling, and which would continue to entrance audiences following the arrival of sound, with advances in filmmaking technology and the amassing of even greater knowledge of what it would be like to walk upon the lunar surface.

163

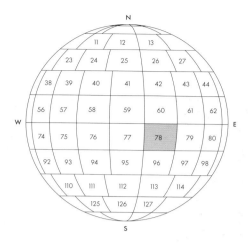

This quadrangle is mostly defined by Moon's central highlands, aside from the portion of Mare Nectaris (Sea of Nectar) to the north of the large crater Theophilus. The terrain in this quadrangle is very rugged, having been shaped and reshaped by multiple periods of bombardment. Theophilus is one of the youngest features in this quadrangle (it is identified here as belonging to the Moon's latest geologic period, but it has since been placed earlier in this chronology based on ejecta samples returned by the crew of Apollo 16). The crater does appear young; it has a well-defined rim, a terraced wall, and a tall central peak. An older crater, though also well-preserved with some signs of wear, is the Delambre crater in the north of this quadrangle. In the Descartes Highlands, on April 21, 1972, at coordinates 8.97S, 15.50E, Apollo 16 became the fifth crewed mission to land successfully on the Moon. The crew used their Lunar Roving Vehicle to visit several craters and other features too small to appear on this map. As this map was drawn in 1968, neither the landing site nor the names of these features (which were not yet adopted) are recorded here.

APOLLO 16	ROCKET	Saturn V
	CREW	J. Young, C. Duke, T. Mattingley II
	LAUNCH DATE	17:54:00, 16 April 1972 (UTC)
	LAUNCH SITE	Cape Kennedy Air Force Station
	PAYLOAD	Casper (CM-113) Orion (LM-11)
	ORBITAL INSERTION	N/A
	LANDING DATE	02:24:00, 21 April 1972 (UTC)
	IMPACT SITE	Descartes Highlands
	LANDING COORDINATES	8.97°S, 15.50°E Ⓐ

SLIM	ROCKET	H-2A
	CREW	N/A
	LAUNCH DATE	23:42:11, 06 September 2023 (UTC)
	LAUNCH SITE	Tanegashima Space Center
	PAYLOAD	Lunar Excursion Vehicles 1 + 2
	LUNAR ORBIT INSERTION	25 December 2023
	LANDING DATE	15:20:00, 19 January 2024 (UTC)
	IMPACT SITE	Near Shioli crater
	LANDING COORDINATES	13.31°S, 25.25°E Ⓑ

FAR SIDE OF THE MOON
Captured by Apollo 16, the far side of the Moon, with Mare Marginis and Mare Smythii visible, is much more cratered than the near side of the Moon.

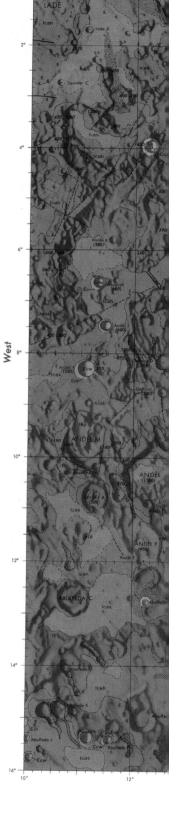

APOLLO 16 LUNAR LANDING: 21 APRIL 1972

PLUM CRATER
John W. Young stands on the rim of Plum crater on 21 April 1972 to collect lunar samples at Station 1 – the Descartes landing site – during EVA 1.

GRAND PRIX
John W. Young demonstrates a lunar roving vehicle's speed in this frame from Charles M. Duke's motion picture, filmed during EVA 1.

OUTHOUSE ROCK
John W. Young and Charles M. Duke return to the Descartes landing site for their final EVA. Here, Charles examines a large boulder at North Ray crater on 23 April 1972.

SHADOW ROCK
John W. Young approaches the site of the permanently shadowed soil to extract a soil sample while Charles M. Duke observes and captures the moment on camera.

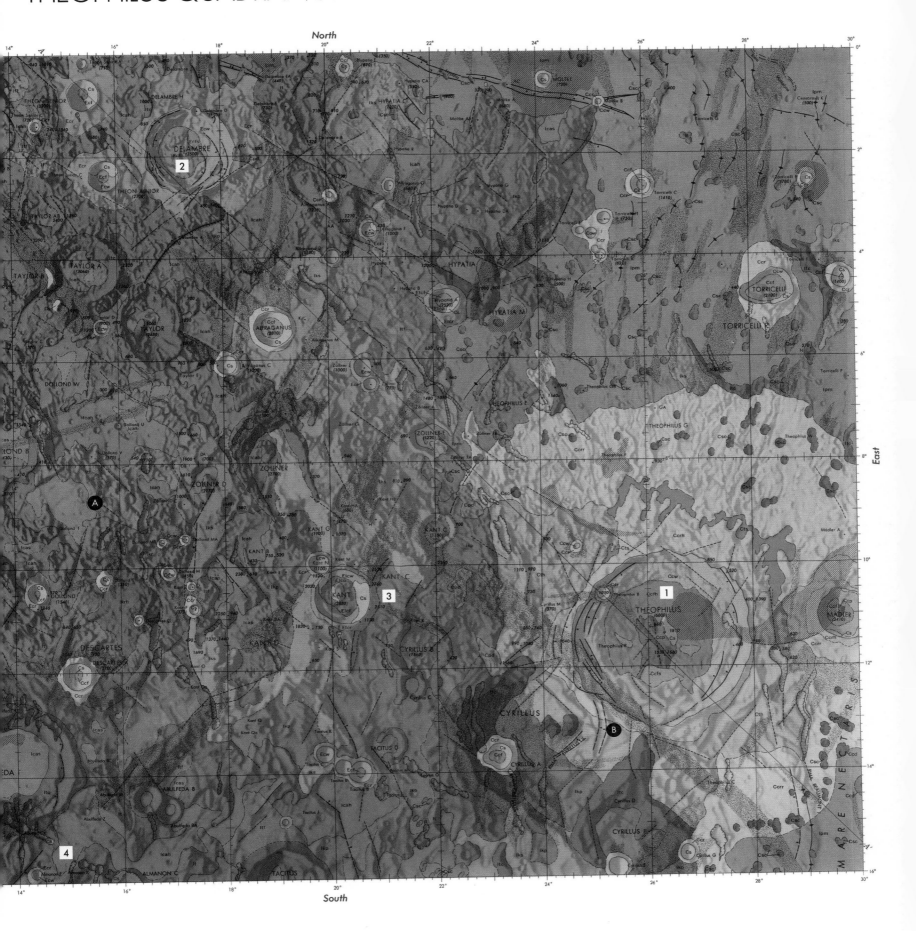

THEOPHILUS QUADRANGLE: KEY CHARACTERISTICS

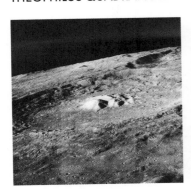

1. THEOPHILUS CRATER
Latitude: 11.45°S / Longitude: 26.28°E
Diameter: 98.6 km / 61 ¼ miles
Depth: 3.2 km / 2 miles
Photograph: Lunar Orbiter III (USA, 1967)

2. DELAMBRE CRATER
Latitude: 1.94°S / Longitude: 17.39°E
Diameter: 51.5 km / 32 miles
Depth: 3.5 km / 2 ¼ miles
Photograph: Lunar Orbiter IV (USA, 1967)

3. KANT CRATER
Latitude: 10.62°S / Longitude: 20.2°E
Diameter: 30.85 km / 19 ¼ miles
Depth: 3.1 km / 2 miles
Photograph: LRO (USA, 2017)

4. CANTENA ABULFEDA
Latitude: 16.6°S / Longitude: 16.7°E
Length: 209.97 km / 130 ½ miles
Photograph: Lunar Orbiter IV (USA, 1967)

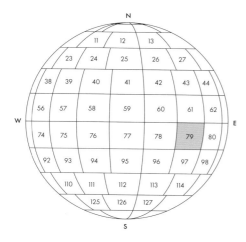

The large impact that formed the Nectaris basin and the concentric upland rings (shown as dotted lines) that surround Mare Nectaris (Sea of Nectar) marks the beginning of one of the earliest periods in the Moon's geologic history, the Nectarian. Many of the features created by this early impact have since been overlaid with craters and materials from subsequent impact events. The mare itself has been filled in with basalt lava. In this quadrangle we see a portion of Mare Nectaris in the south and Mare Fecunditatis (Sea of Fertility) in the north. Between Mare Nectaris and Mare Fecunditatis, within the heavily cratered Montes Pyrenaeus (shown here as Montes Pyrenaei), is the crater Colombo after which this quadrangle is named. Colombo is an old crater, perhaps one of the oldest prominent craters in this quadrangle. Its features are degraded and small craters overlay its surface. Smaller younger craters have formed more recently, such as the simple craters Messier and Messier A in the north, which show visible rays.

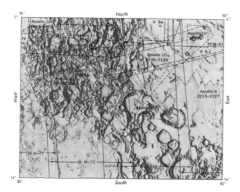

LUNAR ORBITER AND APOLLO PHOTOGRAPHIC
COVERAGE OF COLOMBO QUADRANGLE

COLOMBO / GOCLENIUS
Consecutive frames show an oblique view of craters Colombo (top left) and Goclenius (centre right) and of the western edge of

Mare Fecunditatis. These images were captured on 24 December 1968 by a 70-mm (2 ¾ in) Hasselblad camera as part of the Apollo 8 mission.

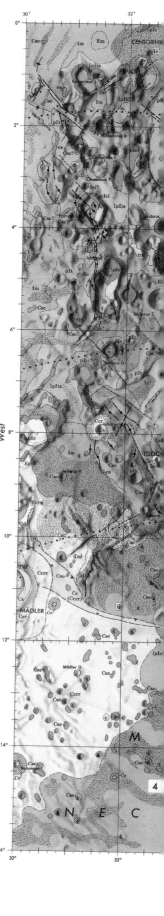

COLOMBO QUADRANGLE: KEY CHARACTERISTICS

1. COLOMBO CRATER
Latitude: 15.26°S / Longitude: 46.02°E
Diameter: 79.02 km / 49 miles
Depth: 2.4 km / 1 ½ miles
Photograph: Apollo 8 (USA, 1968)

2. MARE FECUNDITATIS
Latitude: 7.83°S / Longitude: 53.67°E
Diameter: 840.35 km / 522 ¼ miles
Photograph: LRO (USA, 2012)

3. MESSIER CRATER
Latitude: 1.9°S / Longitude: 47.65°E
Diameter: 13.8 km / 8 ½ miles
Depth: 1,300 m / 4,265 ft 1 in
Photograph: Apollo 11 (USA, 1969)

4. MARE NECTARIS
Latitude: 15.2°S / Longitude: 34.6°E
Diameter: 339.4 km / 211 miles
Photograph: LRO (USA, 2014)

5. DAGUERRE CRATER
Latitude: 11.91°S / Longitude: 33.61°E
Diameter: 45.8 km / 28 ½ miles
Photograph: Apollo 16 (USA, 1972)

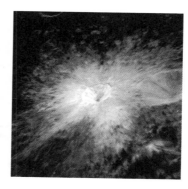

6. CENSORINUS CRATER
Latitude: 0.42°S / Longitude: 32.69°E
Diameter: 4.1 km / 2 ½ miles
Depth: 400 m / 1,312 ft 4 in
Photograph: Apollo 16 (USA, 1972)

7. ISIDORUS CRATER
Latitude: 8°S / Longitude: 33.5°E
Diameter: 41.4 km / 25 ¾ miles
Depth: 1.6 km / 1 mile
Photograph: Apollo 16 (USA, 1972)

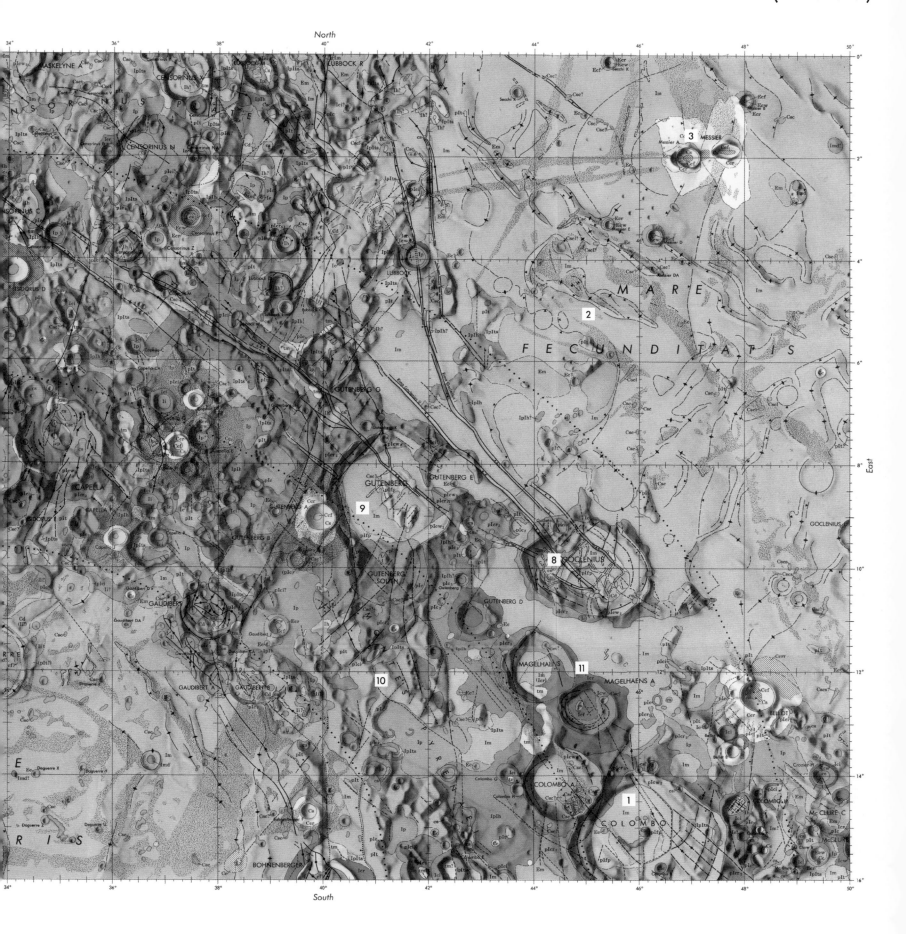

North

East

South

34° 36° 38° 40° 42° 44° 46° 48° 50°

MARE FECUNDITATIS

8. GOCLENIUS CRATER
Latitude: 10.05°S / Longitude: 45.03°E
Diameter: 73.04 km / 45 ½ miles
Depth: 1.5 km / 1 mile
Photograph: Apollo 11 (USA, 1969)

9. GUTENBERG CRATER
Latitude: 8.61°S / Longitude: 41.25°E
Diameter: 70.65 km / 44 miles
Depth: 2.3 km / 1 ½ miles
Photograph: Apollo 11 (USA, 1969)

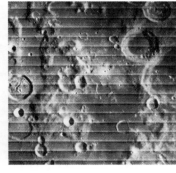

10. MONTES PYRANAEUS
Latitude: 14.05°S / Longitude: 41.51°E
Diameter: 251.33 km / 156 ¼ miles
Max Height: 3 km / 1 ¾ miles
Photograph: Lunar Orbiter IV (USA, 1967)

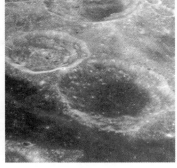

11. MAGELHAENS CRATER
Latitude: 12°S / Longitude: 44.07°E
Diameter: 37.2 km / 23 miles
Depth: 2 km / 1 ¼ miles
Photograph: Apollo 16 (USA, 1972)

SURREALISTS AND THE MOON

Melanie Vandenbrouck

Myth

Lunar

↱
Man Ray's *Le Monde* forms part of a portfolio of photogravures produced in 1931 for electricity supplier CPDE. By 1930, electricity production had become a symbol of modernity, and moonlight had given way to streetlights in Paris.

↓
The brainchild of Spanish filmmaker Luis Buñuel and his friend the painter Salvador Dalí, *Un Chien Andalou* (1929) is a surrealist masterpiece of the silent film era. The scene of the sliced eyeball reportedly came to Buñuel in a dream. It associates the full Moon with the expression of the darkest repressed desires.

↱↱
Helen Lundeberg's 1935 *Cosmicide* is one of several works by the American painter to include a depiction of the Moon. The title, unusual format of the painting, disparate association of objects and crimson tones lend a disquieting undertone to the scene.

↳
British painter Paul Nash was an official war artist during the First and Second World Wars. Painted in 1940–41, *Totes Meer* (Dead Sea) shows a graveyard of enemy planes on a site near Oxford. Bathed in moonlight, the vast expanse of wreckage looks like a turbulent sea.

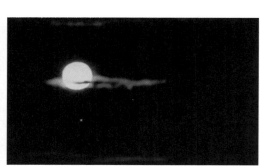

hen in 1931 Man Ray (1890–1976) was commissioned to create a portfolio of photogravures to celebrate the advent of electricity in French homes, he juxtaposed a photograph of the Moon with the photogram of a light switch, setting side-by-side the celestial and human-made. For millennia, the Moon had dictated humans' night-time activities (or lack thereof); it was, quite literally, a lightbulb in the sky, which, before the advent of gas and then electric lights, illuminated travellers' journeys or presided over September harvest. In Man Ray's image, switch and wire are a stark white against the blackness of space, incandescent even by comparison to a pale, partly illuminated Moon. Had the Moon lost its spark (*éclat*) with the advent of modernity? Not in the slightest in the work of his contemporaries, particularly those associated (at times briefly) with surrealism.

Among them was the British painter Paul Nash (1889–1946), who showed a sustained engagement with the Moon throughout his career and his formal experimentation. Executed in black or dark blue ink and wash, Nash's nocturnes from the early 1910s betray the influence of William Blake in their monochromatic approach and romantic feel. The Moon bestows on these landscapes their muted light and dreamy atmosphere, whether as a crescent sliver in its early phase or as a luminous orb. By contrast, in *Voyages of the Moon* (1934–37, Tate), Nash interpreted ceiling lamps reflected in a restaurant's mirror as a sphere moving across a geometricized space of intersecting planes. Nash's distinctive modernism is here marked equally by surrealist figuration and an abstract simplicity of forms, exemplifying the spirit of Unit One, the short-lived avant-garde group he had founded in 1933. Nash suffered post-traumatic stress disorder following the First World War, and must have met the advent of the Second with anguish. As in the previous conflict, he was appointed official war

artist, this time attached to the Air Ministry. *Totes Meer* (German for 'Dead Sea', 1940–41, Tate) is Nash's vision of Cowley Dump, a graveyard of wrecked enemy aircraft, as a sea of shattered fuselages, their jagged, broken shapes crisscrossing the Oxfordshire landscape like the exhumed bones of antediluvian species. The spectral carcasses are bleached by the Moon's bleak light, and Nash described how, 'by moonlight, this waning moon, one could swear they began to move and twist and turn as they did in the air'.[1] Nash's final years were notable for seasonal scenes that endowed the cycles of life with a transcendental quality, staging an omnipresent Moon hovering above rolling hills and clumps of trees.

While Nash excelled in depicting moonlight's transformation of the landscape, the surrealists revelled in the Moon's effects on the human psyche. The opening scene of Spanish filmmaker Luis Buñuel's (1900–1983) silent film *Un Chien Andalou* (1929) sees a man methodically sharpening a razor blade before stepping on a balcony to gaze at

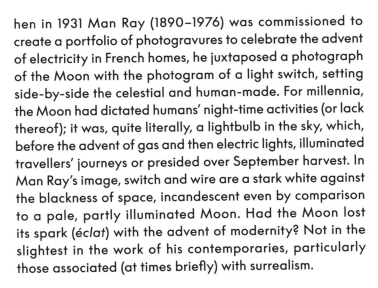

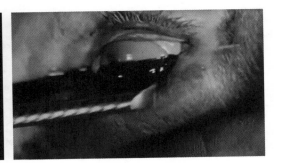

Melanie Vandenbrouck

the sky. As clouds cut across the Moon's dazzling sphere, he grabs a woman by the hair to slash her eyeball from which a thick, translucent substance oozes. Is the Moon witness to a heinous crime, or the purveyor of sadistic dreams? (To the viewer's relief, the heroine reappears unmutilated throughout the film.) The sequence had in fact come to Buñuel in a dream, sparking the decision to create this seminal film in which free association and dreamlike scenes speak of the surrealists' fascination for the subconscious, repressed desires and the uncanny. Long (if erroneously) associated with beliefs that its cycles alter human behaviours and its light when full triggers insanity – or lunacy – the Moon is no stranger to irrational thought. Indeed, the Italian dictator Benito Mussolini feared its powers so profoundly that he would refuse to sleep near a window, for fear its light might fall on his face.

Key to surrealist practice was the uncanny, which could be conveyed in the peculiar association of objects, unusual contexts, enigmatic titles, or mysterious undertones. Belgian painter René Magritte (1898–1967) deployed all these devices in his work, in a trademark hyper-realistic style. In *À la Rencontre du plaisir* (Towards Pleasure) from 1962 a man in a bowler hat shares centre stage with the full Moon.[2] Magritte saw in the man in a bowler hat the banality of human existence, in its anonymous representation of a bourgeois middle class. (Though old-fashioned by the 1960s, the bowler hat had been an omnipresent headgear in European cities in the 1920s and 40s, when Magritte started using the motif.) Yet, crowned with a magnificent full Moon, the scene is imbued in mystery. The man occupies a liminal space between indoors and outdoors (the scene is framed by a curtain to the left, and a tree to the right), between an open field and the forest, between reality and the unknown, darkness mysteriously shrouded in a veil of misty moonlight. What 'pleasure' is this man awaiting to meet is open to speculation; is this the splendour of the full Moon or something more illicit, hidden from view, concealed by the man himself and the night-time world?

With cubist limpidity, the Austrian Wolfgang Paalen's (1905–1959) *Cadran Lunaire (Cycladic Head)* of 1935 evokes the way in which shadow and light meet on the Moon, partitioning its face into black and white. The title recalls the Moon's use for timekeeping – *cadran lunaire* means 'moon dial' – while its reference to Cycladic art (prehistoric Greek sculpture, for which Paalen nourished a passion) alludes to humanity's timeless fascination for the Moon and its relationship to the human form. At once, Paalen reminds us that for thousands of years people have observed the

Moon and seen themselves on its face. Equally rich in meaning, the work of the American artist Joseph Cornell (1903–1972) merges an interest in scientific phenomena and the history of astronomy with the exploration of the natural and divine.[3] In 1936 he created the first of many 'Soap Bubble Sets' (Wadsworth Atheneum, Hartford, Connecticut) as a shadow box laid out with a seemingly disparate arrangement of objects: an egg in a wine glass; a golden doll's head; four white cylinders; flat glass discs; a clay pipe; and, as if blown out of the pipe, a 19th-century map of the Moon. In this cabinet of curiosity, combining scientific map and ordinary objects (some familiar from the surrealist repertoire, others with autobiographical resonance), pipe and glass represent the transience of life, while the childhood pastime of blowing soap bubbles takes on celestial significance. With Cornell, viewers are at once on Earth and in the heavens, straddling the natural and spiritual realms.

For the surrealists, the Moon could also personify the psychic, feminine forces of nature, powerfully so in the work of women subverting surrealism's patriarchal structure and its consideration of the female body as object of desire. A few years before she found herself naturally orbiting the surrealist inner circle, in 1932–33, the teenage Leonora Carrington (1917–2011) depicted in watercolour the Moon as a levitating female figure in 'exotic' attire, and also three *Sisters of the Moon: Fantasia, Diana* and *Indovina Zingara* (all private collection). While reminiscent of the book illustrations of Aubrey Beardsley, they would herald her later, highly singular paintings peopled with fantastic creatures, anthropomorphic or hybrid, often acting under the aegis of the Moon. By 1940 and after many vicissitudes Carrington fled wartime Europe for Mexico, where she joined a circle of émigré artists that included Paalen. The following year, another exiled European, the Spanish painter Remedios Varo (1908–1963) moved to Mexico. At long last, Carrington had met her Moon sister.[4] Together, the two friends studied alchemy, the esoteric and the occult, subjects traditionally associated with the Moon. They influenced each other's otherworldly art, rich in mystical symbolism and in which women, often characterized by an androgynous femininity, take centre stage. In *Papilla Estelar* (Celestial Pablum, 1958, FEMSA Collection, Monterrey), a woman in Varo's own image sits in a small edifice above the clouds. Through the roof, a funnel sucks stars into a mill as she grinds these to spoon-feed stardust to a caged Moon. Absorbing light from the stars, the slim crescent Moon shines brilliantly to illuminate the room. In performing her alchemical ritual, Varo's heroine harvests the cosmos

←
A geographical map of the Moon forms the backdrop for the American Joseph Cornell's *Untitled (Soap Bubble Set)* (1936). Part cabinet of curiosity, part surrealist conglomeration of objects, this artwork conflates scientific language with autobiographical elements and childlike wonder. The Moon is made to look like it is a bubble blowing out of a clay pipe.

↳
Exiled to the United States in 1941, German-born artist Max Ernst created the plaster model of *Moonmad* (1944, cast in bronze in 1956) while staying on the coast of Long Island, New York. This figure was made from an assemblage of casts of utensils and automobile parts, their shapes recalling the phases of the Moon.

↱
Remedios Varo was a Spanish artist exiled to Mexico in 1941 whose surrealist style was inspired by metaphysics, alchemical practices and the occult. She painted *Nacer de nuevo* (To Be Born Again) in 1960. The fantastical setting, in which nature and human-made structures merge into each other, and the association of womanhood and the cosmos, are typical of her work.

↱↱
Remedios Varo's *Papilla Estelar* (Celestial Pablum) (1958) is one of several paintings by Varo featuring an ensnared Moon. Sat on the top floor of a celestial tower, a lone woman alchemist feeds the Moon with stardust. Meticulously but also loosely executed, the painting radiates with cosmic luminosity.

to command the Moon. By contrast, in *Nacer de Nuevo* (To Be Reborn, 1960, Weinstein Gallery, San Francisco), a semi-naked woman in rags breaks through a parchment-like wall into a chamber where a goblet of water reflects the Moon from a pond-like surface in the grassy ceiling above. Opposing perspectives coalesce. Bare tree branches stretch through the windows, while shoots emanating from a knot in the wooden table's grain hang limply over the edge. If the meaning of the painting is open to interpretation, the Moon, femininity and nature are clearly associated.

The Moon also holds a key place in the work of American painter Helen Lundeberg (1908–1999) who developed what she called New Classicism (later termed Post Surrealism), in which she distinguished herself from European surrealism by rejecting free association and the unconscious, instead preferring carefully planned compositions conducive to guided contemplation. In her *Cosmicide* (1935, Sheldon Museum of Art) the Moon occupies the narrow top of the trapezoidal painting, while the middle is devoted to a flowering plant, presented twice, in full and as a cross-section. The title and the motif of a tool bending a nail lend sinister undertones to the painting. Together, these motifs evoke the cycle of life and death. It prefigures her later *Biological Fantasy* (1946) in its association of the Moon with plant growth. To these plant elements she adds a female figure, suggesting her acceptance of the long-held association of the Moon with fertility. That Lundeberg chose to present herself, in a 1944 self-portrait at the easel (Zimmerli Art Museum at Rutgers University), using a sphere to aid the painting of a moonlit scene, confirms the significance of the Moon in her art.

In summer 1943, the exiled German Max Ernst (1891–1976) and his American lover Dorothea Tanning (1910–2012) stayed in Arizona in the southwestern USA. There, the two artists were struck by the primeval qualities of the arid landscape, its eeriness all the more emphasized in moonlight. Celestial discs – Sun or Moon – had been a recurring motif in Ernst's work, and from this trip resulted new nocturnal scenes. The following summer, while staying in Grand River, Long Island, Ernst rekindled a sculptural

practice he had experimented with ten years before in France, creating anthropomorphic plaster sculptures by assembling casts of everyday objects. From this second period of experimentation came *Moonmad* (plaster, 1944, cast in bronze, 1956), a figure born out of kitchen utensils and automobile parts, in which the curved forms that make up its body recall different states or phases of the Moon: crescent, waning or waxing, full. These interconnecting shapes lend the figure a particular dynamism, not unlike that of the Hopi kachina figures Ernst had started avidly collecting during his first trip through the US southwest in 1941. The kachinas are spirits that embody concepts and elements of the cosmos and natural world. Perhaps inspired by his 1943 stay in Arizona, Ernst was moved to place his plaster figure outdoors in a field, to see it under the illumination of the full Moon.[5]

For the surrealists, in whose work the boundaries between dream and reality remained fluid and permeable, the oneiric otherworldliness with which moonlight drapes the world lends itself to spiritual, metaphysical and mystical exploration. They were moonstruck – or 'moonmad'.

In this nocturnal scene, enigmatically titled *À la Rencontre du plaisir* (Towards Pleasure) from 1962, René Magritte's iconic man in a bowler hat turns his back to us to assume the position of the viewer. Is he gazing at the Moon, lost in reverie, or contemplating an unrealized deed, the promise of pleasure in the title?

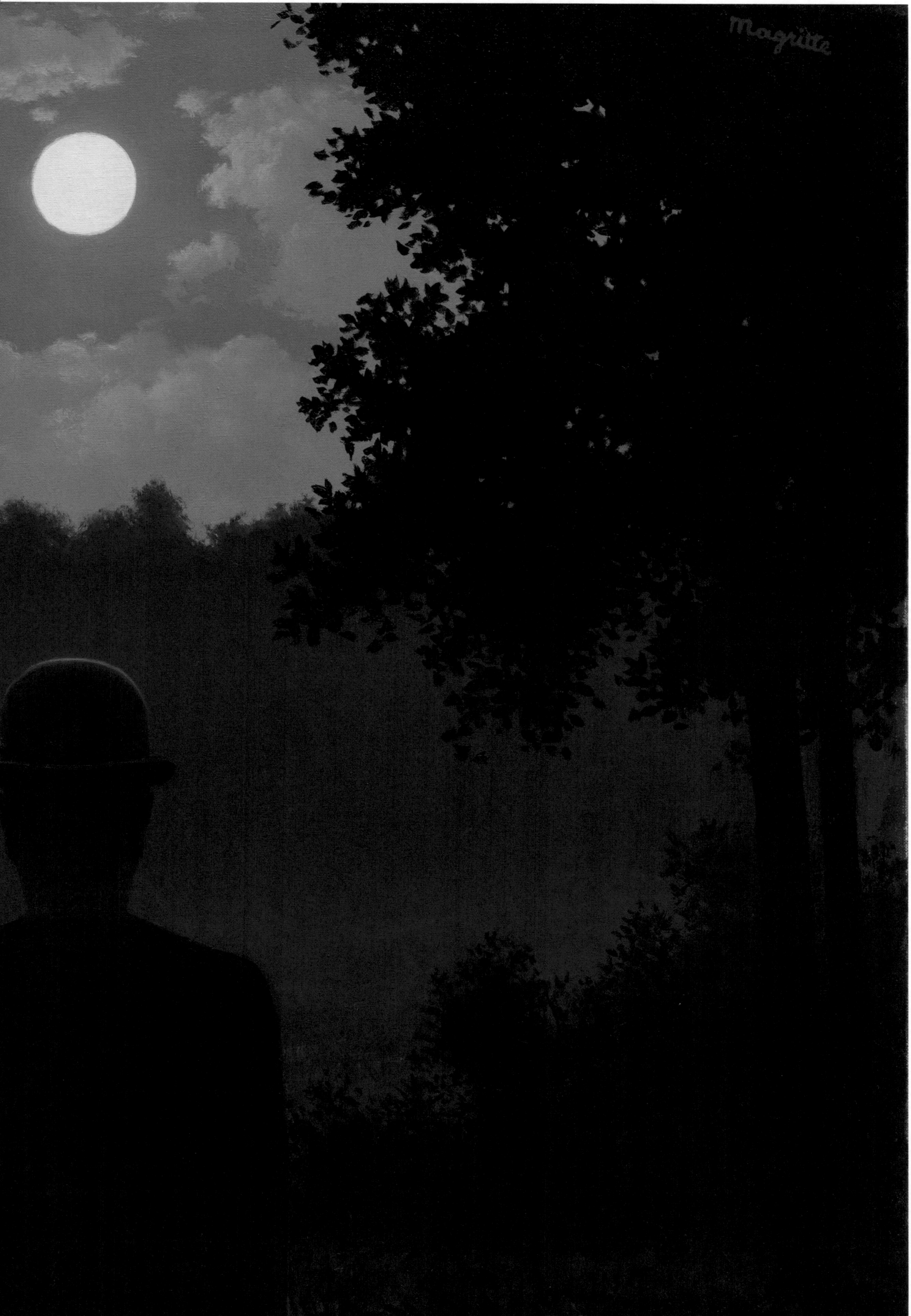
Magritte

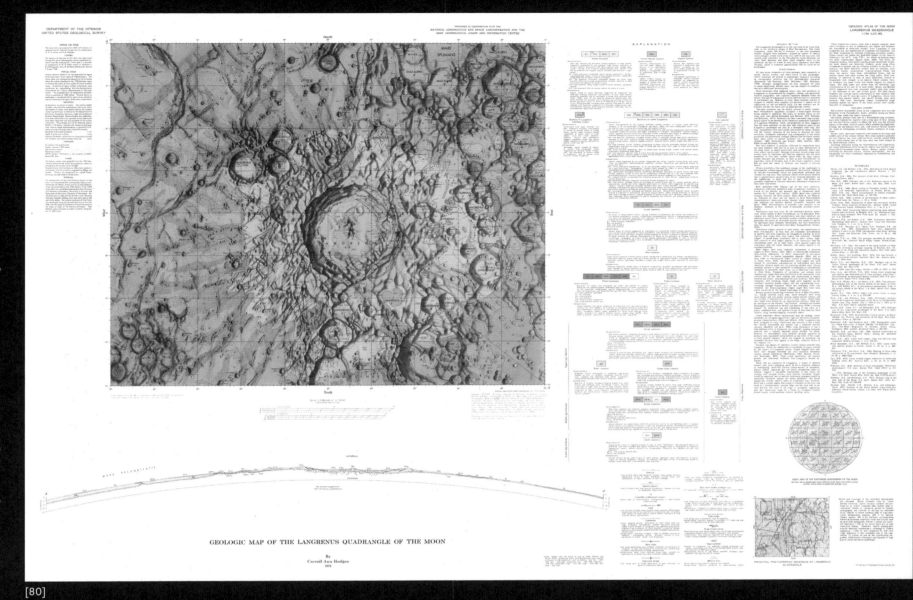

[80]

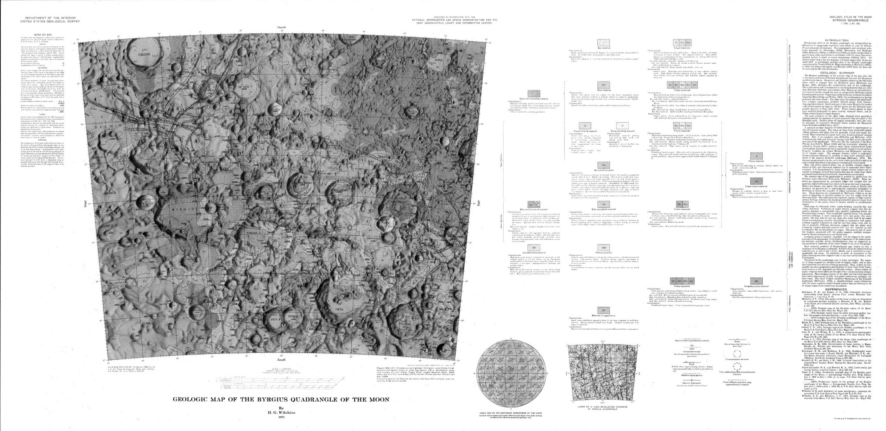

[92]

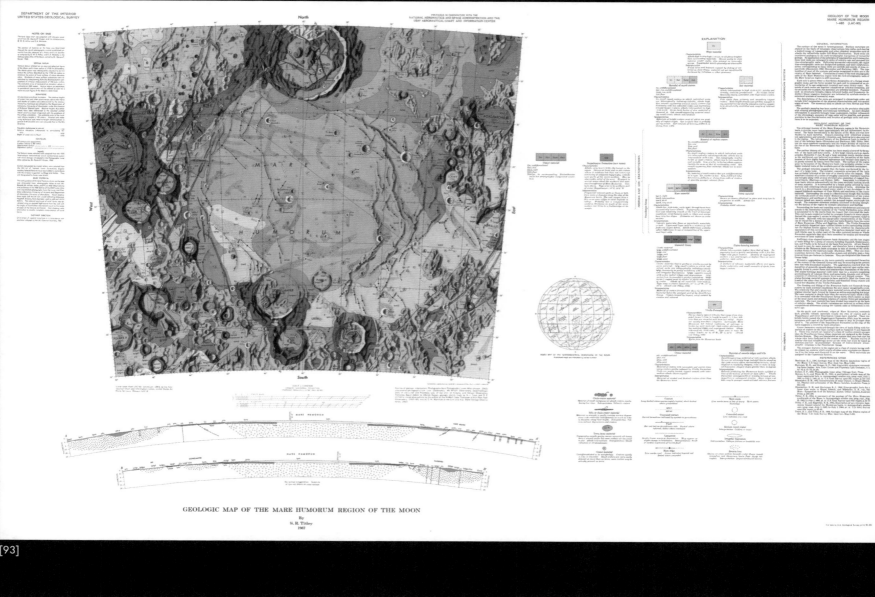

GEOLOGIC MAP OF THE MARE HUMORUM REGION OF THE MOON
By
S.R. Titley
1967

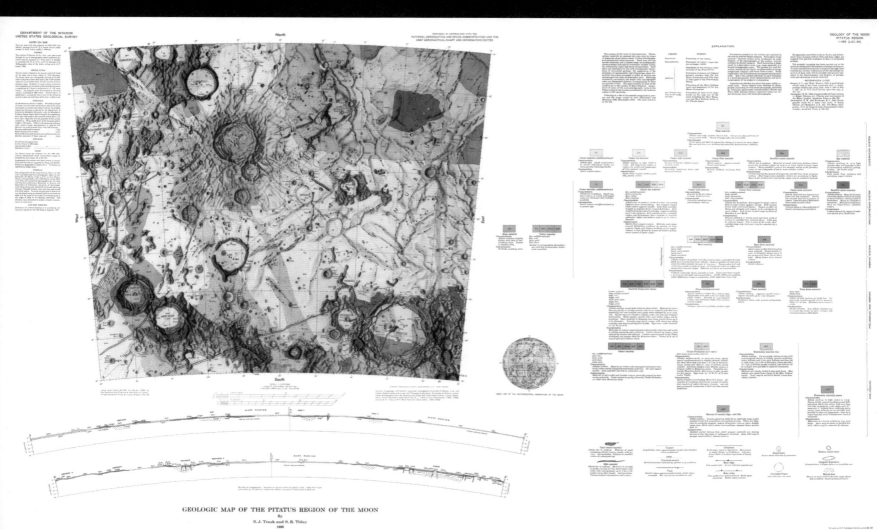

GEOLOGIC MAP OF THE PITATUS REGION OF THE MOON
By
N.J. Trask and S.R. Titley
1966

GEOLOGIC ATLAS OF THE MOON

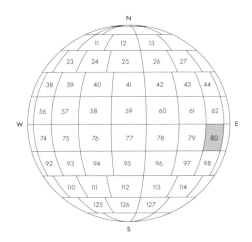

The crater Langrenus dominates this quadrangle. To the crater's east and southwest, rolling terra and impact craters of various sizes and ages occupy much of the map. West of the crater are the lava plains of Mare Fecunditatis (Sea of Fertility). Smaller maria are found in the northeast, as part of a belt of irregular maria. The terrae visible in this region may have once been part of the Fecunditatis basin's ring structure, since battered by impacts over billions of years. The highlands have probably been covered with materials thrown out by large impacts to the north. Blocks of this material may also account for the overlapping secondary craters seen to the southeast of Langrenus. Langrenus itself was only excavated during the Copernican, the Moon's latest geologic period. In the quadrangle's north, at coordinates 0.51°S, 56.36°E, the words 'LUNA 16' mark the 24 September 1970 landing site of the Soviet Union's Luna 16 robotic probe. The probe was the first robotic mission to return lunar soil samples to Earth. In 2007, China's Chang'e-1 spacecraft was deliberately crashed into Mare Fecunditatis.

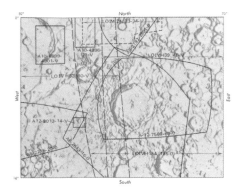

PRINCIPAL PHOTOGRAPHIC COVERAGE
OF LANGRENUS QUADRANGLE

LUNA 16 DRILL CORE
Sample from Mare Fecunditatis retrieved by a robotic drill on the Luna 16 mission. It measures 36 cm / 14 ⅛ in in length and weighs 101 g / 3 ½ oz. Only two sections and one sample rock were shared with NASA.

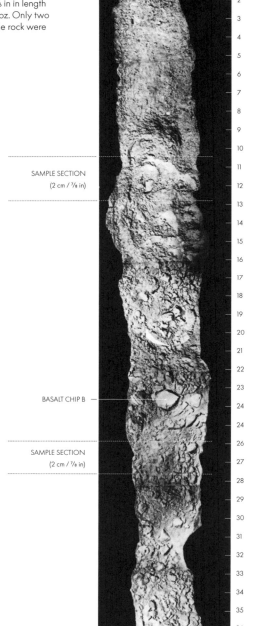

CM
1
2
3
4
5
6
7
8
9
10
11

SAMPLE SECTION
(2 cm / ⅞ in)

12
13
14
15
16
17
18
19
20
21
22
23

BASALT CHIP B

24
24
25
26

SAMPLE SECTION
(2 cm / ⅞ in)

27
28
29
30
31
32
33
34
35
36

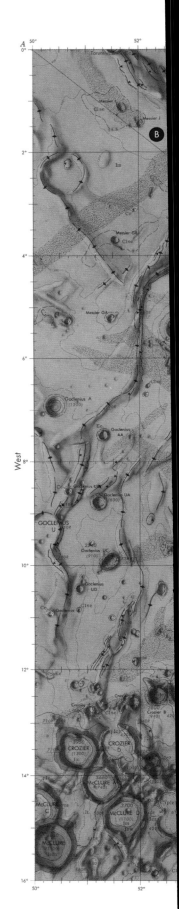

LUNA 16	ROCKET	Proton-K + Blok D
	CREW	N/A
	LAUNCH DATE	13:25:52, 12 September 1970 (UTC)
	LAUNCH SITE	Tyuratam (Baikonur Cosmodrome)
	PAYLOAD	Remote Arm for Sample Collection
	LUNAR ORBIT INSERTION	17 September 1970
	LANDING DATE	05:18:00, 20 September 1970 (UTC)
	LANDING SITE	Mare Fecunditatis
	LANDING COORDINATES	0.51°S, 56.36°E Ⓐ

CHANG'E 1	ROCKET	Chang Zheng 3A
	CREW	N/A
	LAUNCH DATE	10:05:04, 24 October 2007 (UTC)
	LAUNCH SITE	Xichang Satellite Launch Center
	PAYLOAD	Stereoscopic CCD Camera
	LUNAR ORBIT INSERTION	05 November 2007
	LANDING DATE	08:13:00, 01 March 2009 (UTC)
	IMPACT SITE	North of Mare Fecunditatis
	LANDING COORDINATES	1.66°S, 52.27°E Ⓑ

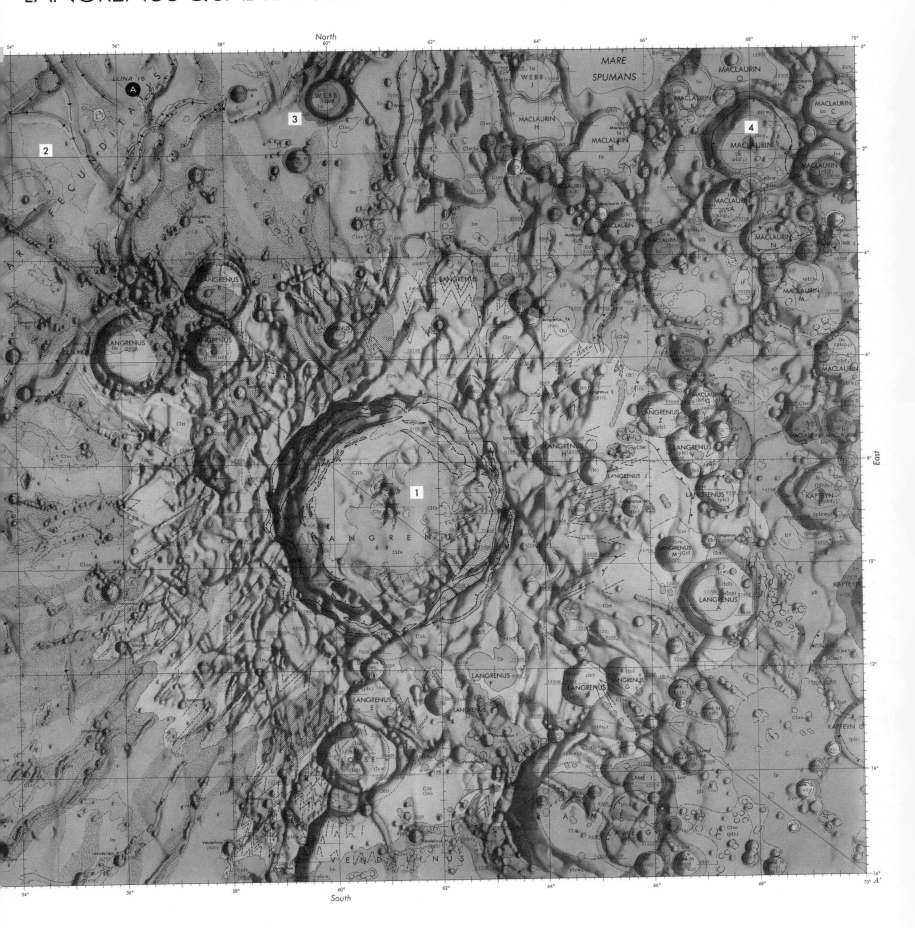

LANGRENUS QUADRANGLE: KEY CHARACTERISTICS

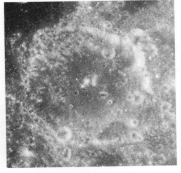

1. LANGRENUS CRATER
Latitude: 8.86°S / Longitude: 61.04°E
Diameter: 132 km / 82 miles
Depth: 2.7 km / 1 ¾ miles
Photograph: Apollo 8 (USA, 1968)

2. MARE FECUNDITATIS
Latitude: 7.83°S / Longitude: 53.67°E
Diameter: 840.35 km / 522 ¼ miles
Photograph: LRO (USA)

3. WEBB CRATER
Latitude: 1°S / Longitude: 60°E
Diameter: 21.41 km / 13 ¼ miles
Depth: 800 m / 2,624 ft 8 in
Photograph: Lunar Orbiter I (USA, 1966)

4. MACLAURIN CRATER
Latitude: 1.92°S / Longitude: 67.99°E
Diameter: 54.33 km / 33 ¾ miles
Depth: 1.5 km / 1 mile
Photograph: Apollo 15 (USA, 1971)

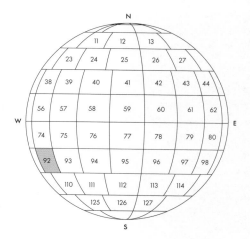

The southwestern half of this quadrangle is dominated by structures and deposits related to the formation of the Orientale basin. The northeastern half is dominated by deposits that are older, including the eroded walls of old craters that have been modified by repeated impacts and the prolonged effects of gravity, as well as in-filling by ejecta materials from other craters. Linear structures that underlay the terra in this region, radial to the Orientale basin, probably formed during the impact that created the basin and were then covered by ejecta from the Orientale impact, although some of the rugged terrain may be the remnants of older structures related to the creation of the Humorum basin. Volcanic domes have since protruded into the terrain, forming rolling plateaus, as have furrows that may once have been vents for more explosive eruptions of volcanic material. Many of the craters in this quadrangle are old, showing extensive wear and erosion. This is the case for the crater Byrgius after which this quadrangle is named. Only a few young craters, such as the bright, rayed crater Byrgius A, which intrudes upon Byrgius' eastern wall and is the brightest feature in this region, evidence more recent impact events.

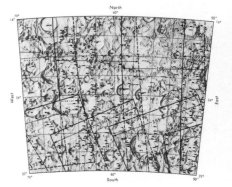

LUNAR ORBITER IV HIGH-RESOLUTION COVERAGE OF BYRGIUS QUADRANGLE

LUNA ORBITER IV MOSAIC
This composite image taken in May 1969 shows Rimae Sirsalis cutting through some craters and winding around others.

BYRGIUS QUADRANGLE: KEY CHARACTERISTICS

1. BYRGIUS CRATER
Latitude: 24.73°S / Longitude: 65.38°W
Diameter: 84.46 km / 52 ½ miles
Depth: 4.6 km / 2 ¾ miles
Photograph: Lunar Orbiter IV (USA, 1967)

2. FOURIER CRATER
Latitude: 30.31°S / Longitude: 53.1°W
Diameter: 51.57 km / 32 miles
Depth: 3.7 km / 2 ¼ miles
Photograph: Lunar Orbiter IV (USA, 1967)

3. PROSPER HENRY CRATER
Latitude: 23.52°S / Longitude: 59°W
Diameter: 41.73 km / 26 miles
Depth: 2 km / 1 ¼ miles
Photograph: Lunar Orbiter IV (USA, 1967)

4. DARWIN CRATER
Latitude: 19.9°S / Longitude: 69.21°W
Diameter: 122.18 km / 75 ¾ miles
Photograph: Lunar Orbiter IV (USA, 1967)

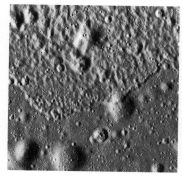

5. CRÜGER CRATER
Latitude: 16.7°S / Longitude: 67°W
Diameter: 46 km / 28 ½ miles
Depth: 500 m / 1,640 ft 6 in
Photograph: LRO mosaic

6. RIMAE ZUPUS
Latitude: 15.46°S / Longitude: 53.76°W
Length: 120 km / 74 ½ miles
Photograph: Lunar Orbiter IV (USA, 1967)

7. CAVENDISH CRATER
Latitude: 24.63°S / Longitude: 53.8°W
Diameter: 52.64 km / 32 ¾ miles
Depth: 2.4 km / 1 ½ miles
Photograph: Lunar Orbiter IV (USA, 1967)

8. VIETA CRATER
Latitude: 29.31°S / Longitude: 58.2°W
Diameter: 87.16 km / 54 ½ miles
Depth: 4.5 km/ 2 ¾ miles
Photograph: Lunar Orbiter IV (USA, 1967)

9. RIMAE SIRSALIS
Latitude: 15°S / Longitude: 61.36°W
Length: 426 km / 264 ¾ miles
Photograph: Lunar Orbiter IV (USA, 1967)

10. DE VICO CRATER
Latitude: 19.71°S / Longitude: 60.32°W
Diameter: 22.13 km / 13 ¾ miles
Photograph: Lunar Orbiter IV (USA, 1967)

11. LAMARK CRATER
Latitude: 23.12°S / Longitude: 70.06°W
Diameter: 114.65 km / 71 ¼ miles
Depth: 2.8 km / 1 ¾ miles
Photograph: Lunar Orbiter IV (USA, 1967)

Michael J. Neufeld

Myth

Lunar

'SELLING' THE MOON IN THE 1950S

↓
Chesley Bonestell's painting of a rocket ship on the lunar surface featured on the front cover of Willy Ley's, *Conquest of Space* (1949). The rocket is inspired by the A-4b, an unsuccessful winged version of the Nazi V-2 ballistic missile. In the background are Bonestell's characteristically rugged lunar mountains. He did not anticipate that a rain of micrometeorites would round off all peaks and crater rims.

↱
Wernher von Braun poses in his NASA Marshall Space Flight Center office in September 1960, two months after his rocket group was transferred from the US Army. Behind him is a painting from an Army Moon mission proposal. On his desk are models of the Juno II launch vehicle (left) and the first stage of the Saturn I rocket.

→
In this Disney publicity photo from the mid-1950s, Wernher von Braun (left), Willy Ley, Walt Disney and Heinz Haber stand in front of a model of von Braun's space station, which appeared in the Moon episode of Disney's *Man in Space* TV series.

↳
Von Braun and Ley illustrate putting a satellite into orbit using a model launch vehicle. This photo was taken in 1955, likely in a US Army office.

↵↵
Cover of *Mechanix Illustrated* (1945). Immediately after the end of the Second World War, Willy Ley published an article about a trip to the Moon. The German V-2 rocket inspired many space advocates such as Ley to conclude that such a trip might be feasible sooner than expected.

↵↵↵
Poster for *Destination Moon* (1950), which won the Academy Award for special effects. Chesley Bonestell painted the movie's backdrops of the Earth from space and of the lunar surface.

↵↵↵↵
On the cover of the first issue of *Collier's* magazine's space series is Bonestell's illustration of von Braun's space shuttle.

to the US military. Von Braun's name appeared episodically in the press, but his public fame really began in 1952, when *Collier's* magazine featured him in a special issue about spaceflight.[2]

Back in September 1945, Ley had published a 'Rocket to the Moon' article in *Mechanix Illustrated*, which led quickly to Ley and Bonestell corresponding. In March 1946, *Life* ran a new Bonestell article, 'Trip to the Moon: Artist Paints Journey by Rocket', including a spectacular painting of Albategnius crater framed in the window of a spaceship. The two soon began to collaborate on a book that would combine Bonestell's paintings with Ley's narrative. *The Conquest of Space* (1949) was a massive success, quickly reprinted and translated into several European languages. It became a milestone in the campaign to 'sell' the idea of spaceflight as feasible, its cover featuring Bonestell's painting of a winged

oon after the Second World War, spaceflight advocates and enthusiasts began to try to convince the wider public of the feasibility of flying to the Moon. The idea of such a trip was not new: writers had penned fantasy and science fiction accounts of lunar voyages for centuries. In the early 20th century, thanks to spaceflight theorists and rocket experimenters, it became possible to imagine a technology that could make these dreams real. But the Second World War was an inflection point. The German V-2 ballistic missile, first launched against western European cities in autumn 1944, was a poor weapon, but it was a revolutionary breakthrough in the size and power of rockets. Suddenly, a lunar vehicle seemed feasible within a few decades. After the war, the V-2's elegant, finned shape became the iconic image of the spaceship.[1]

Three men were at the forefront of popularizing Moon flights: space artist Chesley Bonestell (1888–1986), science writer Willy Ley (1906–1969) and rocket engineer Wernher von Braun (1912–1977). Hollywood special effects artist Bonestell became famous first. In late May 1944, *Life* magazine ran five of his pictures of Saturn as viewed from its moons. Bonestell's paintings birthed a new American space art, one that combined astronomical knowledge, American landscape painting and frontier rhetoric. That same month, the German émigré writer Ley, one of the few veterans of the Weimar spaceflight movement to flee the Nazis, published a book, *Rockets*. It made him into one of the leading journalists on a topic still mainly associated with the space operas of Buck Rogers and Flash Gordon. Von Braun's fame came later. Ley knew him from the Berlin rocket group, but the engineer only arrived in the United States in September 1945, one of the first 'Nazi scientists' imported to help transfer German science and technology

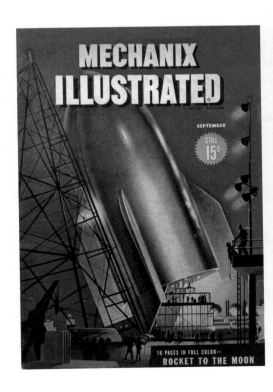

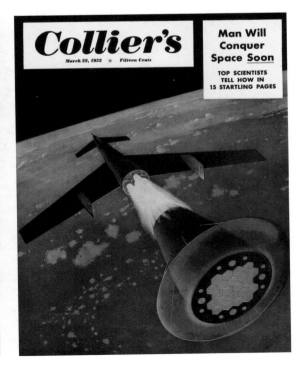

Walt Disney opened his first theme park, Disneyland, in Anaheim, California, in 1955. The Tomorrowland section included a Rocket to the Moon ride, shown here in August 1963. Visitors first passed through a simulated space station with a view of Earth, then took their seats for an imagined trip to the Moon. Ley and von Braun were advisors.

V-2-like ship sitting in a mountainous lunar landscape with space-suited crew members on the surface. While the book was about a lot more than the Moon, a human expedition there always excited public interest because it seemed like the most reasonable near-term possibility.[3]

Before the book appeared, Hollywood producer George Pal hired Chesley Bonestell for a movie based on a novel by science fiction writer Robert Heinlein. The artist, who had produced the matte paintings for *Citizen Kane* and other major motion pictures, was the natural candidate to provide the scenic backdrops for *Destination Moon*. Released in late June 1950, it too was a hit and won a special effects Oscar. The moonship also looked like a V-2 (without wings) but was powered by a nuclear rocket engine, reflecting American faith in everything atomic after the Second World War.

The next big breakthrough was the *Collier's* series, which arose out of a US Air Force space conference in late 1951. The journalist and editor Cornelius Ryan (1902–1974) met Wernher von Braun and other leading scientists there. Ryan became a spaceflight convert and helped organize a special issue with articles by von Braun, Ley and several other advocates. Bonestell's art was again central, but was supplemented by illustrations from Fred Freeman and Rolf Klep. The 22 March 1952 edition of the magazine, which featured von Braun's space station concept, was such a hit that it spun off a book in the same format as *Conquest of*

Space. *Collier's* then devoted parts of two autumn issues to von Braun's vision of a Moon expedition, which broke with the V-2 archetype: his bulbous ships were designed to fly only in a vacuum. Those issues became another book, *Conquest of the Moon*. *Collier's* followed with more space articles in 1953 and 1954, primarily by von Braun. The last was on his imagined Mars expedition – his spare-time focus since 1947, as he thought that the feasibility of a Moon trip was 'too easy' to demonstrate.[4]

Just as the series was finishing, Walt Disney (1901–1966) was trying to figure out what to do with the Tomorrowland section of a new theme park and television show. A deputy, Ward Kimball, pointed to *Collier's* and spaceflight as a theme. Their first hire was Ley, who pushed them to employ von Braun. The two consulted on the Rocket to the Moon ride at Disneyland and became on-screen explainers for programmes that aired in March and December 1955. The second covered von Braun's space station and a trip around the Moon. A third, about Mars, appeared in December 1957, two months after the USSR launched Sputnik (the first artificial Earth satellite).[5]

In March 1957, Bonestell unveiled his ultimate Moon painting, a huge mural for the Boston Science Museum's planetarium. The Smithsonian's National Air and Space Museum now exhibits it. The mural was a fitting capstone to a campaign to convince the public that a Moon flight was not only feasible, but imminent.

↑
Bonestell painted this illustration of an orbital scout version of von Braun's lunar ship for the fall 1952 issue of *Collier's* magazine, which focused on 'The Conquest of the Moon'. Von Braun and Ley detailed the

German rocket engineer's grandiose plan for a massive expedition to the lunar surface. These ships would have landing gear and two of them would take the crew to and from Earth's orbit.

↓
Bonestell's massive *Lunar Landscape* – 3 metres (10 feet) high and 12 metres (40 feet) long – was unveiled in the Boston planetarium in March 1957, six months before Sputnik. It was taken down in 1970,

after Surveyor and Apollo lunar photos made its craggy mountains look inaccurate. This image shows the restored mural at the Smithsonian's National Air and Space Museum after it was installed in 2021.

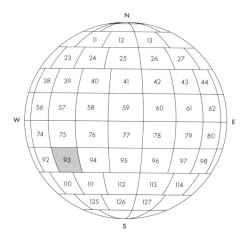

The smooth lava plains of Mare Humorum (Sea of Moisture) occupy the centre of this otherwise rugged and heavily cratered quadrangle. At the northern edge of this mare is the large crater Gassendi. The rim of this crater is heavily worn, and it has been filled in by lava, with only the rim and some central peaks remaining visible. A system of rilles crisscrosses the crater's interior lava plain. At the mare's southern edge are other notable craters, including the partially eroded and filled Doppelmayer and Lee craters, the mostly erased Puiseux crater, as well as the relatively young and well-preserved Vitello crater. The mare and several of the older craters are overlaid by ray material ejected from the younger craters in this region. The cratered areas are also shown here as containing mixtures of pyroclastic volcanic materials and ejecta. Several of the craters on the eastern and western edges of the mare show extensive modification, including Agatharchides in the east and Mersenius in the west, which are both filled and spotted with small craters.

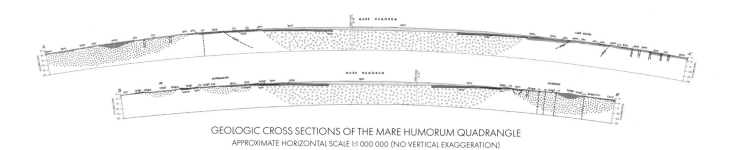

GEOLOGIC CROSS SECTIONS OF THE MARE HUMORUM QUADRANGLE
APPROXIMATE HORIZONTAL SCALE 1:1 000 000 (NO VERTICAL EXAGGERATION)

MARE HUMORUM QUADRANGLE: KEY CHARACTERISTICS

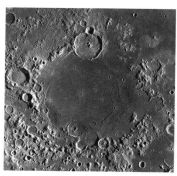

1. MARE HUMORUM
Latitude: 24.5°S / Longitude: 38.57°W
Diameter: 419.67 km / 260 ¾ miles
Photograph: LRO (USA, 2016)

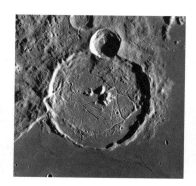

2. GASSENDI CRATER
Latitude: 17.55°S / Longitude: 39.96°W
Diameter: 111.4 km / 69 ¼ miles
Depth: 1.9 km / 1 ¼ miles
Photograph: LRO (USA, 2014)

3. VITELLO CRATER
Latitude: 30.42°S / Longitude: 37.55°W
Diameter: 42.5 km / 26 ½ miles
Depth: 1.7 km / 1 mile
Photograph: Lunar Orbiter IV (USA, 1967)

4. AGATHARCHIDES CRATER
Latitude: 19.85°S / Longitude: 32.03°W
Diameter: 52 km / 32 ¼ miles
Depth: 1,200 m / 3,937 ft
Photograph: Lunar Orbiter IV (USA, 1967)

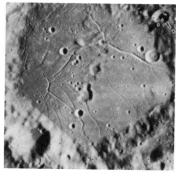

5. MERSENIUS CRATER
Latitude: 21.5°S / Longitude: 49.34°W
Diameter: 84.46 km / 52 ½ miles
Depth: 2.3 km / 1 ½ miles
Photograph: Lunar Orbiter IV (USA, 1967)

6. DOPPELMAYER CRATER
Latitude: 28.48°S / Longitude: 41.5°W
Diameter: 65.08 km / 40 ½ miles
Depth: 1.6 km / 1 mile
Photograph: Lunar Orbiter IV (USA, 1967)

7. LEE CRATER
Latitude: 30.66°S / Longitude: 40.76°W
Diameter: 41.17 km / 25 ½ miles
Depth: 1,300 m / 4,265 ft 1 in
Photograph: Lunar Orbiter IV (USA, 1967)

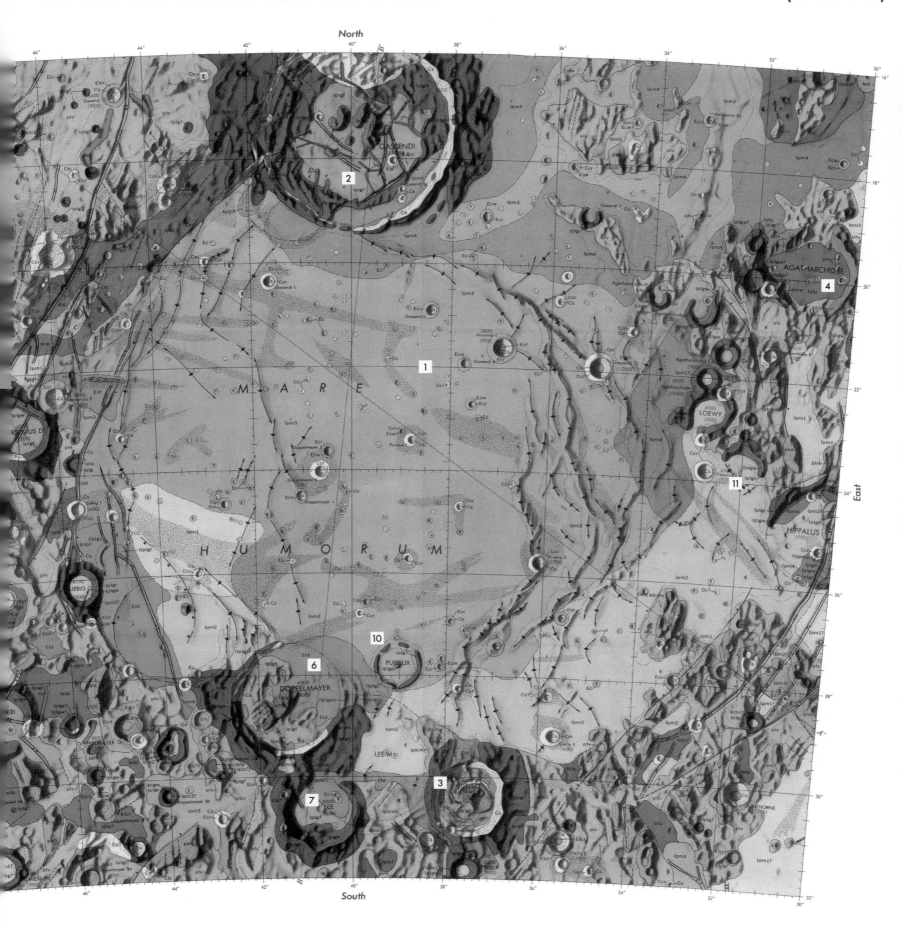

8. LIEBIG CRATER
Latitude: 24.35°S / Longitude: 48.3°W
Diameter: 38.96 km / 24 ¼ miles
Depth: 1.5 km / 1 mile
Photograph: Lunar Orbiter IV (USA, 1967)

9. RIMA PALMIERI
Latitude: 27.83°S / Longitude: 47.17°W
Diameter: 27.13 km / 16 ¾ miles
Photograph: Lunar Orbiter IV (USA, 1967)

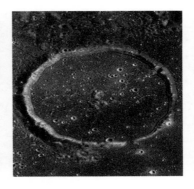

10. PUISEUX CRATER
Latitude: 27.82°S / Longitude: 39.2°W
Diameter: 24.95 km / 15 ½ miles
Depth: 400 m / 1,312 ft 4 in
Photograph: LRO (USA, 2015)

11. LOEWY CRATER
Latitude: 22.69°S / Longitude: 32.85°W
Diameter: 22.45 km / 14 miles
Depth: 1,100 m / 3,608 ft 11 in
Photograph: Lunar Orbiter IV (USA, 1967)

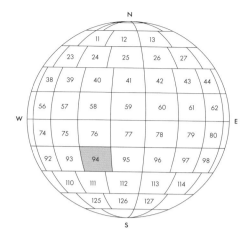

Pitatus is an ancient impact crater in Mare Nubium (Sea of Clouds). Though not the Moon's oldest crater, it was excavated during the Moon's early geologic history. Its rim has been heavily worn and its interior, long ago flooded when lava welled up from its floor, almost joins the lava plains that surround the crater. A system of rilles intersects with and roughly outlines Pitatus' walls. The nearby Bullialdus crater is, by contrast, nearly pristine. It has a high rim, terraced inner walls, mountainous central peaks and a large ejecta blanket. Bullialdus is surrounded by several older craters, including the Campanus and Mercator crater pair in the southwest of the quadrangle, both of which have smooth and dark interiors similar to the surrounding mare. These craters are similar in age to Pitatus. To Bullialdus' north are additional old craters Lubiniezky and Opelt, of which very little remains other than traces of the rims. Other craters in this quadrangle, such as Wolf, have almost entirely lost their characteristic features, having been completely flooded and significantly eroded. The otherwise smooth mare of this quadrangle is interrupted by occasional crater chain systems, smooth ridges and hills.

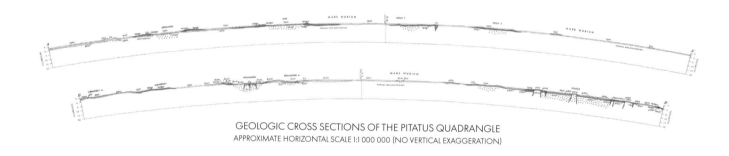

GEOLOGIC CROSS SECTIONS OF THE PITATUS QUADRANGLE
APPROXIMATE HORIZONTAL SCALE 1:1 000 000 (NO VERTICAL EXAGGERATION)

PITATUS QUADRANGLE: KEY CHARACTERISTICS

1. PITATUS CRATER
Latitude: 29.88°S / Longitude: 13.53°W
Diameter: 100.63 km / 62 ½ miles
Depth: 900 m / 2,952 ft 9 in
Photograph: Lunar Orbiter IV (USA, 1967)

2. MARE NUBIUM
Latitude: 20.59°S / Longitude: 17.29°W
Diameter: 714.5 km / 444 miles
Photograph: Apollo 16 (USA, 1972)

3. BULLIALDUS CRATER
Latitude: 20.75°S / Longitude: 22.26°W
Diameter: 60.72 km / 37 ¾ miles
Depth: 3.5 km / 2 ¼ miles
Photograph: Apollo 16 (USA, 1972)

4. CAMPANUS CRATER
Latitude: 28.04°S / Longitude: 27.9°W
Diameter: 46.41 km / 28 ¾ miles
Depth: 2.1 km / 1 ¼ miles
Photograph: Lunar Orbiter IV (USA, 1967)

5. RIMA HESIODUS
Latitude: 30.54°S / Longitude: 21.85°W
Diameter: 251.46 km / 156 ¼ miles
Depth: 1,100 m / 3,608 ft 11 in
Photographed from Earth (UK, 2014)

6. RIMAE HIPPALUS
Latitude: 25.6°S / Longitude: 29.36°W
Diameter: 266 km / 165 ¼ miles
Photographed from Earth
(Germany, 2020)

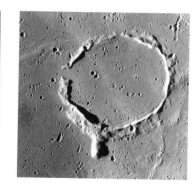

7. KIES CRATER
Latitude: 26.31°S / Longitude: 22.63°W
Diameter: 45.54 km / 28 ¼ miles
Depth: 400 m / 1,312 ft 4 in
Photograph: LRO (USA, 2010)

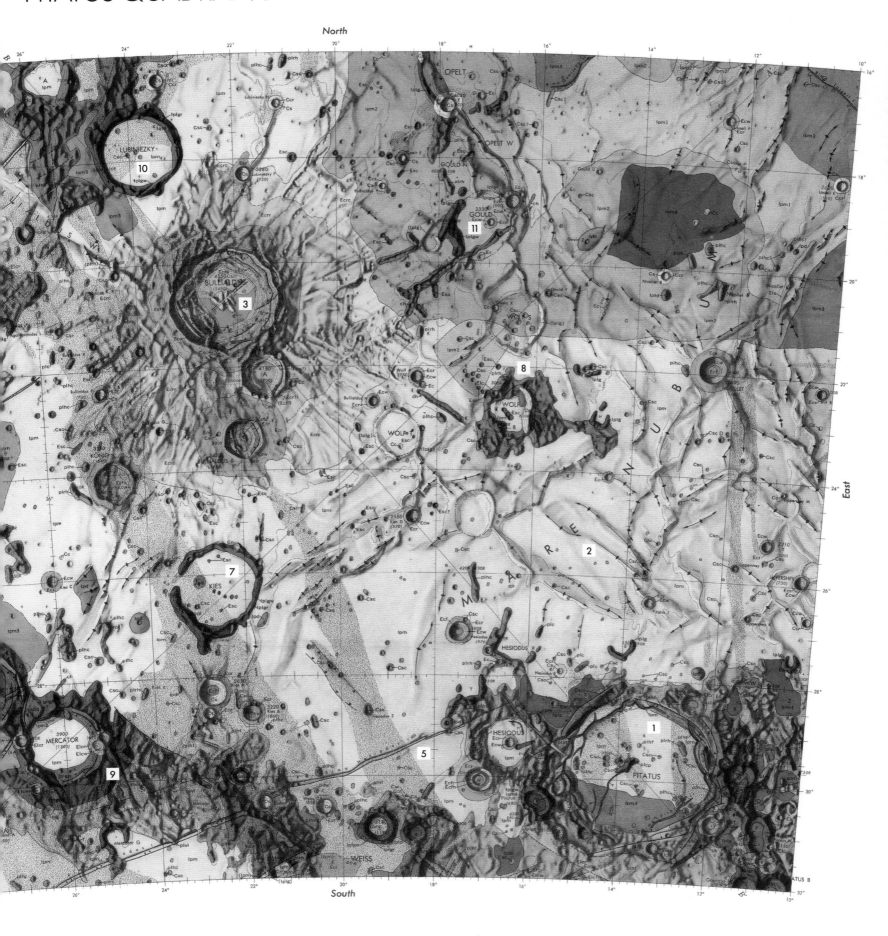

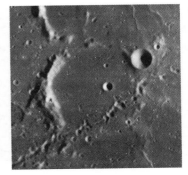

8. WOLF CRATER
Latitude: 22.79°S / Longitude: 16.63°W
Diameter: 25.74 km / 16 miles
Photograph: LRO (USA, 2019)

9. MERCATOR CRATER
Latitude: 29.25°S / Longitude: 26.11°W
Diameter: 46.32 km / 28 ¾ miles
Depth: 1.8 km / 1 mile
Photograph: Lunar Orbiter (USA, 1967)

10. LUBINIEZKY CRATER
Latitude: 17.88°S / Longitude: 23.89°W
Diameter: 42.97 km / 26 ¾ miles
Depth: 800 m / 2,624 ft 8 in
Photograph: Lunar Orbiter IV (USA, 1967)

11. GOULD CRATER
Latitude: 19.26°S / Longitude: 17.25°W
Diameter: 32.99 km / 20 ½ miles
Photograph: Lunar Orbiter IV (USA, 1967)

THE FIRST MOON RACE — LUNA VERSUS PIONEER

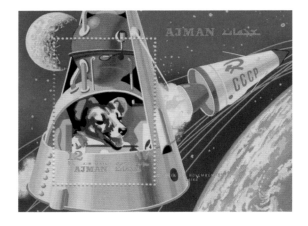

he Moon race of the 1960s is well remembered. Less so is an earlier mad scramble by the USSR and the USA to be the first to send a probe there. Starting less than a year after the USSR launched the Sputnik satellite on 4 October 1957, this competition resulted in several major space firsts – almost all of them by the Soviets.

There is little evidence of forethought in the robotic exploration of the Moon. Pre-Sputnik, space advocates talked about landing people on the surface, both because they were obsessed with the idea and because it was what excited the public. The great Soviet rocket engineer, Sergei Pavlovich Korolev (1907–1966), the 'Chief Designer' of the space programme, formulated plans in 1957 for more powerful boosters and lunar and planetary spacecraft. But there would be no funding until after the propaganda coups of two Sputniks in the autumn. (The second, carrying Laika the dog, orbited on 3 November.)[1] In the USA, all Moon plans were reactive – a response to a perceived national humiliation.

Three weeks after Sputnik, William Pickering (1910–1966), director of the Jet Propulsion Laboratory (JPL) in Pasadena, California, suggested leapfrogging the Soviets by sending a probe to the Moon. Caltech-run JPL was contracted to the US Army for rocket and missile research and was aligned with the Army Ballistic Missile Agency in Huntsville, Alabama, whose technical director was Wernher von Braun. The two agencies were collaborating on an Earth satellite programme as a backup to the Navy's Vanguard. But Red Socks, as the lunar project was nicknamed, was premature; American space policy was in turmoil. The Army faced competition from its chief rival in missiles and space, the US Air Force (USAF). The USAF ballistic missile programme, run by General Bernard Schriever (1910–2005), soon proposed putting a bigger spacecraft in lunar orbit.[2]

Korolev had a distinct advantage: he received rapid translations from the American press, whereas all his launches were secret and the USA had only limited intelligence about Soviet plans. Late in January 1958, Korolev sent Communist Party leader Nikita Khrushchev (1894–1971) a programme designed to beat the Americans

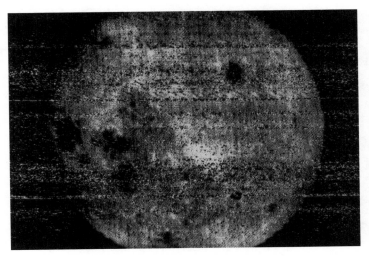

and fell back, but gathered even better data about the Earth's radiation belts than Pioneer 1. Two days earlier, Korolev's group had launched their third impactor. The R-7 worked this time, but the upper stage did not fire.[5]

The new year finally brought success. On 4 January 1959, the Soviet spacecraft *Mechta* (Dream), later retroactively labelled Luna 1, missed the Moon by 5,965 kilometres (3,707 miles), but became the first human-made object to escape Earth and orbit the Sun. Korolev's team was surprised that their failure was hailed internationally as a great achievement. The USA achieved the same result in early March when the Army's Pioneer 4 passed by the Moon at 60,000 kilometres (37,282 miles) and went into solar orbit. An interlude followed. Korolev's team pivoted to ICBM tests and the initial Pioneer project was finished.[6]

A new Soviet launch attempt failed in June, but three months later the impactor programme finally worked. Luna 2 crashed into the Moon on 14 September, two days after taking off from Kazakhstan. It was new, an international first, immediately followed by another. On 4 October, Korolev's next rocket also worked, carrying a more sophisticated, far-side-imaging spacecraft. Luna 3 made a loop around the Moon, automatically oriented itself and took twenty-nine pictures with a television camera projecting onto film. The film was developed on board and scanned. After difficulties getting a good signal, Korolev's team received seventeen usable images. They were low-resolution and noisy, but showed that the Moon's 'far' side was almost all highland terrain, with very few dark maria (so-called 'seas': actually low-lying lava plains). It was one of the greatest achievements of the space age.[7]

Luna 3 was the effective end of the first Moon race. The USA attempted to send three more sophisticated Pioneer Lunar Orbiters in 1959–60, but all the launches failed. Two more Soviet attempts to photograph the far side were also lost to booster failures.[8] It would be 1964 before the next successful lunar mission: NASA's camera-carrying Ranger 7 impactor. But great things had been accomplished during that first mad dash.

to the Moon. In America, the Army finally orbited the first US satellite on 31 January (two previous Vanguard attempts had failed). Three weeks later, President Dwight Eisenhower (1890–1969) created the Advanced Research Projects Agency (ARPA) to rationalize a military space programme riven by interservice rivalry. It would also be a caretaker for projects that would go to a new civilian agency, which became NASA. ARPA immediately approved both the Air Force and Army Moon projects, ostensibly united under the rubric Pioneer.

In late March, Korolev got the go-ahead. The first objective would be to crash a spacecraft into the Moon to establish a Soviet first, followed by photographing the far side, which had been hidden from view for all human history.[3]

On 17 August 1958, the USAF's first Pioneer attempt blew up spectacularly during launch. Half a world away, Korolev's engineers were preparing their first impactor probe launch, but he told them to stand down after the American failure. More preparation did not help; on 23 September, the Soviet booster broke up in the air. Mounting an upper stage on the R-7 intercontinental ballistic missile, which was needed to reach lunar velocity, produced structural vibrations that tore the rocket apart. Determined to win, the Soviets made interim fixes and prepared a new booster in case the USAF succeeded with their next attempt. The Americans' Pioneer 1 (the earlier failure never got a name), launched on 11 October, just missed the necessary velocity and fell back to Earth after rising some 114,750 kilometres (71,300 miles). That smashed all previous altitude records and was a rare space propaganda coup for the USA. Korolev then started his rocket, but it failed exactly like the first. A more thorough fix was needed.[4]

General Schriever's contractors prepared the last of the three authorized USAF attempts. Pioneer 2 ascended on 8 November 1958, but the rocket's third stage failed and the spacecraft only made it to 1,550 kilometres (963 miles). It was the Army's turn. On 6 December, von Braun and Pickering's Pioneer 3, a tiny probe meant simply to fly by the Moon, rose about 107,500 kilometres (66,800 miles)

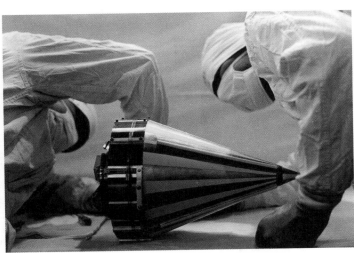

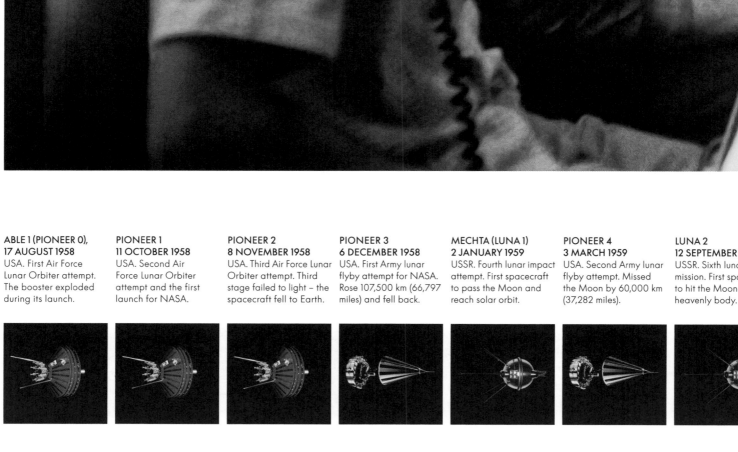

**ABLE 1 (PIONEER 0),
17 AUGUST 1958**
USA. First Air Force
Lunar Orbiter attempt.
The booster exploded
during its launch.

**PIONEER 1
11 OCTOBER 1958**
USA. Second Air
Force Lunar Orbiter
attempt and the first
launch for NASA.

**PIONEER 2
8 NOVEMBER 1958**
USA. Third Air Force Lunar
Orbiter attempt. Third
stage failed to light – the
spacecraft fell to Earth.

**PIONEER 3
6 DECEMBER 1958**
USA. First Army lunar
flyby attempt for NASA.
Rose 107,500 km (66,797
miles) and fell back.

**MECHTA (LUNA 1)
2 JANUARY 1959**
USSR. Fourth lunar impact
attempt. First spacecraft
to pass the Moon and
reach solar orbit.

**PIONEER 4
3 MARCH 1959**
USA. Second Army lunar
flyby attempt. Missed
the Moon by 60,000 km
(37,282 miles).

**LUNA 2
12 SEPTEMBER 1959**
USSR. Sixth lunar impact
mission. First spacecraft
to hit the Moon or any
heavenly body.

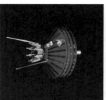

LUNA 3
4 OCTOBER 1959
USSR. First lunar far-side photography attempt. Returned seventeen images of the far side.

PIONEER P-3
26 NOVEMBER 1959
USA. First NASA attempt with Atlas-Able rocket and a larger Pioneer Lunar Orbiter – launch failure.

PIONEER P-30
25 SEPTEMBER 1960
USA. Second NASA attempt with a larger Pioneer Lunar Orbiter – launch failure.

PIONEER P-31
15 DECEMBER 1960
USA. Third NASA attempt with a larger Pioneer Lunar Orbiter – launch failure.

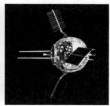
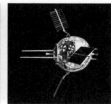
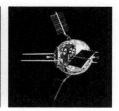

↑
The Cape Canaveral launch crew during Pioneer 4's firing in March 1959. On the upper right are Wernher von Braun and Maj. Gen. John Barclay of the Army Ballistic Missile Agency, which was responsible for the Juno II rocket.

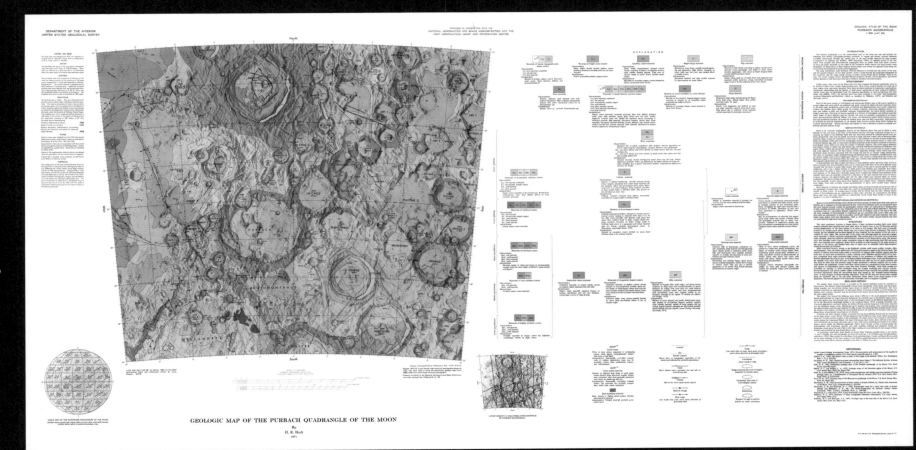

GEOLOGIC MAP OF THE PURBACH QUADRANGLE OF THE MOON
By
H. E. Holt
1971

[95]

GEOLOGIC MAP OF THE RUPES ALTAI QUADRANGLE OF THE MOON
By
L. C. Rowan
1971

[96]

95
Facsimile:
*Geologic Map of the Purbach
Quadrangle of the Moon*, 1974,
by H. E. Holt

96
Facsimile:
*Geologic Map of the Rupes Altai
Quadrangle of the Moon*, 1971,
by L. C. Rowan

97
Facsimile:
*Geologic Map of the Fracastorius
Quadrangle of the Moon*, 1972,
by Desiree E. Stuart-Alexander
and Rowland W. Tabor

98
Facsimile:
*Geologic Map of the Petavius
Quadrangle of the Moon*, 1973,
by Carroll Ann Hodges

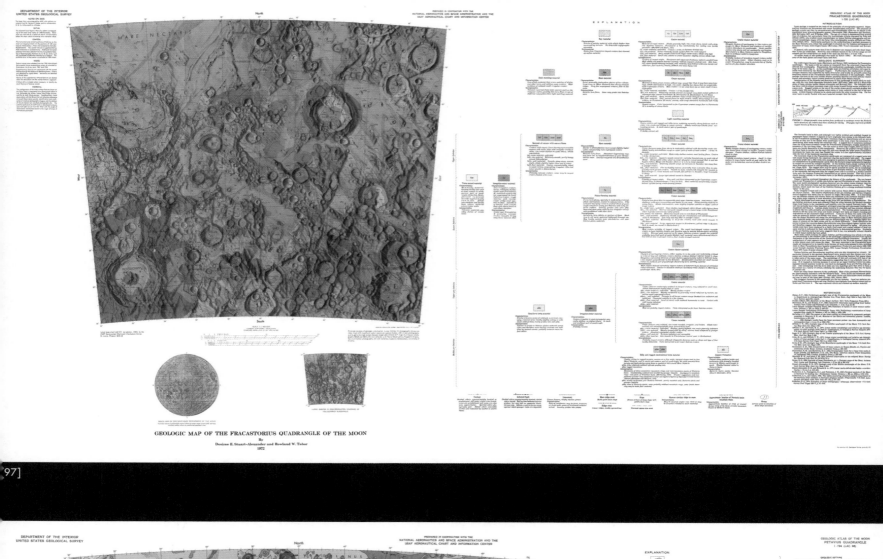

GEOLOGIC MAP OF THE FRACASTORIUS QUADRANGLE OF THE MOON
By
Desiree E. Stuart-Alexander and Rowland W. Tabor
1972

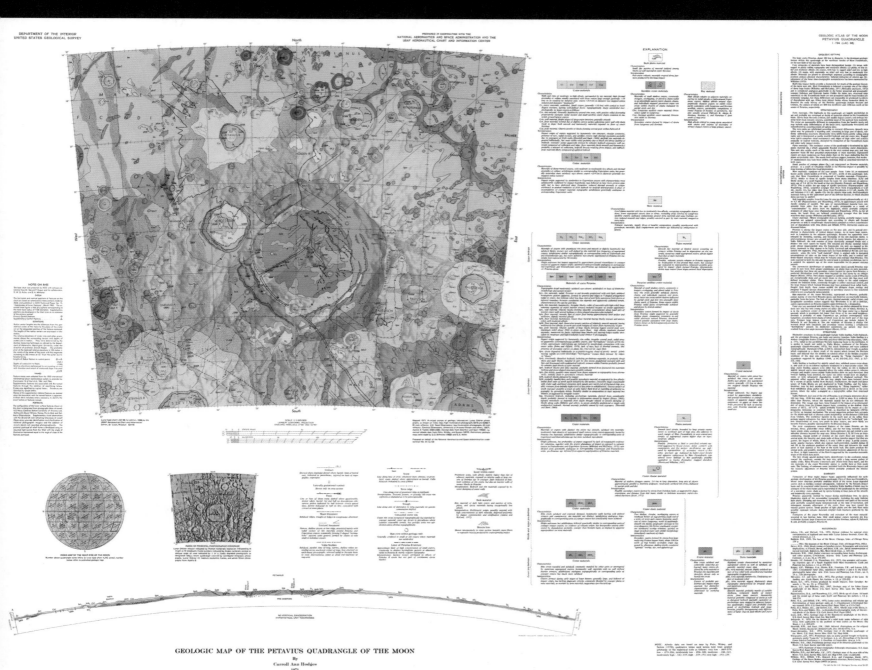

GEOLOGIC MAP OF THE PETAVIUS QUADRANGLE OF THE MOON
By
Carroll Ann Hodges
1973

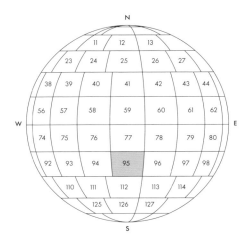

The remnants of large craters that broke and blended the surface of the lunar crust form the rugged ancient features of the southern highlands. These old craters, such as the crater Purbach after which this quadrangle is named, are heavily worn and subdued. Many of the features in the northern portion of this quadrangle were first laid down by the impact that formed the Imbrium basin to the north. Ejecta from the impact created bowl-shaped secondary craters. Periods of cratering, volcanism and tectonic adjustments continued. The formation of the crater Arzachel was the dominant impact event during this early history. A large ejecta blanket and some resulting volcanism transformed the rugged terrain into rounded hills. Some terra in the west was buried by the lava flood that created the plains of Mare Nubium (Sea of Clouds). The large crater Werner was created some time in the middle of the geologic history of this region. Most recently, the impact that created Tycho crater to the southwest of this region covered this quadrangle in ejecta and produced groups of young secondary impact craters. Tectonic faulting in this region created visible scarps such as Rupes Recta.

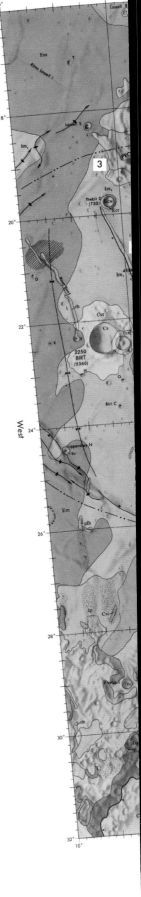

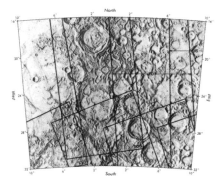

LUNAR ORBITER IV HIGH-RESOLUTION COVERAGE
OF PURBACH QUADRANGLE

THE STRAIGHT WALL (RUPUS RECTA)
Oblique view of Rupes Recta with Birt crater and Rima Birt to the right, captured during the Apollo 16 mission in April 1972.

PURBACH QUADRANGLE: KEY CHARACTERISTICS

1. PURBACH CRATER
Latitude: 25.51°S / Longitude: 2.03°W
Diameter: 114.97 km / 71 ½ miles
Depth: 3 km / 1 ¾ miles
Photograph: Lunar Orbiter IV (USA, 1967)

2. REGIOMONTANUS
Latitude: 28.28°S / Longitude: 1.09°W
Diameter: 126.64 km / 78 ¾ miles
Depth: 1.7 km / 1 mile
Photograph: Lunar Orbiter IV (USA, 1967)

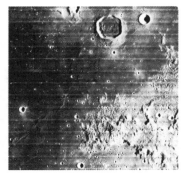

3. PROMONTORIUM TAENARIUM
Latitude: 18.63°S / Longitude: 7.34°W
Length: 70 km / 43 ½ miles
Height: 700 m / 2,296 ft 7 in
Photograph: Lunar Orbiter IV (USA, 1967)

4. ARZACHEL CRATER
Latitude: 18.26°S / Longitude: 1.93°W
Diameter: 97 km / 60 ¼ miles
Depth: 3.6 km / 2 ¼ miles
Photograph: Apollo 16 (USA, 1972)

5. LA CAILLE CRATER
Latitude: 23.68°S / Longitude: 1.08°E
Diameter: 67.22 km / 41 ¾ miles
Depth: 2.8 km / 1 ¾ miles
Photograph: Lunar Orbiter IV (USA, 1967)

6. WERNER CRATER
Latitude: 28.03°S / Longitude: 3.29°E
Diameter: 70.59 km / 43 ¾ miles
Depth: 4.2 km / 2 ½ miles
Photograph: Lunar Orbiter IV (USA, 1967)

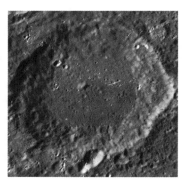

7. ALIACENSIS CRATER
Latitude: 30.6°S / Longitude: 5.13°E
Diameter: 79.65 km / 49 ½ miles
Depth: 3.7 km / 2 ½ miles
Photograph: LRO (USA, 2018)

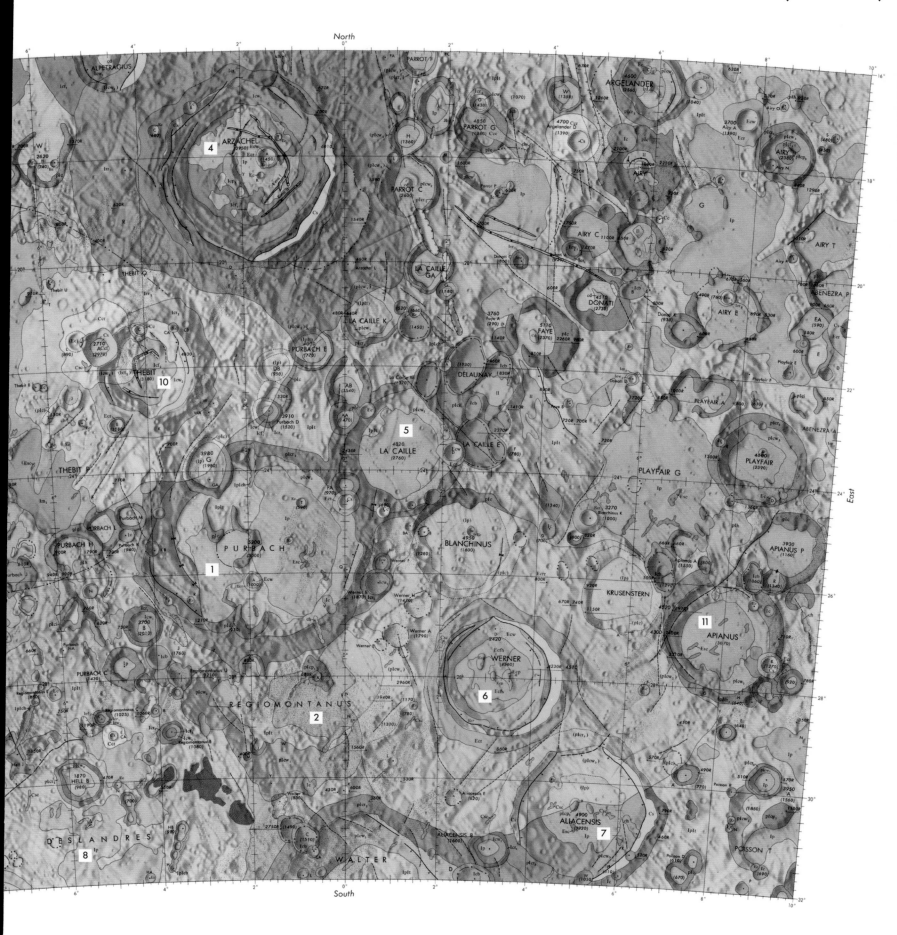

North

South

West

East

8. DESLANDRES
Latitude: 32.55°S / Longitude: 5.57°W
Diameter: 227 km / 141 miles
Photograph: Lunar Orbiter IV (USA, 1967)

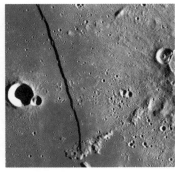

9. RUPES RECTA
Latitude: 21.67°S / Longitude: 7.7°W
Length: 110 km / 68 ¼ miles
Photograph: LRO (USA, 2015)

10. THEBIT CRATER
LLatitude: 22.01°S / Longitude: 4.02°W
Diameter: 54.64 km / 34 miles
Depth: 3.3 km / 2 miles
Photograph: Lunar Orbiter IV (USA, 1967)

11. APIANUS CRATER
Latitude: 26.96°S / Longitude: 7.87°E
Diameter: 63.44 km / 39 ½ miles
Depth: 2.08 km / 1 ¼ miles
Photograph: Lunar Orbiter IV (USA, 1967)

This quadrangle is named for the Rupes Altai cliffs that run diagonally across this region of the Moon's south-central highlands. The escarpment runs from roughly the centre north to the southeast corner of this map. The terrain in these highlands is rugged and chaotic, showing a complicated history of bombardment and alteration, with craters of varying sizes and age peppering the surface. One of the larger craters, Sacrobosco, is drastically altered. Multiple smaller, younger craters lie within Sacrobosco and overlay its degraded rim and wall. There are twenty-three craters counted as satellite craters to Sacrobosco. The area to the west of Rupes Altai is the Nectaris basin. There are fewer craters on this side of the scarp, and the area is characterized by benches and shallowly filled troughs that run parallel to the cliffs, as well as exposed mare materials. Nonetheless, there are still craters to the west of the scarp, including the ancient crater Catharina (here spelled Catherina), which has also been significantly altered by subsequent impacts. Bright rays and satellite craters associated with the younger craters Tycho to the south and Theophilus to the east cover much of the quadrangle.

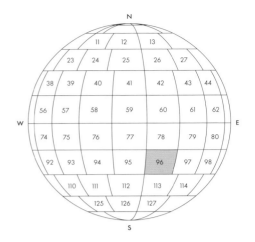

LUNAR ORBITER PHOTOGRAPHIC COVERAGE
OF RUPES ALTAI QUADRANGLE

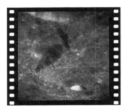

RUPES ALTAI
Near-consecutive frames from the Apollo 16 mission taken on 23 April 1972, showing oblique views of Rupes Altai with decreasing magnification.

RUPES ALTAI QUADRANGLE: KEY CHARACTERISTICS

1. RUPES ALTAI
Latitude: 24.32°S / Longitude: 23.12°E
Diameter: Length: 427 km / 265 ¼ miles
Photograph: LRO (USA, 2011)

2. SACROBOSCO CRATER
Latitude: 23.75°S / Longitude: 16.64°E
Diameter: 97.67 km / 60 ¾ miles
Depth: 2.8 km / 1 ¾ miles
Photograph: Lunar Orbiter IV (USA, 1967)

3. CATHERINA CRATER
Latitude: 17.98°S / Longitude: 23.55°E
Diameter: 98.77 km / 61 ¼ miles
Depth: 3.1 km / 2 miles
Photograph: Lunar Orbiter IV (USA, 1967)

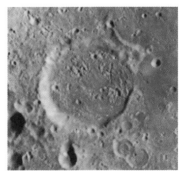

4. BEAUMONT CRATER
Latitude: 18.08°S / Longitude: 28.82°E
Diameter: 50.69 km / 31 ½ miles
Depth: 1.7 km / 1 mile
Photograph: Lunar Orbiter IV (USA, 1967)

5. AZOPHI CRATER
Latitude: 21°S / Longitude: 11.9°E
Diameter: 43.2 km / 26 ¾ miles
Depth: 3.7 km / 2 ¼ miles
Photographed from Earth (USA, 2004)

6. PONTANUS CRATER
Latitude: 28.42°S / Longitude: 14.36°E
Diameter: 55.66 km / 34 ½ miles
Depth: 2.1 km / 1 ¼ miles
Photograph: Lunar Orbiter IV (USA, 1967)

7. GEBER CRATER
Latitude: 19.46°S / Longitude: 13.85°E
Diameter: 44.68 km / 27 ¾ miles
Depth: 3.5 km / 2 ¼ miles
Photograph: LRO (USA, 2017)

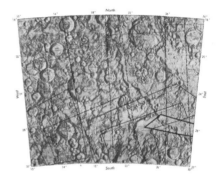

8. POISSON CRATER
Latitude: 30.34°S / Longitude: 10.56°E
Diameter: 41.4 km / 25 ½ miles
Depth: 2 km / 1 ¼ miles
Photograph: Lunar Orbiter IV (USA, 1967)

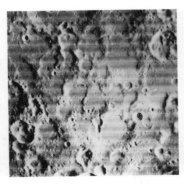

9. WILKINS CRATER
Latitude: 29.58°S / Longitude: 19.58°E
Diameter: 59.44 km / 37 miles
Depth: 1,200 m / 3,937 ft
Photograph: Lunar Orbiter IV (USA, 1967)

10. ROTHMANN CRATER
Latitude: 30.81°S / Longitude: 27.7°E
Diameter: 41.67 km / 26 miles
Depth: 4.2 km / 2 ½ miles
Photograph: Lunar Orbiter IV (USA, 1967)

11. POLYBIUS CRATER
Latitude: 22.46°S / Longitude: 25.63°E
Diameter: 40.81 km / 25 ¼ miles
Depth: 2.1 km / 1 ¼ miles
Photograph: Lunar Orbiter IV (USA, 1967)

THE CINEMATIC MOON

→
Stills *from Luna* (Moon) (1965), directed by Pavel Klushantsev. The film begins with a factual presentation of what is known about the Moon, before speculating on the possibility of travel to it.

→→
Still from Ron Howard's 1995 drama *Apollo 13*. In this film an aborted lunar mission is reconfigured as a victory of human ingenuity and resilience in the face of insurmountable odds.

→→→
Still from Roland Emmerich's fantastical 2022 blockbuster *Moonfall*. It features a dazzling climactic sequence that reveals the Earth's satellite to be a vast space-craft with a spectacularly designed interior.

↳
In this scene from Stanley Kubrick's 1968 space opus, *2001: A Space Odyssey*, a group of scientists faces a mysterious, obsidian-like monolith buried beneath the Moon's surface.

↳↳
Still from *Ad Astra*. In James Gray's brooding 2019 film, the Moon has become populated by gangs of pirates, setting the scene for a thrilling Moon buggy chase sequence.

↳
The genres of these film posters run the gamut from aspirational and existential dramas to thrillers and horrors. But each of these films, employing the Moon as a character or a backdrop, captures a combination of our wonder, awe and trepidation of this neighbouring world.

he first lunar landing, on 20 July 1969, was not only a momentous achievement and truly global event, but also forever transformed the way in which people would view representations of the Moon. There was no longer any need to imagine what its surface was like; humans now had first-hand knowledge of its texture, its thin exosphere and its low gravitational pull.

NASA had invested heavily in creating camera lenses to capture pin-sharp images from the six Moon landings and multiple lunar missions. They obtained so much footage, in fact, that it has taken years for much of it to be developed. As recently as 2019, in the documentary *Apollo 11* (released to coincide with the fiftieth anniversary of the first lunar landing) audiences witnessed previously unseen footage. That film was composed solely of archive material edited into a cohesive narrative arc tracing the mission's journey to and from the Moon. It followed the earlier *In the Shadow of the Moon* (2007), which also featured footage never seen before, alongside interviews with all the surviving astronauts who made the journey to the Moon or its orbit, save for Neil Armstrong, who remained quiet on the subject after the post-landing tour had concluded. But arguably the finest of all the Moon-landing documentaries remains Al Reinert's *For All Mankind* (1989), which combines footage from all the missions into one single journey, accompanied by the unnamed (but collectively credited) voices of those involved, alongside a sublime soundtrack by Brian Eno and Daniel Lanois. It perfectly captures the majesty of both the Moon and Earth from space, and NASA's extraordinary accomplishment.

An earlier Russian documentary, *Luna* (1965), saw scientists offering ways in which future colonization of the Moon could benefit humanity, with a final section dramatizing a lunar landing by cosmonauts. It was no longer screened after the success of Apollo 11. More recently, the Russian mockumentary *First Men on the Moon* (2005) employs

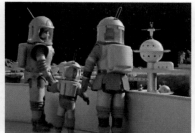

'found' footage purporting to show that Russians first travelled there in the 1930s. In the USA, Craig Baldwin's satirical *Mock Up on Mu* (2008) uses archive footage to suggest that, in 2019, L. Ron Hubbard was living in a Scientology theme park and rehab centre on the Moon. *Lunopolis* (2010) and *Apollo 18* (2011) also played with the 'found footage' genre. The former offers up a time-travel conspiracy theory, while the latter, drawing on the actual cancelled final mission to the Moon, makes effective use of horror tropes.

More conventional narrative features have presented a variety of takes on the lunar missions. *La La Land* (2016) director Damien Chazelle's *First Man* (2018) is a moving portrait of Neil Armstrong, culminating in the first moonwalk. Richard Linklater's *Apollo 10½: A Space Age Childhood* (2022) presents the historic event through the eyes of an impressionable young boy, while Australian comedy *The Dish* (2000) dramatizes the role played by the Parkes Observatory, in New South Wales, in relaying images of the Moon landing around the world. All three films capture the excitement and tension leading up to the landing and the hopes of a safe return.

The potential for disaster surrounding the Moon missions was conveyed by Robert Altman in *Countdown* (1967) and John Sturges in *Marooned* (1969). But Ron Howard's *Apollo 13* (1995), a record of that mission's near-fatal outcome when systems failed aboard the spacecraft, proved a winning account of success wrested from the jaws of disaster – the astronauts' safe return to Earth, in the face of extraordinary adversity.[1]

Considering how laughably inaccurate some fictional, pre-Apollo 11 accounts of the Moon's environment were, it is a testament to Stanley Kubrick's directorial meticulousness and scientific curiosity that the lunar scenes in *2001: A Space Odyssey* (1968), which was released before humankind set foot on the Moon, still look so impressive. The film spends only a fraction of its running time on the Moon, but it still ranks as one of the finest evocations of how a colony on the satellite's surface might look. Only Duncan Jones' ingenious, low-budget drama *Moon* (2009) and James Gray's *Ad Astra* (2019) present a comparative portrait of human existence on the Moon. *Ad Astra*, like *2001*, features only one lunar segment, before Brad Pitt's astronaut ventures deeper into the solar system, but it is the only film to feature an inventively staged action sequence – a chase in Moon buggies across the rocky landscape.

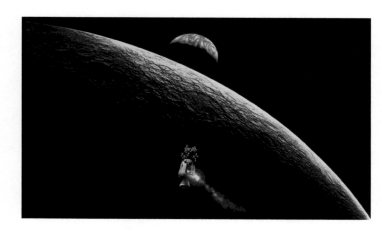

Moon unfolds mostly inside a base occupied by Sam Rockwell's lone mining operative. But when it does venture outside, the film makes the most of the desolate surface, a reflection of its protagonist's loneliness. It is a long way from Aardman Animation's early Wallace and Gromit short *A Grand Day Out* (1989), which playfully embraces the notion of a Moon made entirely of cheese.

Cinema's interest in the Moon has rarely abated. Some films have used the lunar surface as a stepping stone for an alien invasion. In *Independence Day* (1996), the dust in which the footprints of the Apollo 11 astronauts are preserved is disturbed as a vast spacecraft flies overhead. The 2016 sequel offered a more expansive prelude to the return of the voracious extra-terrestrials. While in *Transformers: Dark of the Moon* (2011), the second sequel in the effects-laden action series, hope for the future of the human race lies in a secret buried deep beneath the lunar surface. However, none of these scenarios equal the jaw-dropping audaciousness of *Iron Sky* (2012), an action comedy about an army of Nazis who fled to the Moon in 1945 and, from their secret base there, have been planning their violent return to Earth.

In *Moon Man* (2022), a Chinese science fiction comedy based on a popular South Korean webcomic, a lone astronaut comes to terms with being the sole surviving human after an asteroid hits the Earth and wipes out the species. While in *Moonfall* (2022), *Independence Day* director Roland Emmerich posits that the Moon is hollow; it was created by humanity's ancestors thousands of years ago to escape the destructiveness of a rogue artificial intelligence.[2] The film may embrace the fantasy of fiction more than the logic of science, but the vision of the Moon it conjures up perfectly encapsulates our continuing sense of wonder regarding our barren, yet endlessly fascinating, neighbour.

An epic drama of adventure and exploration

Space Station One: your first step in an Odyssey that will take you to the Moon, the planets and the distant stars.

2001: a space odyssey

MGM PRESENTS A STANLEY KUBRICK PRODUCTION

CINERAMA® Super Panavision® and Metrocolor

1965, USSR

1967, USA

1969, USA

1989, USA

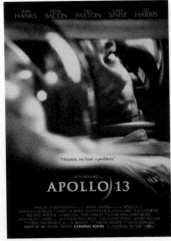

1995, USA

1996, USA

2000, AUSTRALIA

2005, RUSSIA

2007, USA

2009, USA

2010, USA

2011, USA

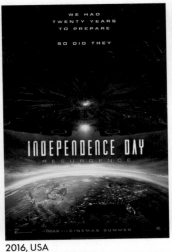

2011, USA

2012, USA

2016, USA

2018, USA

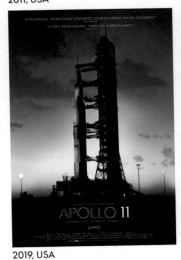

2019, USA

2019, USA

2022, CHINA

2022, USA

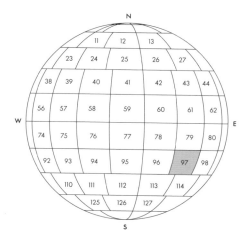

This quadrangle is dominated by the multi-ringed Nectaris basin, which extends across four quadrangles. Fracastorius quadrangle contains three rings of the basin, and many of its oldest features resulted from the impact that formed the basin. Rings of jagged massifs grade into hilly terrain created by ejecta from the Nectaris impact. Two of the largest craters in this quadrangle are the eponymous Fracastorius crater and the Piccolomini crater. Fracastorius is noticeably degraded, speaking to its relative old age, while Piccolomini is well preserved but still old when compared to the brighter craters in this region. The youngest material in this quadrangle is the ray material extending from young distant craters Tycho and Stevinus A (not located in this quadrangle).

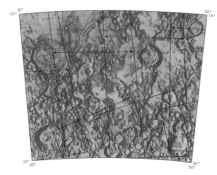

LUNAR ORBITER IV HIGH-RESOLUTION COVERAGE OF FRACASTORIUS QUADRANGLE

APOLLO 8: AS8-13-2243
South view of Frascatorius crater alongside Daguerre and Rosse craters with Mare Nectaris in the foreground. Captured in December 1968 by Apollo 8's Hasselblad camera.

APOLLO 11: AS11-42-6235
South view of Frascatorius crater alongside Beaumont, Daguerre and Rosse craters with Mare Nectaris in the foreground. Captured in July 1969 by Apollo 11's Hasselblad camera.

FRACASTORIUS QUADRANGLE: KEY CHARACTERISTICS

1. FRACASTORIUS CRATER
Latitude: 21.36°S / Longitude: 33.07°E
Diameter: 120.58 km / 75 miles
Photograph: Lunar Orbiter IV (USA, 1967)

2. BOHNENBERGER CRATER
Latitude: 16.24°S / Longitude: 40.06°E
Diameter: 31.74 km / 19 ¾ miles
Depth: 1,100 m / 3,608 ft 11 in
Photograph: LRO (USA, 2011)

3. PICCOLOMINI CRATER
Latitude: 29.7°S / Longitude: 32.2°E
Diameter: 87.58 km / 54 ½ miles
Depth: 4.5 km / 2 ¾ miles
Photograph: Lunar Orbiter IV (USA, 1967)

4. SANTBECH CRATER
Latitude: 21°S / Longitude: 44.06°E
Diameter: 62.24 km / 38 ¾ miles
Depth: 4.5 km / 2 ¾ miles
Photograph: Lunar Orbiter IV (USA, 1967)

5. COOK CRATER
Latitude: 17.5°S / Longitude: 48.8°E
Diameter: 45.16 km / 28 miles
Depth: 1,200 m / 3,937 ft
Photograph: Lunar Orbiter IV (USA, 1967)

6. REICHENBACH CRATER
Latitude: 30.48°S / Longitude: 47.95°E
Diameter: 64.85 km / 40 ¼ miles
Depth: 4 km / 2 ½ miles
Photograph: LRO (USA, 2011)

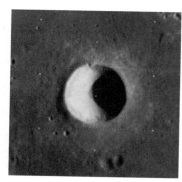

7. ROSSE CRATER
Latitude: 17.95°S / Longitude: 34.98°E
Diameter: 11.43 km / 7 miles
Depth: 2.4 km / 1 ½ miles
Photograph: Lunar Orbiter IV (USA, 1967)

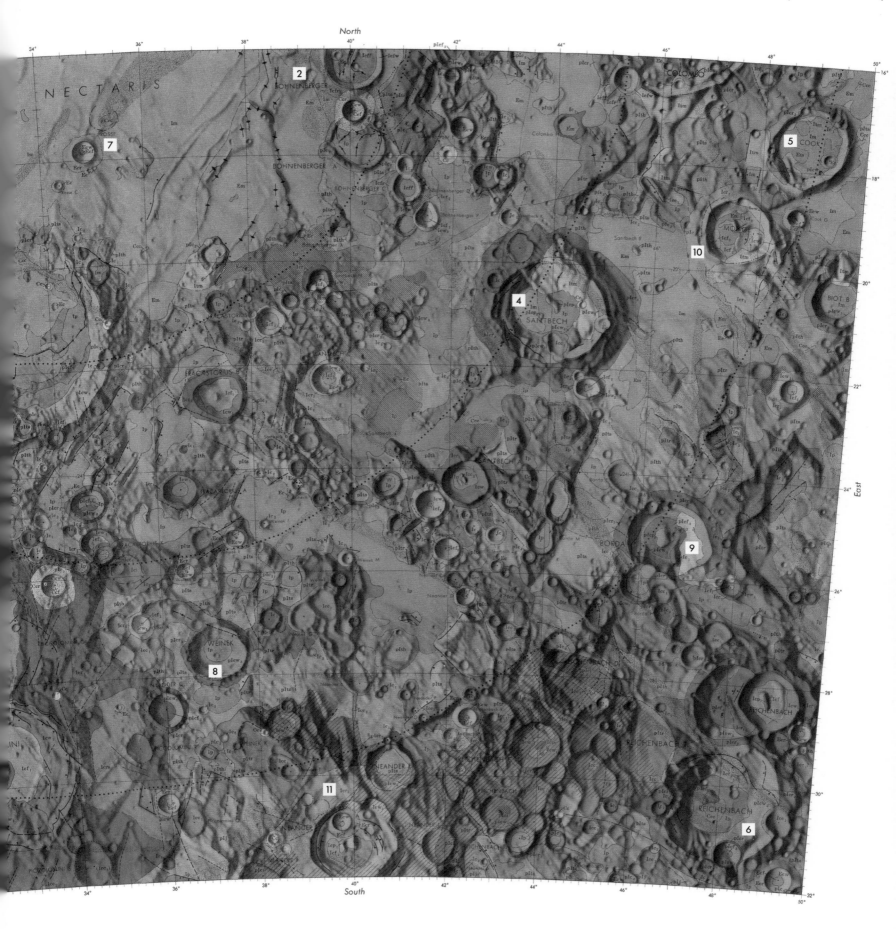

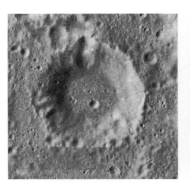

8. WEINECK CRATER
Latitude: 27.57°S / Longitude: 37.06°E
Diameter: 32.01 km / 20 miles
Depth: 3.4 km / 2 miles
Photograph: LRO (USA, 2011)

9. BORDA CRATER
Latitude: 25.2°S / Longitude: 46.52°E
Diameter: 45.4 km / 28 miles
Depth: 2.9 km / 1 ¾ miles
Photograph: Lunar Orbiter IV (USA, 1967)

10. MONGE CRATER
Latitude: 19.24°S / Longitude: 47.54°E
Diameter: 36.6 km / 22 ¾ miles
Depth: 2.6 km / 1 ½ miles
Photograph: Lunar Orbiter IV (USA, 1967)

11. NEANDER CRATER
Latitude: 31.35°S / Longitude: 39.88°E
Diameter: 49.22 km / 30 ½ miles
Depth: 3.4 km / 2 miles
Photograph: LRO (USA, 2011)

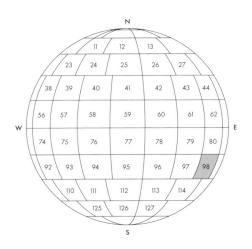

The large crater Petavius is the dominant geologic feature in this quadrangle, resting southeast of the border of the Mare Fecunditatis (Sea of Fertility) lava plains. The terra of this region's highlands is probably originally formed from materials related to the impact event that created the Fecunditatis basin, one of the oldest multi-ring basins on the Moon. The formation of other nearby basins introduced such features as the Vallis Snellius valley and secondary crater chains. The original features have since been covered by more recent ejecta and volcanic materials. The terrain ranges from rugged to gently rolling. The impact that created Petavius blanketed a large portion of this terrain before it and the surrounding terrain were overlain with material from smaller, younger craters such as Wrottesley, Stevinus, Petavius B and Palitzsch B.

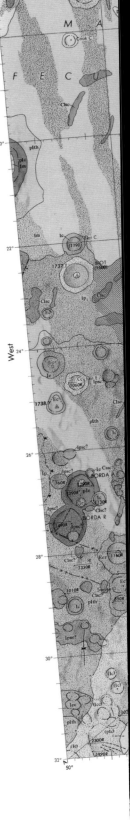

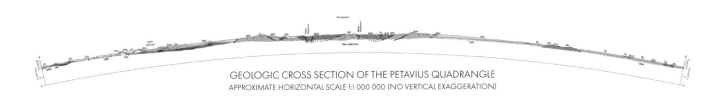

GEOLOGIC CROSS SECTION OF THE PETAVIUS QUADRANGLE
APPROXIMATE HORIZONTAL SCALE 1:1 000 000 (NO VERTICAL EXAGGERATION)

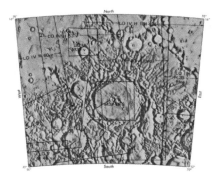

INDEX OF PRINCIPAL PHOTOGRAPHIC COVERAGE
OF PETAVIUS QUADRANGLE

PETAVIUS: IMPACT MELT CHANNEL / OUTCROP

Above right: Evidence of a frozen melt pond of rock, is visible at the base of this photograph, taken on 20 December 2009 after an impact cratering event caused a melt channel from the rim of Petavius B crater. Below right: An outcrop is exposed by a large fracture in the central peak of the Petavius crater in this photograph taken on 18 November 2012.

PETAVIUS QUADRANGLE: KEY CHARACTERISTICS

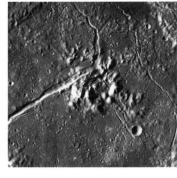

1. PETAVIUS CRATER
Latitude: 25.39°S / Longitude: 60.78°E
Diameter: 184.06 km / 114 ¼ miles
Depth: 3.4 km / 2 miles
Photograph: Lunar Orbiter IV (USA, 1967)

2. RIMA PETAVIUS
Latitude: 25.23°S / Longitude: 60.48°E
Length: 80 km / 49 ¾ miles
Photograph: LRO (USA, 2012)

3. VENDELINUS
Latitude: 16.46°S / Longitude: 61.55°E
Diameter: 141.21 km / 87 ¾ miles
Depth: 2.6 km / 1 ½ miles
Photograph: Lunar Orbiter IV (USA, 1967)

4. HOLDEN CRATER
Latitude: 19.2°S / Longitude: 62.53°E
Diameter: 47.6 km / 29 ½ miles
Depth: 4 km / 2 ½ miles
Photograph: Lunar Orbiter IV (USA, 1967)

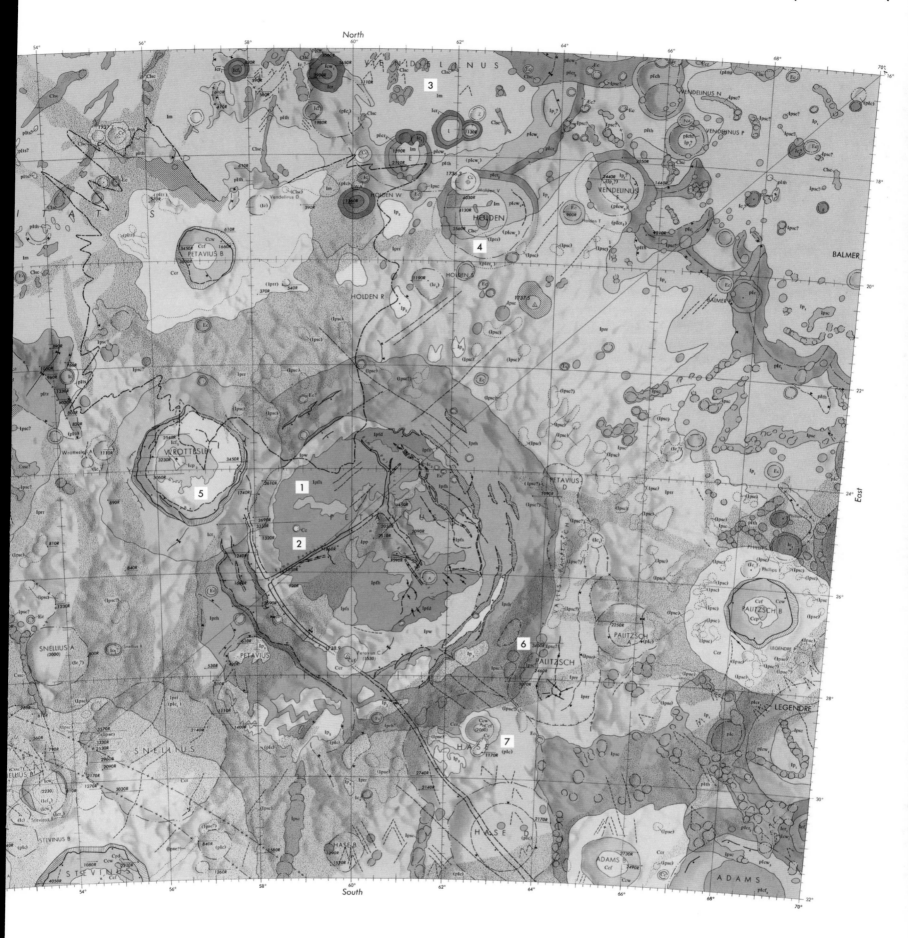

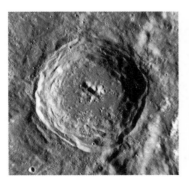

5. WROTTESLEY CRATER
Latitude: 23.9°S / Longitude: 56.62°E
Diameter: 58.38 km / 36 ¼ miles
Depth: 2.3 km / 1 ½ miles
Photograph: LRO (USA, 2014)

6. PALITZSCH CRATER
Latitude: 28.02°S / Longitude: 64.39°E
Diameter: 41.87 km / 26 miles
Photograph: Lunar Orbiter IV (USA, 1967)

7. HASE CRATER
Latitude: 29.37°S / Longitude: 62.68°E
Diameter: 82.08 km / 51 miles
Depth: 2.5 km / 1 ½ miles
Photograph: LRO (USA, 2014)

8. MARE FECUNDITATIS
Latitude: 7.83°S / Longitude: 53.67°E
Diameter: 840.35 km / 522 ¼ miles
Photograph: Apollo 8 (USA)

FLYBYS, LANDERS AND ORBITERS

↑
The first image of the Earth as seen from the Moon was captured by NASA's Lunar Orbiter I on 23 August 1966 for later transmission.

↳
Les Whaley reviews a Surveyor I photomosaic. All Surveyor mosaics of the lunar landscape were assembled by hand, employing the expert eyes of NASA employees between 1966 and 1968.

↱
NASA's Surveyor landers carried a television camera to methodically image its landing site. The footage was transmitted to Earth, converted into still imagery and assembled into mosaics.

↳
The view of the Earth from the Moon provided a new perspective on home, a fact reinforced in 1968 by the *Earthrise* image captured by the Apollo 8 crew.

n 31 July 1964, NASA's Ranger 7 impactor crash-landed as planned in the Mare Nubium (Sea of Clouds). The spacecraft carried an array of television cameras and returned over 4,000 images of the Moon during its descent – the first close-up images of the lunar surface collected by either the USA or USSR. The area of its crash site was later named Mare Cognitum (Sea that has Become Known), in honour of the detailed images. Rangers 8 and 9 in early 1965 were equally successful and returned even more imagery of the Moon. Ranger 8 impacted about 24 kilometres (15 miles) from its planned landing site in Mare Tranquillitatis (Sea of Tranquillity) and Ranger 9 impacted in the lunar highlands.

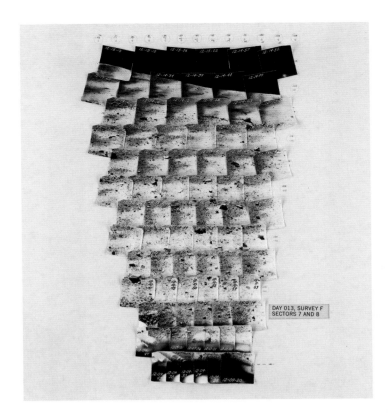

DAY 013, SURVEY F
SECTORS 7 AND 8

These dramatic successes spoke to the increasingly sophisticated robotic capabilities of the US space programme. With the first Moon race having been defined as a mad dash to reach the Moon with a robotic spacecraft (see p. 188 [The First Moon Race]), US robotic missions were now able to prioritize the needs of the upcoming crewed Apollo missions – exploring potential landing sites and characterizing the properties of the Moon's surface to determine the feasibility of landing and operating there. These missions set the tone for the decades of robotic lunar exploration that followed the Apollo era.

The USSR likewise saw more successes during the 1960s. They continued to lose spacecraft, such as the landers Lunas 5 and 6, but found success with the Zond 3 flyby, which returned images of the Moon's far side. Two more landers, Lunas 7 and 8, failed before Luna 9 successfully landed on the Moon on 3 February 1966 – the first successful soft landing on another world. This achievement was followed two months later by Luna 10, the first spacecraft to orbit another world.

Beginning in May 1966, NASA sent a series of Surveyor soft-landers to explore the Moon's surface. Remarkably, all seven of the Surveyors reached the Moon and only two failed to land successfully. The Surveyors carried varying instrumentation, but all had a camera system with which they could survey their surroundings. These images were stitched together into mosaics that geologists could use to study the lunar surface. Mechanical scoops and a scientific instrument on some of the landers allowed geologists to test the physical and chemical properties of the lunar dust and rocks.

A new suite of Lunar Orbiters collected imagery in the service of mapping and developing a global understanding of the Moon. Between 1966 and 1967, NASA sent five Lunar Orbiters to perform the first systematic photographic survey of the lunar surface. Every Orbiter mission was a success, and by the end of the programme 99 per cent of the Moon had been photographed with a resolution of 100 metres (328 feet) or better. This included coverage of twenty candidate landing sites for Apollo. On 23 August 1966 Lunar Orbiter 1 even captured the first image of the Earth as seen from the Moon – a feat that would later be duplicated (to greater fanfare) by the crew of Apollo 8.

Soviet robotic achievements on the Moon continued, even as the emphasis of US lunar missions turned to human explorers. The USSR continued to send orbiters, flybys and landers to the Moon. In 1970 the robotic lander Luna

16 collected dust and rock samples and returned them to Earth for analysis – the first robotic sample return. Later that same year, Luna 17 delivered the first robotic lunar rover, Lunokhod 1. Two additional successful sample-return missions and the Lunokhod 2 rover followed.

The 1976 Luna 24 sample-return mission marked the end of this era of early lunar exploration. No nation sent spacecraft there again until the 1990s. The US has sent multiple successful missions since that time, including Clementine, Lunar Prospector and the Lunar Reconnaissance Orbiter. The Lunar Reconnaissance Orbiter (in operation since 2009) can be singled out for returning some of the most stunning images of the lunar surface ever collected by a robotic spacecraft, including some in which the Apollo landing sites, the equipment abandoned there, and the tracks from the Lunar Roving Vehicles can be seen.[1]

Many other nations have joined in lunar exploration. The European Space Agency in 2003 sent the Small Mission for Advanced Research in Technology (SMART) spacecraft into lunar orbit. China has sent seven successful missions to the Moon, including the 2020 Chang'e 5 mission that returned lunar samples for laboratory analysis on Earth. In 2024 China's Chang'e-6 spent two days collecting the first samples ever from the Moon's far side before carrying them back to Earth. Japan had three successful flights to the Moon, including Hiten and Kaguya. India has had two successes at the Moon, including the short-lived 2023 Vikram lander with its Pragyan rover.[2]

The Moon remains a hard target. Many missions, such as the 2023 Russian Luna 25 lander, still fail. Commercial attempts to land on the surface – including SpaceIL and Israel Aerospace Industries' Beresheet lander (2019) and the Japanese company ispace's Hakuto-R Mission 1 (2023) – have so far achieved only partial success.[3] Still, a new era of privately funded robotic exploration may be on the lunar horizon.

In January 2024 a third commercial attempt was made to land on the Moon, this time by the American company Astrobotic. Shortly after a successful launch, the spacecraft experienced a leak in its propulsion system that made the planned lunar landing impossible.[4] On 22 February 2024, a fourth commercial lander, built by Intuitive Machines, became the first US spacecraft to land on the Moon in the 21st century. The IM-1 lander, nicknamed Odysseus, survived a rough landing that broke one of its struts and left the spacecraft laying on its side – an outcome that limited the operation of some of the instruments it carried.[5] The information gathered by the Astrobotic and Intuitive Machines teams will help them prepare for future missions, including an upcoming contract to deliver a new NASA rover to the Moon. NASA is relying on commercial lunar delivery missions to prepare for the eventual return of humans to the Moon's surface.[6]

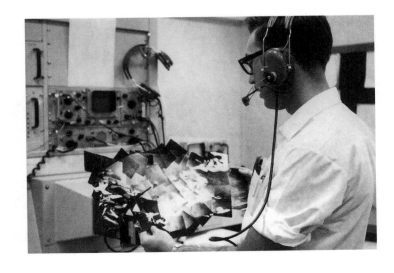

Matthew Shindell

Matter

Lunar

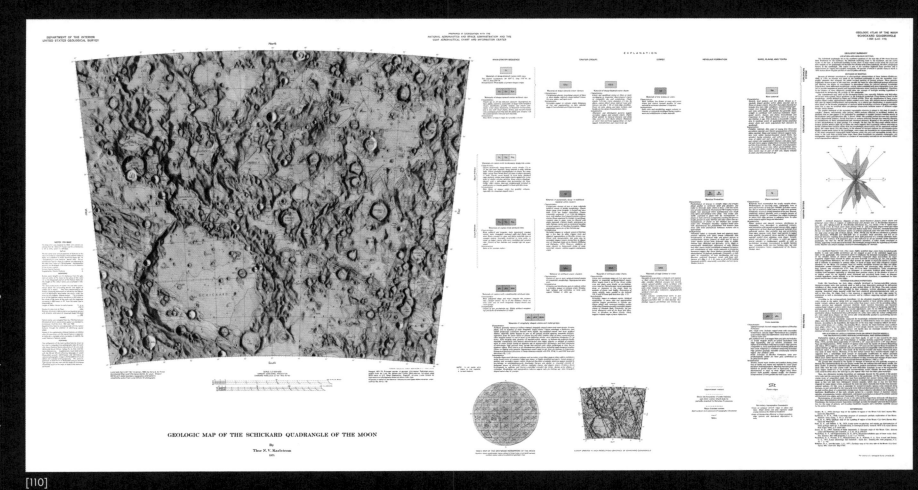

[110]

[111]

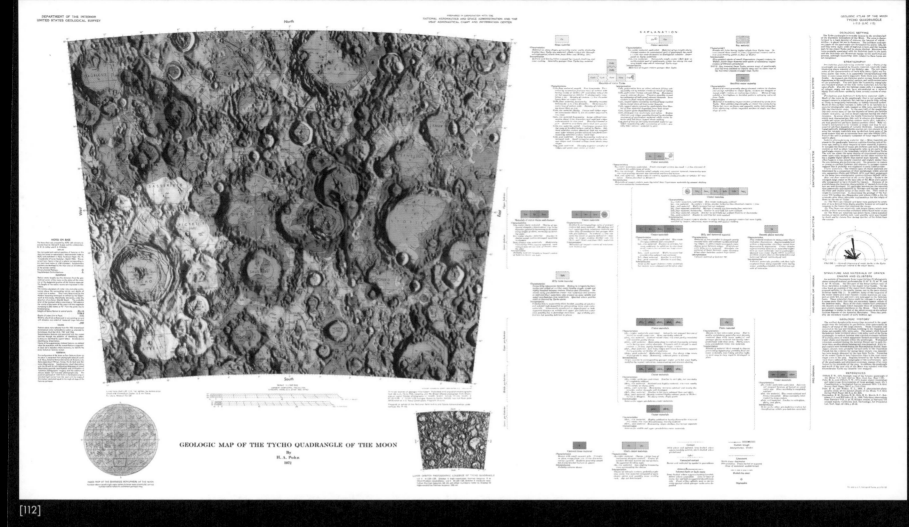

GEOLOGIC MAP OF THE TYCHO QUADRANGLE OF THE MOON
By
H. A. Pohn
1972

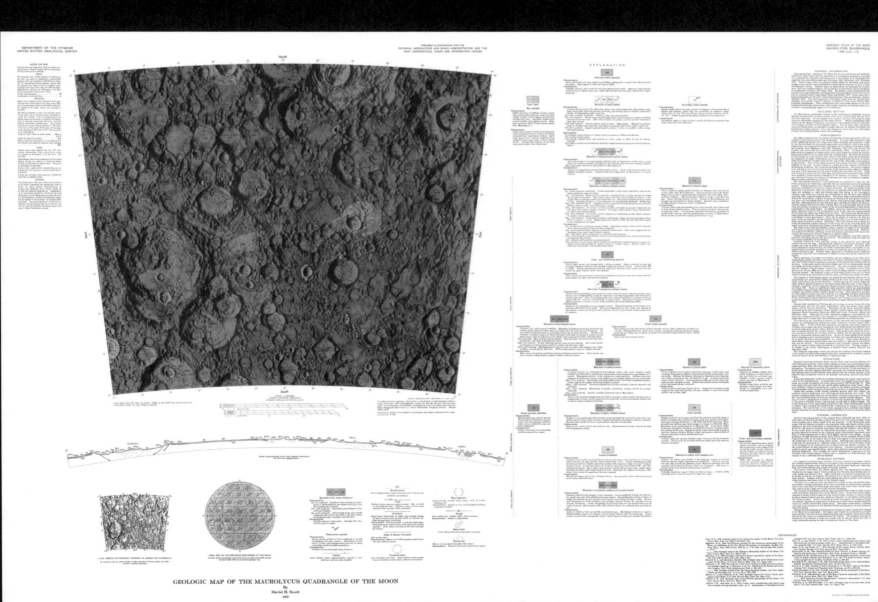

GEOLOGIC MAP OF THE MAUROLYCUS QUADRANGLE OF THE MOON
By
David H. Scott
1972

[LAC-110]

I-823

Schickard Quadrangle

Maps

Lunar

212

GEOLOGIC ATLAS OF THE MOON

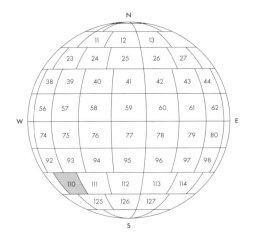

A chain of craters bows along the south and west margins of this quadrangle, including the very large crater Schickard that dominates this rugged portion of the Moon's southern highlands. The craters, ridges and plains found in these highlands were shaped very early in the Moon's geologic history. Very few young features are seen in this region. With a few exceptions, the craters here are modified and subdued, many filled with smooth lava floors. Schickard crater has been transformed into a walled plain, with a worn rim that has been altered by smaller impact craters. The portion of the crater's floor in the southwest was not covered in lava, and so maintains some rougher terrain. Some of the smaller bowl-shaped craters and crater chains and clusters in this quadrangle do still have sharp rims and higher reflectivity, making them stand out as brighter than their surroundings. On 3 September 2006, the European Space Agency deliberately crashed the robotic SMART-1 orbiter into the small unmarked mare in the northeast of this quadrangle (this region was named Lacus Excellentiae [the Lake of Excellence] after this map was made).

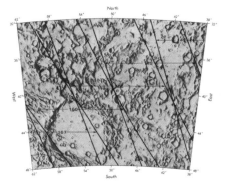

LUNAR ORBITER HI-RESOLUTION COVERAGE
OF SCHICKARD QUADRANGLE

THE LUNAR SOUTH POLE
Right: Composite image of the lunar south pole taken during the lunar southern summer by the SMART-1 satellite between December 2005 and March 2006.

THE LUNAR NORTH POLE
Below right: Composite image of the lunar north pole during differing phases of the SMART-1 mission, taken between May 2005 and February 2006.

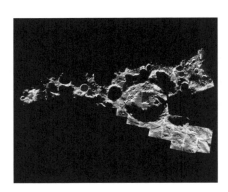

SMART-1	ROCKET	Ariane 5G (no. V162) (L516)
	CREW	N/A
	LAUNCH DATE	23:14:46, 27 September 2003 (UTC)
	LAUNCH SITE	ELA-3, Centre Spatial Guyanais
	PAYLOAD	Advanced Moon Micro-Imager Experiment (AMIE)
	LUNAR ORBIT INSERTION	15 November 2004
	LANDING DATE	05:42:22, 03 September 2006 (UTC)
	IMPACT SITE	Lacus Excellentaie
	LANDING COORDINATES	46.20 °W, 34.30°S (A)

SCHICKARD QUADRANGLE: KEY CHARACTERISTICS

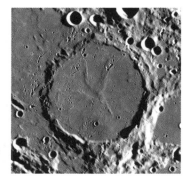

1. SCHICKARD CRATER
Latitude: 48.0°N / Longitude: 20.0°W
Diameter: 90 km / 56 miles
Height: 1.8 km / 1 mile
Photograph: Lunar Orbiter IV (USA, 1967)

2. LACROIX CRATER
Latitude: 37.93°S / Longitude: 59.2°W
Diameter: 36.07 km / 22 ½ miles
Depth: 1.6 km / 1 mile
Photograph: Lunar Orbiter IV (USA, 1967)

3. WARGENTIN CRATER
Latitude: 49.53°S / Longitude: 60.44°W
Diameter: 84.69 km / 52 ½ miles
Photograph: LRO (USA, 2016)

4. DREBBEL CRATER
Latitude: 40.93°S / Longitude: 49.12°W
Diameter: 30.23 km / 18 ¾ miles
Depth: 2.5 km / 1 ½ miles
Photograph: Lunar Orbiter IV (USA, 1967)

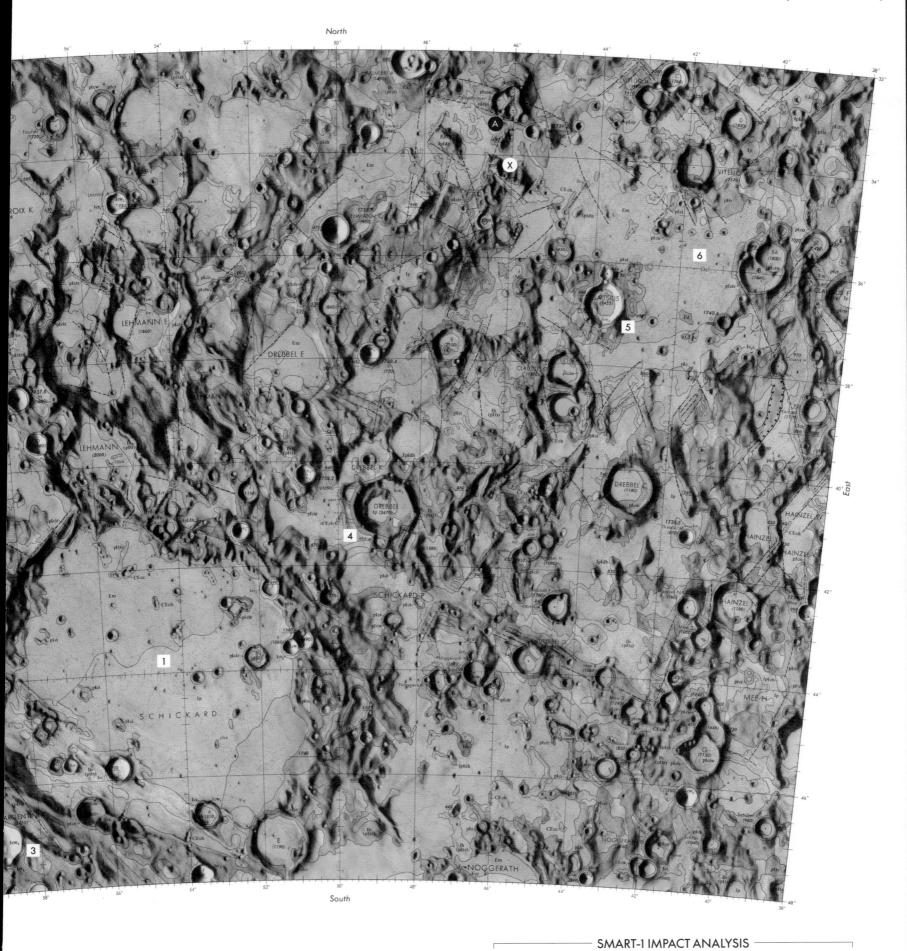

North

South

SMART-1 IMPACT ANALYSIS

5. CLAUSIUS CRATER
Latitude: 36.9°S / Longitude: 43.93°W
Diameter: 24.2 km / 15 miles
Depth: 2.5 km / 1 ½ miles
Photograph: Lunar Orbiter IV (USA, 1967)

6. LACUS EXCELLENTIAIE
Latitude: 35.65°S / Longitude: 43.58°W
Diameter: 197.74 km / 122 ¾ miles
Photograph: Lunar Orbiter IV (USA, 1967)

X. SMART-1 IMPACT LOCATION (UPDATED)
Surface impact occurred at approximately
05:42:21 CEST on 3 September 2006,
34.262°S and 46.193°W.

SMART-1 IMPACT FLASH (UPDATED)
Infrared image of Smart-1 Impact flash
taken by the Canada-France-Hawaii
telescope on 3 September 2006.

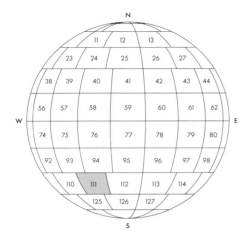

This quadrangle straddles a mare–highlands boundary and contains pitted terra, patches of mare, plains, large craters and massifs. This region has seen the influence of large nearby impact events from almost every direction. The impact that formed the Nubian basin created troughs and rises and deposited ejecta material and craters. More cratering followed, and then the impact that formed the Humorum basin produced new troughs and ridges to crisscross those of Nubium, and deposited new ejecta over the quadrangle, along with fresh craters and crater chains. Ejecta from the formation of the Imbrian basin created new clusters of large craters, small pits and furrows. The impact that formed the Orientale basin deposited a thin layer of ejecta on the region and excavated additional secondary craters. Volcanism then transformed crater floors and troughs into smooth plains. Then, for billions of years, not many large geologic events happened here. Faulting formed the network of rilles that defines Rimae Ramsden, and a few fresh impacts excavated young craters. Ray materials from the young crater Tycho (not located in this quadrangle) lie across the floor of the walled-plane of Wilhelm crater, for which this quadrangle is named.

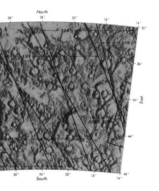

LUNAR ORBITER PHOTOGRAPHIC COVERAGE OF WILHELM QUADRANGLE

MARSH OF EPIDEMICS
Palus Epidemiarum with craters Campanus (above centre), Mercator (above right) and Capuanus (below) also in the frame. Captured by Lunar Orbiter IV, May 1967.

WILHELM QUADRANGLE: KEY CHARACTERISTICS

1. WILHELM CRATER
Latitude: 43.21°S / Longitude: 20.94°W
Diameter: 100.83 km / 62 ¾ miles
Depth: 3 km / 1 ¾ miles
Photograph: Lunar Orbiter IV (USA, 1967)

2. CAPAUNUS CRATER
Latitude: 34.09°S / Longitude: 26.73°W
Diameter: 59.69 km / 37 miles
Photograph: Lunar Orbiter IV (USA, 1967)

3. RIMAE RAMSDEN
Latitude: 32.93°S / Longitude: 31.32°W
Length: 108 km / 67 miles
Photographed from Earth (Netherlands, 2019)

4. HEINSIUS CRATER
Latitude: 39.48°S / Longitude: 17.82°W
Diameter: 64.87 km / 40 ¼ miles
Depth: 2.7 km / 1 ¾ miles
Photograph: LRO (USA, 2010)

5. MONTANARI CRATER
Latitude: 45.83°S / Longitude: 20.76°W
Diameter: 77.05 km / 47 ¾ miles
Depth: 2 km / 1 ¼ miles
Photograph: Lunar Orbiter IV (USA, 1967)

6. WURZELBAUER CRATER
Latitude: 34.04°S / Longitude: 16.06°W
Diameter: 86.77 km / 54 miles
Depth: 2.2 km / 1 ¼ miles
Photograph: Lunar Orbiter IV (USA, 1967)

7. BROWN CRATER
Latitude: 46.53°S / Longitude: 17.99°W
Diameter: 34.03 km / 21 ¼ miles
Depth: 2.3 km / 1 ½ miles
Photograph: Lunar Orbiter IV (USA, 1967)

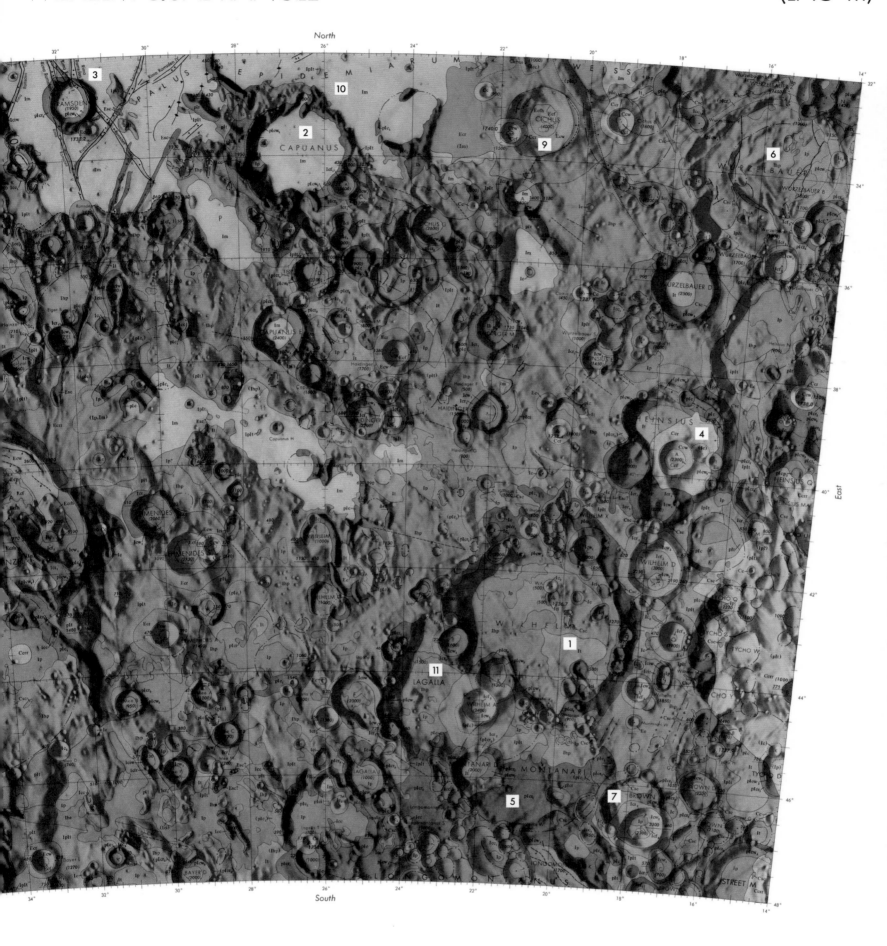

8. HAINZEL CRATER
Latitude: 41.23°S / Longitude: 33.52°W
Diameter: 70.56 km / 43 ¾ miles
Depth: 3 km / 1 ¾ miles
Photograph: Lunar Orbiter IV (USA, 1967)

9. CICHUS CRATER
Latitude: 33.29°S / Longitude: 21.18°W
Diameter: 39.18 km / 24 ¾ miles
Depth: 2.8 km / 1 ¾ miles
Photograph: Lunar Orbiter IV (USA, 1967)

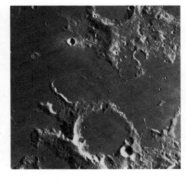

10. PALUS EPIDEMIARUM
Latitude: 32°S / Longitude: 27.54°W
Diameter: 300.38 km / 186 ¾ miles
Photograph: Lunar Orbiter IV (USA, 1967)

11. LAGALLA CRATER
Latitude: 44.48°S / Longitude: 22.36°W
Diameter: 88.78 km / 55 miles
Depth: 1,300 m / 4,265 ft 1 in
Photograph: Lunar Orbiter IV (USA, 1967)

THE FAR SIDE OF THE MOON

→
An artist's representation
of the first-ever view of the
far side of the Moon as seen
by the Soviet Union's Luna
spacecraft on 4 October
1959. The dark patch in
the upper right is Mare
Moscoviense.

↱
A portion of the far side
of the Moon photographed
byApollo 16 in 1972. Mare
Smythii and Mare Marginis
are visible here from the lunar
near side. The north pole of
the Moon is pointing towards
the bottom-right corner.

↱↱
The far side of the Moon
photographed by NASA's
Lunar Reconnaissance
Orbiter Camera (LROC)
in 2011. The dark mare
in the northwest is Mare
Moscoviense. The darker
region near the bottom
of the image is the South
Pole Aitken basin.

ecause the Moon's rotation means that the same face is always visible from Earth, the concept of a 'dark' side has taken root in the human imagination; the top internet search result for 'dark side of the Moon' is Pink Floyd's eighth studio album. In fact, the far side of the Moon receives just as much sunlight as the near side, but this does not necessarily make it a cheerful place. For an astronaut, the far side of the Moon is *radio* dark. Without a line of sight to Earth, communication is impossible without a satellite relay. Apollo 11 command module pilot Michael Collins (1930–2021), expressing this sense of distance and isolation, said of his time passing over the lunar far side: 'I am alone now, truly alone, and absolutely isolated from any known life. I am it. If a count were taken, the score would be three billion plus two over on the other side of the moon, and one plus God knows what on this side'.[1] The far side of the Moon will never be darker than the near side, but it will forever be a more difficult, lonely place for humans to explore.

Lunar scientists tend to think about the Moon in two ways: geographically (divided into the near side and far side) and geologically. The Moon's terrains are broadly defined, geologically, as either maria (low-lying regions filled with basalt, smooth and dark) or highlands (bright, rough, hilly regions). Humans have traditionally seen a 'Man in the Moon' (or another creature) in the near side's pattern of light and dark. For this author, the smooth, dark areas of the maria evoke a soccer player with a Mare Serenitatis-shaped head, kicking a Mare Crisium soccer ball, with the 'feet' of Maria Nectaris and Fecunditatis. Lunar geology and geography coincide when the Moon is considered as a whole: the near side is dominated by maria, and the far side by highland terrains.

Humans have known about the Moon since they first looked at the night sky. Only relatively recently, however, have scientists begun to understand Earth's satellite, because of the work of important individuals who laid the foundations for what would become the field of lunar and planetary science. Grove K. Gilbert (1843–1918), a legendary geologist, began applying his physical geological techniques to the Moon in the 1880s. In the 1950s, scientists such as Harold Urey (1893–1981) and Gerard Kuiper (1905–1973) explored the Moon from a geochemical and cosmochemical perspective, using meteorites and ground-based telescopic observations. They were trying to answer questions about the Moon's composition, interior structure and origin. In the early 1960s Eugene Shoemaker (1928–1997) and others, including Robert J. Hackman (1923–1980), began applying geological mapping techniques to unravel the stratigraphy (or geological history) of the lunar near side, wherever imagery from telescopes and, later, spacecraft allowed such investigations.

The far side of the Moon was finally observed by the Soviets' Luna 3 mission on 7 October 1959 (having launched on 4 October). Scientists had expected the far side to look similar to the near side seen from Earth: dark maria interspersed with bright highlands. Luna 3 showed just enough detail to reveal that the far side was almost exclusively made up of highland terrains. In the late 1950s, the diversity of surfaces of planets and their moons was yet to be explored by robotic spacecraft. Predicting that the lunar far side would be dominated by highlands interspersed with a few maria, and would look distinctly different from the near side, would have been pure conjecture. No one could have guessed that the Moon would host such a striking dichotomy.

The far side highlands are home to the Moon's topographic extremes, with impact craters carved by billions of years of bombardment of the surface by meteorites. The highest points on the Moon are on the far side. They reach up to 10 kilometres (about 6 miles) high near the northern rim of the colossal South Pole Aitken (SPA) basin, which stretches for 2,500 kilometres (1,553 miles), more than halfway across the Moon's approximately 3,375-kilometre (2,097-mile) diameter. The lowest elevations are found at Antoniadi crater (9 kilometres or about 5.6 miles deep) located within the SPA basin. The presence of this large basin was not confirmed until the US Lunar Orbiter programme in the mid- to late 1960s. It is deep enough (around 8 kilometres or 5 miles) that it should be filled with thick flows of mare basalts sourced from deep inside the Moon. The far side is not without maria, but these maria lying in the floors of

craters are many times smaller than those on the near side. One such far side mare basalt-filled crater is Tsiolkovsky. It was imaged in detail by Apollo 15, and Apollo 17 astronaut Harrison 'Jack' Schmitt (b. 1935) – the only Apollo astronaut who was a trained geologist – was so interested in Tsiolkovsky crater that he urged NASA to make it his landing site, facilitated by a communication satellite relay.

Extremes in topography are reflected in the thickness of the lunar crust. On Earth, the crust beneath the continents is thicker than that beneath the oceans, because thicker crust is required to uphold land with hills and mountains. Similarly, the lunar crust of the highlands is, on average, much thicker than that of the near side maria; this was confirmed by a set of twin spacecraft of the GRAIL mission (2011–12), named Ebb and Flow. In addition to differences in crustal thickness, lunar highlands are compositionally different, made up of a lighter, lower-density rock called anorthosite that is depleted in iron, unlike the iron-rich basalts of the lunar maria, which are denser and darker.

South Pole Aitken (SPA) basin, lying within the far side highlands of the Moon, is now a leading focus in lunar exploration. SPA was the target of a landed rover mission from the China National Space Administration in 2019. It has been the objective for numerous proposed mission concepts, including rovers that would fetch samples across large distances for the astronauts that NASA plans to land on the Moon with its Artemis programme. Targeted exploration of the SPA basin could reveal the missing pieces of the history of meteorite bombardment, providing the most precise geological chronometer ever for the Moon. Additionally, as the largest basin on the Moon, and one of the oldest, SPA may have probed deep enough into the Moon's interior that it can reveal critical details about the Moon's interior structure, and the evolution of differentiated planets and moons. SPA also, being at a lunar pole, is host to numerous locations called 'permanently shadowed regions' (PSRs) which are the true dark places on the Moon. Scientists believe these PSRs may host large deposits of water ice, which could help maintain a sustained human presence on the lunar surface. SPA basin is one of the most exciting, and perhaps geologically significant, places on the Moon left to be studied in detail, and up to now its location on the far side has hampered exploration by humans and robots alike. Answering pressing questions about lunar chronologies and interior evolutions would have important implications across the field of planetary geology far beyond the Earth–Moon system, reaching into the depths of our solar system, and perhaps, one day, planetary systems beyond our own.

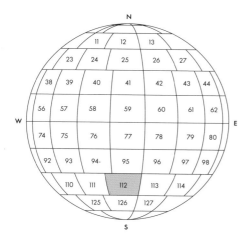

The relatively young Tycho crater that gives this quadrangle its name is one of the most recognizable craters on the Moon's surface. Because of its size and the fact that it is one of the Moon's brightest features, the crater and its surrounding ejecta blanket and rays stand out to observers from Earth. Tycho's surroundings in the western part of this quadrangle are heavily cratered. The northern and eastern parts of the quadrangle are dominated by plains and hilly terra. Many of the larger craters in this quadrangle formed early in its history. This was followed by the creation of plains and hilly terra by volcanism. Nearby impacts and the materials ejected from them then created crater chains and clusters in this region. And finally the creation of Tycho crater blanketed materials across the southwestern portion of the quadrangle and created many secondary craters in the area. Not shown are the rays from Tycho that cover this quadrangle and extend to several other nearby quadrangles. On 10 January 1968, the USA's Surveyor 7 mission successfully soft-landed on Tycho's ejecta blanket at coordinates 40.97°S, 11.44°W, here marked with the words 'SURVEYOR VII LANDING SITE'.

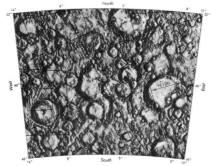

LUNAR ORBITER PHOTOGRAPHIC COVERAGE
OF TYCHO QUADRANGLE

SURVEYOR 7	ROCKET	Atlas Centaur
	CREW	N/A
	LAUNCH DATE	06:30:00, 07 January 1968 (UTC)
	LAUNCH SITE	Cape Kennedy Air Force Station
	PAYLOAD	TV Camera, Alpha-Scattering Instrument
	LUNAR ORBIT INSERTION	N/A
	LANDING DATE	01:05:36, 10 January 1968 (UTC)
	LANDING SITE	Tycho crater
	LANDING COORDINATES	40.97°S, 11.44°W Ⓐ

SURVEYOR 7 PANORAMAS
Above: 100 black-and-white photographs – mounted and composited – of the lunar surface taken by Surveyor 7 on day 19 of the Lunar Surface Survey, 19 January 1968. Below: A panorama of the lunar highlands composed of Surveyor 7 printed photographic mosaics that were digitally reconstructed and reassembled.

TYCHO QUADRANGLE: KEY CHARACTERISTICS

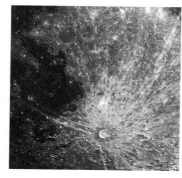

1. TYCHO CRATER
Latitude: 43.3°S / Longitude: 11.22°W
Diameter: 85.29 km / 53 miles
Depth: 4.7 km / 3 miles
Photograph: LRO (USA, 2015)

2. TYCHO RAY PATTERNS
Latitude: 43.3°S / Longitude: 11.22°W
Max Ray Length: 1,500 km / 932 miles
Photographed from Earth (USA, 2020)

3. TYCHO EJECTA DEPOSITS
The impact that formed Tycho splashed lava north of the crater. Polygonal fractures and small pits formed in the flow when it cooled.
Photograph: LRO mission (USA, 2011)

4. STÖFLER CRATER
Latitude: 41.24°S / Longitude: 5.93°E
Diameter: 129.87 km / 80 ¾ miles
Depth: 2.8 km / 1 ¾ miles
Photograph: LRO (USA, 2010)

North

South

East

5. SASSURE CRATER
Latitude: 43.38°S / Longitude: 3.88°W
Diameter: 54.56 km / 34 miles
Depth: 1.8 km / 1 mile
Photograph: Lunar Orbiter IV (USA, 1967)

6. GAURICUS CRATER
Latitude: 33.91°S / Longitude: 12.74°W
Diameter: 79.64 km / 49 ½ miles
Depth: 2.7 km / 1 ¾ miles
Photograph: Lunar Orbiter IV (USA, 1967)

7. STREET CRATER
Latitude: 46.58°S / Longitude: 10.74°W
Diameter: 58.52 km / 36 ¼ miles
Depth: 1.5 km / 1 mile
Photograph: Lunar Orbiter IV (USA, 1967)

8. WALTER CRATER
Latitude: 33.25°S / Longitude: 0.62°E
Diameter: 134.23 km / 83 ½ miles
Depth: 4.1 km / 2 ½ miles
Photograph: Lunar Orbiter IV (USA, 1967)

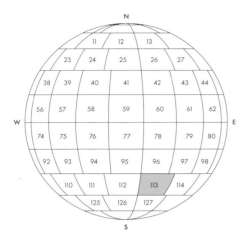

This quadrangle, located in the Moon's southeastern highlands, is dominated by densely cratered terrain. Some of the craters here, such as Zagut, are among the most ancient on the Moon. The oldest unit on this map, in the eastern part of the quadrangle, is the Janssen Formation. This is characterized by hilly, hummocky and rolling terrain with linear ridges running north–northeast, radial to the Nectaris basin, and was understood to be composed of material ejected from the Nectaris basin during impact. It is also possible to discern the outlines of craters buried by this ejecta blanket. Smooth and pitted plains material fills some of the craters and depressions in the hilly terrain. The youngest craters in this region are the smallest, some in clusters, and probably resulted from secondary impacts from the formation of the young craters Tycho and Stevinus (not in this quadrangle).

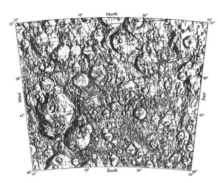

LUNAR ORBITER PHOTOGRAPHIC COVERAGE
OF MAUROLYCUS QUADRANGLE

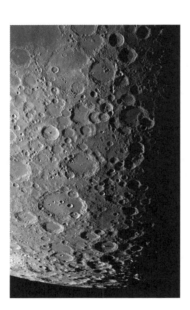

SOUTHERN HIGHLANDS
A composite image of the Moon's surface created from images taken by NASA's Lucy spacecraft. The heavily cratered southern highlands are visible at the bottom of the frame, 16 October 2022.

MAUROLYCUS QUADRANGLE: KEY CHARACTERISTICS

1. MAUROLYCUS CRATER
Latitude: 41.77°S / Longitude: 13.92°E
Diameter: 115.35 km / 71 ¾ miles
Depth: 4.7 km / 3 miles
Photograph: Lunar Orbiter IV (USA, 1967)

2. ZAGUT CRATER
Latitude: 31.94°S / Longitude: 21.89°E
Diameter: 78.92 km / 49 miles
Depth: 3.2 km / 2 miles
Photograph: Lunar Orbiter IV (USA, 1967)

3. JANSSEN FORMATION
Latitude: 44.96°S / Longitude: 40.82°E
Diameter: 200.65 km / 124 ¾ miles
Depth: 2.9 km / 1 ¾ miles
Photograph: LRO (USA, 2014)

4. BAROCIUS CRATER
Latitude: 44.98°S / Longitude: 16.81°E
Diameter: 82.72 km / 51 ½ miles
Depth: 3.5 km / 2 ¼ miles
Photograph: Lunar Orbiter IV (USA, 1967)

5. SPALLANZANI
Latitude: 46.38°S / Longitude: 24.73°E
Diameter: 30.86 km / 19 ½ miles
Depth: 1.5 km / 1 mile
Photograph: LRO (USA, 2010)

6. RABBI LEVI CRATER
Latitude: 34.78°S / Longitude: 23.46°E
Diameter: 82.44 km / 51 ¼ miles
Depth: 3.5 km / 2 ¼ miles
Photograph: Lunar Orbiter IV (USA, 1967)

7. GEMMA FRISIUS CRATER
Latitude: 34.33°S / Longitude: 13.37°E
Diameter: 88.54 km / 55 miles
Depth: 4.7 km / 3 miles
Photograph: Lunar Orbiter IV (USA, 1967)

8. RICCIUS CRATER
Latitude: 37.02°S / Longitude: 26.43°E
Diameter: 71.79 km / 44 ½ miles
Depth: 1.8 km / 1 mile
Photograph: Lunar Orbiter IV (USA, 1967)

9. BUCH CRATER
Latitude: 38.9°S / Longitude: 17.68°E
Diameter: 51.31 km / 32 miles
Depth: 1,400 m / 4,593 ft 2 in
Photograph: Lunar Orbiter IV (USA, 1967)

10. NICOLAI CRATER
Latitude: 42.47°S / Longitude: 25.87°E
Diameter: 40.54 km / 25 ¼ miles
Depth: 1.8 km / 1 mile
Photograph: Lunar Orbiter IV (USA, 1967)

11. STIBORIUS CRATER
Latitude: 34.5°S / Longitude: 32°E
Diameter: 43.76 km / 27 ¼ miles
Depth: 3.7 km / 2 ¼ miles
Photograph: Lunar Orbiter IV (USA, 1967)

THE APOLLO PROGRAMME

Matter

Lunar

↱
Astronauts Charles 'Pete' Conrad Jr (left) and Alan L. Bean (right) simulate the photographic documentation of lunar rock samples during training for the Apollo 12 lunar landing mission at the Kennedy Space Center in October 1969.

↳
A close-up photograph of an astronaut's shoe and shoeprint in the lunar oil, photographed by Buzz Aldrin with a 70-mm (2 ¾ in) lunar surface camera during the Apollo 11 EVA in July 1969.

↳
Astronauts in training for the eleven crewed Apollo missions conducted between 1968 and 1972 in the volcanic terrain of northern Arizona and the waters of the Gulf of Mexico. Activities ranged from geology classes to driving the lunar roving vehicle and from learning to walk in zero gravity to hours spent in spacecraft simulators. Ahead of a mission, crews would train for fourteen hours a day, six days a week. For every hour of a mission, astronauts spent hundreds of hours in training.

As the Apollo 11 spacecraft rounded the far side of the Moon and regained contact with Mission Control in Houston, Texas, astronaut Michael Collins reported 'a geologist up here would just go crazy.' After a three-day journey from Earth, the crew revelled in seeing the geological features they had studied for years, up close and three-dimensional.[1] The next day, on 20 July 1969, Neil Armstrong (1930–2012) and Buzz Aldrin (b. 1930) became the first humans to land on another celestial body and took their first steps on the Moon. Throughout their two-and-a-half-hour moonwalk, they balanced symbolic gestures and ceremonies with scientific activities. The crew set up a seismometer to measure moonquakes, collected over 21 kilograms (46 pounds) of lunar samples – the first ever to be brought back to Earth – and deployed other experiments. The data collected on Apollo 11 and the subsequent Apollo missions transformed our understanding of the evolution and composition of the Moon, as well as the solar system.

Eight years earlier, on 25 May 1961, when American President John F. Kennedy (1917–1963) proposed 'landing a man on the Moon and returning him safely to Earth' in an address to a Joint Session of Congress, he did not foresee Apollo's resulting advances in lunar geology, or even science. For Kennedy, 'no single space project in this period will be more impressive to mankind' than a human setting foot on the lunar surface.[2] He saw Project Apollo as part of his larger geopolitical strategy. The feat would be 'impressive' to the world, demonstrating American values, technical expertise, industrial capacity and national strength. In doing so, the Kennedy administration argued, Apollo might influence the political trajectory of newly independent nations and encourage alignment with the USA instead of the Soviet Union. As historian Matthew Hersch noted, 'To many of its critics in the science community, Apollo was a "mission looking for a science" rather than "science looking for a mission".'[3] Even though geopolitics drove the programme, Apollo's impact was far broader and deeper than Kennedy's original target.[4]

The Apollo programme initiated a national mobilization that drew on a workforce in the hundreds of thousands. NASA staff, contractors and subcontractors worked around the clock – late into the evenings and at weekends – to meet the deadline Kennedy set. Apollo became the most expensive civilian engineering project in US history, consuming over 4 per cent of the federal budget by the mid-1960s.[5] Before the decade was out, NASA developed the Saturn V, a liquid-fuel rocket that produced 7.5 million pounds (over 3.4 million kilograms) of thrust, and a three-part spacecraft that consisted of: a command module, the

astronauts' home and control section; the service module, which housed the consumables and support systems; and a lunar module that ferried two astronauts from lunar orbit to and from the surface of the Moon. Beyond major hardware, Apollo required a new guidance and navigation system, new managerial practices, a global communications infrastructure (including ground stations that spanned the Earth), flight simulation, and extensive training of the astronaut corps. The geology training, in particular, was so extensive that it was said to be equivalent to a master's degree.[6]

In 1959, NASA recruited military test pilots for its first class of astronauts. Accustomed to risk-taking and high-performance flight hardware, the Mercury Seven, as the astronauts became known, proved that the USA could put humans in orbit and return them to Earth alive and well. When interviewed about their motivation for joining the astronaut corps, none of the Mercury Seven mentioned science.[7] But with each successive astronaut class, technical and scientific expertise weighed heavier in NASA's selection process. Although relatively homogeneous in terms of gender, race, age and religion, the new additions to the astronaut corps who would fly the Apollo missions included experienced pilots with advanced degrees in engineering and science. In 1965, NASA explicitly sought scientists for lunar missions. One of these recruits – Harrison 'Jack' Schmitt – had earned a PhD in geology and became the first scientist to traverse the Moon on the Apollo 17 mission in 1972.[8]

In 1963, with the mission to the Moon firmly set, the astronauts began receiving geology training. None of the first twenty-nine Apollo astronauts had any previous educational experience in geology, so geoscientists with the US Geological Survey (USGS) and NASA developed an extensive multi-phase curriculum in the subject. USGS instructors accompanied the astronauts from Arizona to Iceland, examining impact craters and volcanic flows. By the final stage of training, astronauts played the 'Moon Game', with crews of two simulating lunar traverses and sample collection. Apollo 17 astronauts alone visited seventeen geology field sites ahead of their mission.[9]

Between 1969 and 1972, crews on six Apollo missions brought over 380 kilograms (842 pounds) of rocks, regolith and core samples back to Earth. The astronauts also deployed instruments that collected data on the Moon's density and temperature as well as the recession of its orbit. The lunar samples and data provided invaluable information on the origin, age and evolution of Earth's nearest neighbour. This science continues today, with numerous new studies conducted with lunar samples and data every year by scientists from around the world.[10]

The Apollo Programme · Teasel Muir-Harmony · Matter · Lunar · 223

APOLLO 7

APOLLO 8

APOLLO 11

APOLLO 12

APOLLO 15

APOLLO 17

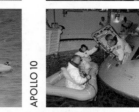

APOLLO 9

APOLLO 10

APOLLO 13

APOLLO 14

APOLLO 16

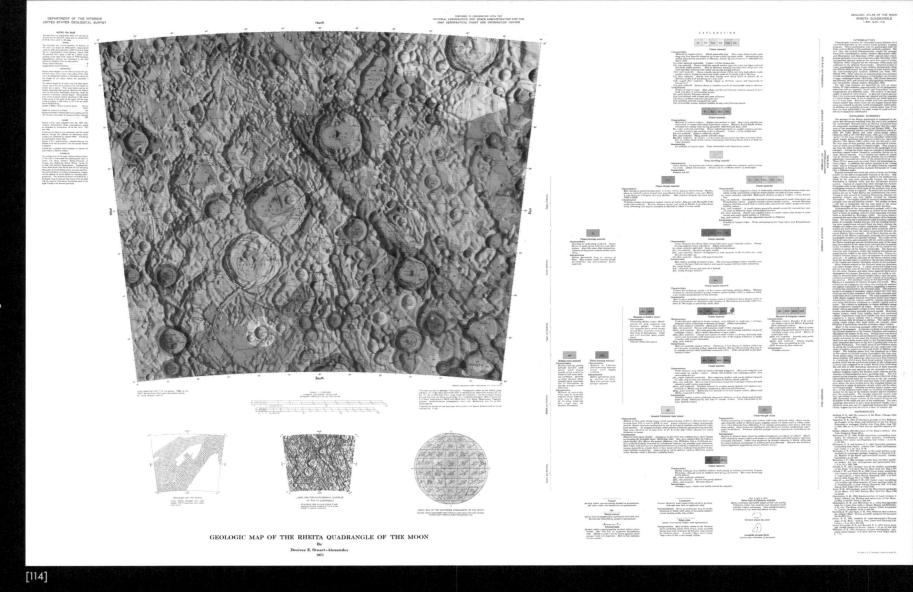

GEOLOGIC MAP OF THE RHEITA QUADRANGLE OF THE MOON
By
Desiree E. Stuart-Alexander
1971

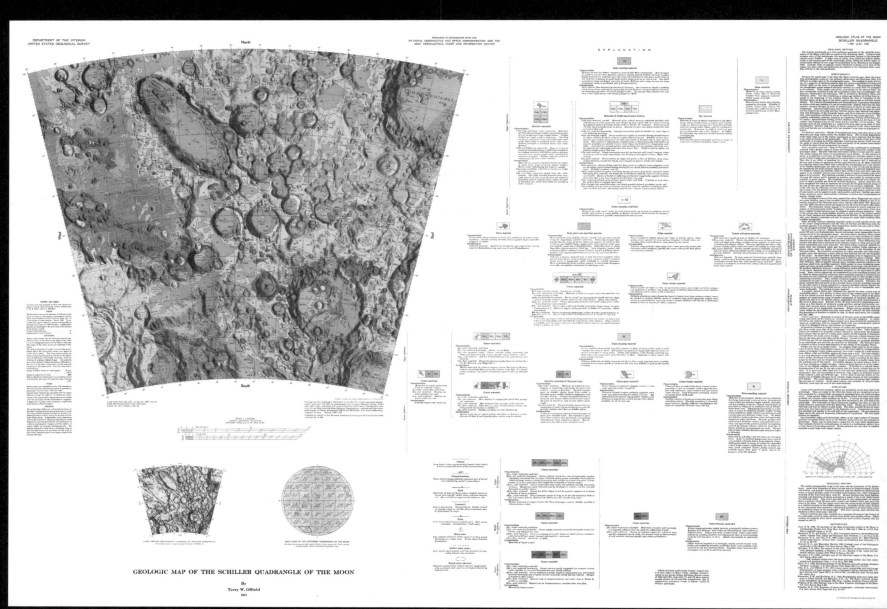

GEOLOGIC MAP OF THE SCHILLER QUADRANGLE OF THE MOON
By
Terry W. Offield
1971

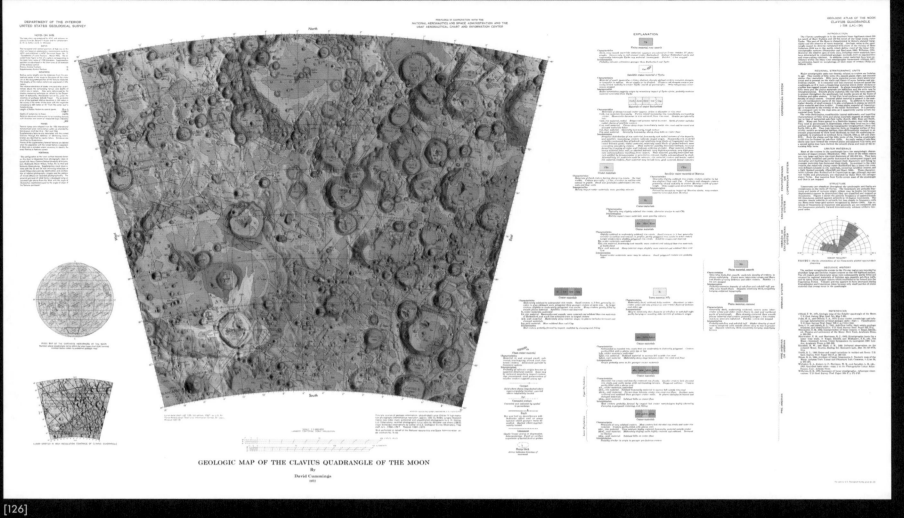

GEOLOGIC MAP OF THE CLAVIUS QUADRANGLE OF THE MOON
By
David Cummings
1972

[126]

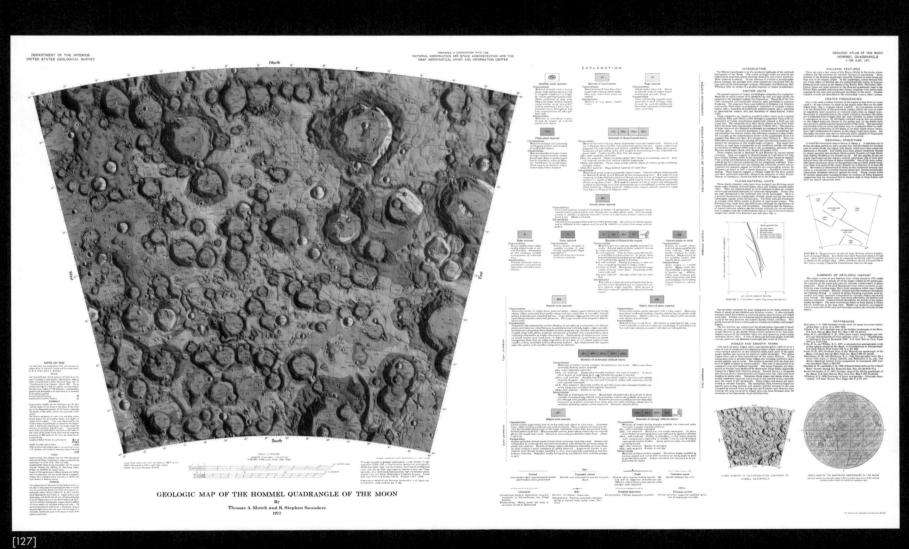

GEOLOGIC MAP OF THE HOMMEL QUADRANGLE OF THE MOON
By
Thomas A. Mutch and R. Stephen Saunders
1972

[127]

114
Facsimile:
Geologic Map of the Rheita Quadrangle of the Moon, 1971, by Desiree E. Stuart-Alexander

125
Facsimile:
Geologic Map of the Schiller Quadrangle of the Moon, 1971, by Terry W. Offield

126
Facsimile:
Geologic Map of the Clavius Quadrangle of the Moon, 1972, by David Cummings

127
Facsimile:
Geologic Map of the Hommel Quadrangle of the Moon, 1972, by Thomas A. Mutch and R. Stephen Saunders

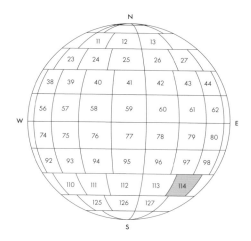

This quadrangle is dominated by features resulting from the formation of the Nectaris basin to its northwest. The Vallis Rheita valley, which shares a name with this quadrangle and a nearby crater, is oriented radially with Mare Nectaris (Sea of Nectar). This and other similar crater-trough valleys are among the oldest features shown on this map. Craters in the region vary in age, and younger volcanic 'plains-forming' material has filled in some crater floors and depressions. Some of the oldest craters are ones with buried rims, and all but the largest craters have been completely obliterated by smaller impacts. The ancient crater Janssen is so worn that it is hard to spot beneath the younger craters that overlay it. The area is partially covered by rolling terra hills that may be composed of material from the ejecta blanket of the Nectaris basin. The crater Rheita remains well defined although multiple small impacts have cratered its rim. Younger craters such as Fabricius and Stevinus stand out. Stevinus crater is one of the youngest prominent features in this quadrangle, having been excavated in the Moon's latest geologic period, the Copernican. On 10 April 1993, Japan's Hiten orbiter was deliberately crashed into this quadrangle, south of Stevinus crater.

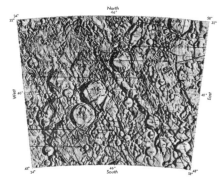

LUNAR ORBITER PHOTOGRAPHIC COVERAGE
OF RHEITA QUADRANGLE

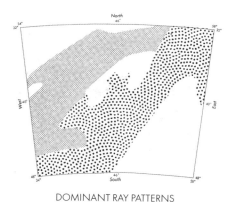

DOMINANT RAY PATTERNS

HITEN LUNAR IMPACT
A sequence of images taken on 10 April 1993 by Hiten's Optical Navigation System as the craft plunged towards the Moon's surface. The white dot marks the impact point.

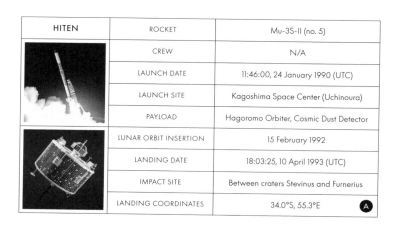

HITEN	ROCKET	Mu-3S-II (no. 5)
	CREW	N/A
	LAUNCH DATE	11:46:00, 24 January 1990 (UTC)
	LAUNCH SITE	Kagoshima Space Center (Uchinoura)
	PAYLOAD	Hagoromo Orbiter, Cosmic Dust Detector
	LUNAR ORBIT INSERTION	15 February 1992
	LANDING DATE	18:03:25, 10 April 1993 (UTC)
	IMPACT SITE	Between craters Stevinus and Furnerius
	LANDING COORDINATES	34.0°S, 55.3°E Ⓐ

RHEITA QUADRANGLE: KEY CHARACTERISTICS

1. RHEITA CRATER
Latitude: 48.0°N / Longitude: 20.0°W
Diameter: 90 km / 56 miles
Height: 1.8 km / 1 mile
Photograph: Lunar Reconnaisance Orbiter

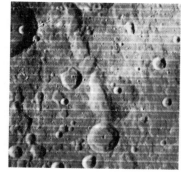

2. VALLIS RHEITA
Latitude: 42.51°S / Longitude: 51.65°E
Length: 450 km / 279 ½ miles
Max Width: 30 km / 18 ¾ miles
Photograph: Lunar Orbiter IV (USA, 1967)

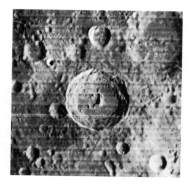

3. STEVINUS CRATER
Latitude: 32.5°S / Longitude: 54.14°E
Diameter: 71.54 km / 44 ½ miles
Depth: 3 km / 1 ¾ miles
Photograph: Lunar Orbiter IV (USA, 1967)

4. FABRICIUS CRATER
Latitude: 42.75°S / Longitude: 41.84°E
Diameter: 78.9 km / 49 miles
Depth: 2.5 km / 1 ½ miles
Photograph: LRO (USA)

North

South

East

5. FURNERIUS CRATER
Latitude: 36°S / Longitude: 60.54°E
Diameter: 135.03 km / 84 miles
Depth: 3.5 km / 2 ¼ miles
Photograph: Lunar Orbiter IV (USA, 1967)

6. METIUS CRATER
Latitude: 40.42°S / Longitude: 43.37°E
Diameter: 83.81 km / 52 miles
Depth: 3 km / 1 ¾ miles
Photograph: Lunar Orbiter IV (USA, 1967)

7. LOCKYER CRATER
Latitude: 46.27°S / Longitude: 36.6°E
Diameter: 35.08 km / 21 ¾ miles
Depth: 3.8 km / 2 ¼ miles
Photograph: LRO (USA)

8. YOUNG CRATER
Latitude: 41.54°S / Longitude: 50.98°E
Diameter: 71.44 km / 44 ½ miles
Depth: 4.2 km / 2 ½ miles
Photograph: Lunar Orbiter IV (USA, 1967)

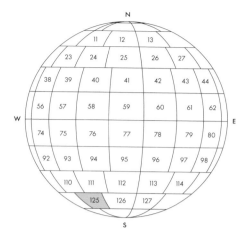

This quadrangle is defined primarily by cratered highland and mountainous terra. The elongated impact crater Schiller is the most prominent feature. At the time this map was made, geologists speculated that the crater might not be an impact crater, but instead a result of volcanic and tectonic activity. Today, however, Schiller is recognized as an impact crater, possibly the result of the fusion of two or more craters that then filled partially with lava. The degraded mountainous terrain is likely among the oldest in this quadrangle. Some of this terra, as well as the rims of craters, has been covered by hummocky volcanic material, possibly arising from vents not identified in this map. Craters of all ages can be found in this quadrangle. Zucchius crater (shown here as Zuchius) is probably the youngest prominent feature in this quadrangle. Its terraced walls, central peaks and visible rays indicate that it was most probably excavated in the Moon's latest geologic period. The young ejecta material from Zucchius overlays the neighbouring Segner crater, which has been significantly degraded.

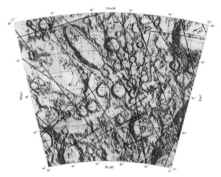

LUNAR ORBITER PHOTOGRAPHIC COVERAGE
OF SCHILLER QUADRANGLE

SCHILLER CRATER AREA
Photograph taken by Lunar Orbiter IV in May 1967 of the elongated Schiller crater (top right), thought to be a fusion of two or more craters and its neighbouring craters.

SCHEINER CRATER AREA
Photograph of the Scheiner crater (centre left) and the Blancanus crater (bottom centre) lying adjacent to the enormous Clavius crater (centre right), taken by Lunar Orbiter IV in May 1967.

SCHILLER QUADRANGLE: KEY CHARACTERISTICS

1. SCHILLER CRATER
Latitude: 51.72°S / Longitude: 39.78°W
Diameter: 179.36 km / 111 ½ miles
Depth: 3.9 km / 2 ½ miles
Photograph: Lunar Orbiter IV (USA, 1967)

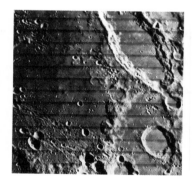

2. SCHILLER-ZUCCHIUS BASIN
Latitude: 56°S / Longitude: 45°W
Diameter: 325 km / 202 miles
Photograph: Lunar Orbiter IV (USA, 1967)

3. LONGOMONTANUS
Latitude: 49.55°S / Longitude: 21.88°W
Diameter: 145.5 km / 90 ½ miles
Depth: 4.5 km / 2 ¾ miles
Photograph: Lunar Orbiter IV (USA, 1967)

4. ROST CRATER
Latitude: 56.42°S / Longitude: 33.84°W
Diameter: 46.85 km / 29 miles
Depth: 2 km / 1 ¼ miles
Photograph: Lunar Orbiter IV (USA, 1967)

5. ZUCCHIUS CRATER
Latitude: 61.38°S / Longitude: 50.65°W
Diameter: 63.18 km / 39 ¼ miles
Depth: 3.2 km / 2 miles
Photograph: Lunar Orbiter IV (USA, 1967)

6. NÖGGERATH CRATER
Latitude: 48.82°S / Longitude: 45.84°W
Diameter: 32.12 km / 20 miles
Depth: 1.5 km / 1 miles
Photograph: Lunar Orbiter IV (USA, 1967)

7. SCHEINER CRATER
Latitude: 60.35°S / Longitude: 27.81°W
Diameter: 110.07 km / 68 ½ miles
Depth: 4.5 km / 2 ¾ miles
Photograph: LRO (USA, 2010)

8. SEGNER CRATER
Latitude: 58.96°S / Longitude: 48.68°W
Diameter: 67.84 km / 42 ¼ miles
Depth: 1,300 m / 4,265 ft 1 in
Photograph: Lunar Orbiter IV (USA, 1967)

9. BAYER CRATER
Latitude: 51.62°S / Longitude: 35.14°W
Diameter: 48.51 km / 30 ¼ miles
Depth: 2 km / 1 ¼ miles
Photograph: Lunar Orbiter IV (USA, 1967)

10. BETTINUS CRATER
Latitude: 63.4°S / Longitude: 45.16°W
Diameter: 71.78 km / 44 ½ miles
Depth: 3.8 km / 2 ¼ miles
Photograph: Lunar Orbiter IV (USA, 1967)

11. BLANCANUS CRATER
Latitude: 63.77°S / Longitude: 21.63°W
Diameter: 105.82 km / 65 ¾ miles
Depth: 3.7 km / 2 ¼ miles
Photograph: Lunar Orbiter IV (USA, 1967)

CONTEMPORARY ARTISTS AND THE MOON

'[T]here is a reason, after all, that some people wish to colonize the moon, and others dance before it as an ancient friend.'

James Baldwin, *No Name in the Street*, 1972

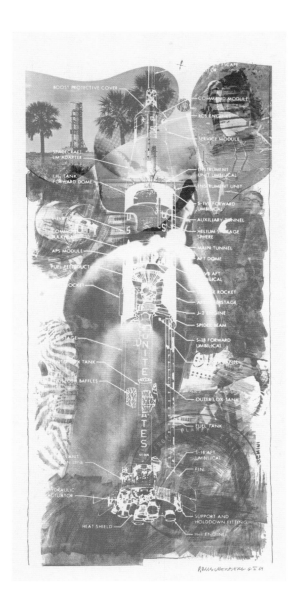

he era-defining years of the Space Race were marked by artworks rhapsodizing the Apollo project. Commissioned by NASA, Robert Rauschenberg (1925–2008) was depicting the events from the inner sanctum at Cape Canaveral.[1] Meanwhile Andy Warhol (1928–1987) had the walls of his New York 'Factory' covered in silver paint and tin foil because it reminded him of space, and the unprecedented events of 'man on the moon' marked him right up to the last months of his life, when he produced his nostalgia-tinged *Moonwalk* edition (1987). But artists also met these unprecedented endeavours more questioningly. Using as a backdrop a composite of Lunar Orbiter 2's reconnaissance picture of the Moon's Copernicus crater (1966) merged with Apollo 8's famous *Earthrise* picture taken on Christmas Eve 1968, Martha Rosler's (b. 1943) *House Beautiful: The Colonies, Frontier* (photomontage, 1966–72) conflates America's expansionist ambitions in outer space with a dreary rendering of its founding myth of 'the next frontier', turning the plains of the Moon into a gridlocked commuting highway of consumerist America. The alien lunarscape is made mundane and domesticated, the Moon's greyness turning into the concrete of block buildings, lunar dust merging with the exhaust fumes of rush hour.

Others still questioned the very *raison d'être* of the astronomically costly Apollo project, its funding raised through taxes, when so much was at stake at home, between the seemingly endless Vietnam War, rampant inequality and racial tensions. This was best summarized in Gil Scott-Heron's (1949–2011) spoken-word poem, 'Whitey on the Moon' (1970), in which he describes the rising cost of living, resulting poverty and marginalization of Black communities, ending with the words: 'Y'know I jus' 'bout had my fill / (of Whitey on the moon) / I think I'll sen' these doctor bills, / Airmail special / (to Whitey on the moon)'. A similar sentiment is powerfully laid out in Faith Ringgold's (1930–2024) *Flag for the Moon: Die N****** (1969) which spells out within the Stars and Stripes the tragic reality of racial discrimination. Others pondered whether something was lost in having finally crossed the threshold of outer space to roam the Moon's pristine

← Robert Rauschenberg's *Sky Garden* is one of thirty-three lithographs from the *Stoned Moon* series (1969–70). Rauschenberg witnessed the Apollo 11 lift off on 16 July 1969. With its vivid colours and attention to the rocket's engineering, this image captures the sense of awe felt as the Saturn V soared into the air.

→ Andy Warhol's 1987 *Moonwalk* evokes long after the event the euphoria of the Moon landing. In pop colours, Warhol conflated two iconic photographs taken on the Moon in July 1969: Buzz Aldrin standing on the barren surface, with Neil Armstrong's silhouette reflected in his gold-plated visor, and Aldrin standing to the right of the American flag.

↓ Martha Rosler's *Frontier* from the series *House Beautiful: The Colonies* (1969–72) is related to her project, *House Beautiful: Bringing the War Home*, in which she protested against the Vietnam War, drawing a parallel between the Space Race and US interventionism.

plains, now littered with space junk and scarred with deep bootprints. Aside from urine bags, stranded rovers and other redundant equipment, the Apollo astronauts left retro-reflectors behind. This was to help gather scientific data, which revealed that the Moon retreats 3.8 centimetres (1 ½ inches) away from the Earth every year. But were we, too, distancing ourselves from the Moon? When James Baldwin (1924–1987) wrote his now oft-quoted passage in *No Name in the Street*, published in 1972, what he meant, of course, was that the quest for the 'next frontier' was linked to an imperialist zeal shown across human history from what we now call the Global North. He was writing at a time when the last Apollo crew was soon to leave the Moon, and the world's appetite for stepping onto lifeless lumps of rock had abated. It was a time of decolonization across the African continent, and the height of the civil rights movement. But he was also touching on another profound truth, which was that the Moon, humanity's constant companion, had not lost its timeless mystery and appeal. Having spent centuries dreaming of

travelling there, and having finally reached its surface, humanity was to rekindle the magic of being under its spell.

In *Earth-Moon-Earth (Moonlight Sonata Reflected from the Surface of the Moon)* (2007), Scottish artist Katie Paterson (b. 1981) used Earth-Moon-Earth technology. Commonly known as E.M.E., or 'moonbounce', it is a kind of transmission used by radio amateurs in which a message is sent from Earth in Morse code, reflected onto the Moon, and received back on Earth. The work beams forth the Morse score of Ludwig van Beethoven's Sonata Op 27.2, which returns from the Moon to be played back on an automated grand piano here on Earth. Popularly known as the *Moonlight Sonata*, Beethoven's piece is one of the most recognizable pieces of classical music, with an achingly romantic mood. Yet reflected back from the Moon, it takes on a distinct, almost haunting resonance, for some of the code is lost in the lunar craters. The mesmerizing melody returns slightly altered, co-authored almost by the Moon, as the missing information translates into intervals or pauses in the score. Was the Moon still reflecting human desires, hopes and dreams?

In other work, Paterson sought to replicate the light of the full Moon in the form of a lifetime's supply of bulbs in *Light bulb to Simulate Moonlight* (2008), and imagined, as part of her series of *Ideas* (2015–ongoing), *A place that exists only in moonlight*. Moonlight is sunlight (and to, a much lesser extent, Earth-light) reflected on the Moon's surface and bouncing back to Earth, its intensity depending on the lunar phase as much as the cloud cover. In music and literature, as well as the visual arts, the transient, ethereal and intangible qualities of moonlight have often been tinged with melancholy and sadness, but also otherworldliness. British artist Darren Almond's (b. 1971) *Fullmoon* series (1998–ongoing) is shot under moonlight, with long exposures of 12 to 30 minutes. The extended opening of the camera's shutter and the photosensitive film within capture the nocturnal landscapes in ways that the naked eye cannot. (Poorly equipped for night-time vision, the human eye perceives in moonlight a world in monochrome blues and greys.) Thus saturated with moonlight, Almond's landscapes seem to be

↓
Mick Namarari Tjapaltjarri painted *Moon Dreaming* for the 1978 film, *Mick and the Moon*, which focuses on the artist. The perspective, as if the viewer were looking down onto the desert, incorporates the cosmos: the Moon is seen above the Earth, with ceremonial objects casting shadows on the landscape.

↗
Richard Long's *Gravity Crescent* was a temporary mural created on the occasion of the artist's 'Circle to Circle' exhibition at Lisson Gallery, London, in May–June 2018. He slathered the gallery wall with mud from the River Avon, to form a crescent Moon. The resulting drips evoke how moonlight bathes the world below.

→
For *Light bulb to Simulate Moonlight* (2008), Katie Paterson worked with scientists, engineers and manufacturer Osram to recreate a lifetime's supply of moonlight. Her work took the form of 289 bulbs approximating the light of the full Moon and lasting 2,000 hours each.

↳
A Partial Eclipse (1987) and *Moon Trail, Tokyo* (1989) both form part of Kikuji Kawada's *The Last Cosmology*. The series conflates events in the sky with the end of the Shōwa era, which concluded with the death of Japanese Emperor Hirohito in 1989.

infused with muted daytime colours, softer shadows and hazy contours. They reveal a different reality.[2] So does *Moon Dreaming* (1978, National Gallery of Australia, Canberra) by First Nations painter from the Pintupi people, Mick Namarari Tjapaltjarri (1926–1998), a leading light of the Papunya Tula Art Movement that developed from 1971. The paintings produced by the movement are invested with deep spiritual meaning and contain stories from the Dreaming. The Dreaming, in First Nations belief, is considered both past – the period when Ancestral Spirits walked the Earth and created life – and present – the spiritual world which exists alongside the living world and in which these Ancestors reside. In characteristic fingertip stippled technique and aerial perspective, *Moon Dreaming* (1978) presents a view of the Earth from above the Moon, looking down onto Australia's Western Desert region. In a colour palette otherwise dominated by ochre, brown and black, the Moon is a luminous orange globe striated in dazzling white, its light bathing the textured land while the darker shapes represent ceremonial objects. The painting forms a pair with Namarari Tjapaltjarri's *Sunrise Chasing Away the Night* (1977–78).[3] Among the First Nations Peoples in Australia, as well as cultures the world over, the Moon is often considered male and the Sun female, or vice versa; in this guise, they are associated with fertility, and sometimes meet during eclipses.

This long-held association between Moon and fertility persists today. A believer in astrology and in the interdependence between people and nature, American artist Kiki Smith (b. 1954) explored in 1993 the relationship between the Moon and the female body. Her *Untitled (Moons)* consists of photocopies of her breasts, lithographed onto hand-made paper and collaged together. The textured paper endows the circular motif of the breasts, a stark white

on the dark expanse left by the photocopying process, with the relief of a rugged terrain, like that of the Moon. The link between the Moon and the female body (see p. 52) originates in the lunar cycle of 29.5 days coinciding with women's fertility cycle (the word 'menstruation' takes its roots from the word 'Moon'). While this link has now been debunked as pure coincidence, there's a widely held belief that maternity wards are busier on nights of the full Moon. Other species, however, are tuned to the lunar cycle: some flowers bloom only in moonlight, while the yearly mass spawning of corals happens right after a full Moon. As Smith highlighted in another work, *Tidal* (1998, artist book, Museum of Modern Art, New York), which combines photogravures of the thirteen full Moons of a calendar year with a photolithograph image of waves, the Moon does govern the oceans.[4] The pull of its gravity dictates the rising and falling of the tides twice a day. And in this, perhaps, resides the Moon's most profound link to life on Earth: not only would the tides have carried the materials of life across oceans, but it is thought that large tidal ranges and the resulting rock pools are responsible for complex life first making it out of the sea.

For millennia, cultures around the world believed in the correlation of astronomical phenomena with earthly events, a belief that has largely subsided with modern science. Eclipses, particularly, were seen as ominous signs. Mingling the personal and collective, Kikuji Kawada's (b. 1933) *Last Cosmology* series (1969–2000) pays homage to this, suggesting that the communal behaviours of Sun, Moon and Earth that occur during eclipses may be seen as astrological omens of the end of the Shōwa era in Japan, with the death of the emperor in 1989 and the close of the 20th century. Portentously brooding, his eclipse pictures are marked by the contrast between the textured darkness of the sky and the blinding luminosity of light piercing from around the disc of the Moon. Kawada was born near the beginning of the Shōwa era, and his youth was marked by the bombing of Hiroshima and Nagasaki and the occupation of Japan. He described the series as 'the cosmology of a changing heart', the symbolist photographs being evocative of his own troubled soul. By contrast, Katie Paterson's *Totality* (2016) provokes joy and wonder: a mirror-ball, faceted with tiles combining almost every documented solar eclipse in over 10,000 images, casts thousands of eclipses on to walls, floors, ceiling and viewers alike. An astronomy enthusiast since a tender age, German photographer Wolfgang Tillmans (b. 1968) is a committed eclipse-chaser. In his two-part video work, *Printing Press Heidelberg Speedmaster XL – Real Time Total Eclipse Nightfall and Exit* (2015), he connects mundane with astronomical. In the first part, Tillmans focuses on the hypnotic movements of the printing press, the colourful inks, the sense of light and depth. In the second, he records the Moon's vast shadow sweeping across the Earth at 100 kilometres (around 62 miles) per hour. Observing how the eclipse affects not only the surroundings but also people at a physiological, but also deeper level, Tillmans expresses that unearthly feeling as the Earth is suddenly plunged into darkness, the birds stop singing and coldness envelops onlookers in such a way that their fingers tingle.

It is as an object moving in the sky that the Moon has had its most considerable influence on human activity, aiding the recording of time through calendars, and the plotting of navigational routes. And yet its motions have also enabled us to make sense of our place within the cosmos. In his series *Revolution*, Japanese photographer Hiroshi Sugimoto (b. 1948) captures the movement of the Moon across the sky as a luminous trail against the dark expanses of sky and sea. By rotating the nocturnal seascape clockwise by 90 degrees, the horizon now vertical, Sugimoto operates a decentring, a perspective shift akin to a vision of the Earth from space. There is no longer an above and a below; yet the series encapsulates the osmotic relationship between Moon and Earth, grounding both within the cosmos.[5] In turn, in his work British land artist Richard Long (b. 1945) grounds the Moon on Earth. He imbues carefully ordered parkland with ancient gravitas, arranging sheets of grey slate in a circle that glistens in sun- and moonlight (*Full Moon Circle*, 2003) or paints the Moon in thick Avon mud on the walls of art galleries (*Gravity Crescent*, 2018). He sets off on long walks invested in lunar meaning, these ephemeral works of art captured in his *Textworks*, part descriptive, part poetic pithy recordings of the journey. A 167-kilometre (104-mile) *Tide Walk* (1992) takes the wandering artist from the English Channel to the Bristol Channel. *Walking to a Lunar Eclipse* (1996) is described as 'a walk of 366 miles in 8 days ending at a midnight total eclipse of the full moon', and *Walking from a Full Moon to a New Moon* as 'fourteen days of waning moonlight in the Sierra Nevada Spain'. In his *Moor Moon* (2009) Dartmoor sites echo lunar sites. Duck's Pool is paired with the Sea of Rains (Mare Imbrium) and Gallaven Mire with the Sea of Cold (Mare Frigoris), mapping the world below with that above: a symmetry equally witty and poetic that not only conjures afresh the vision of the Moon as a mirror of the Earth, but also reflects on a sense of place, the meaning – and arbitrariness – of naming geographical features. In *Space Race* (2018–20), the work of Haitian-Swiss artist Sasha Huber (b. 1975), the practice of naming lunar sites is scrutinized to reveal that some eighty places on Earth, the Moon and Mars have been named after the eminent Swiss glaciologist Louis Agassiz (1807–1873), who was also a notorious and influential proponent of polygenism

and developed pseudo-scientific theories to justify racism and white supremacy. Beyond questioning who we seek to honour and memorialize through the act of naming, Huber questions whether such naming of lunar sites, imprinting them with Earth's darker history, perpetuates acts of colonization and racial violence in outer space.[6]

The Moon is writ large in humanity's history, just as it is in the work of Ghanaian artist El Anatsui (b. 1944). The son of kente weavers, Anatsui is best-known for his use of humble found objects – bottle tops and commercial packaging – to create monumental metallic sheets that he likens to cloth. The sheets themselves evoke a sense of place, African identity, and universality. The grid-like composition of his 1995 *Earth-Moon Connexions* (paint on wood, National Museum of African Art, Smithsonian Institution) conjures the plotting of celestial objects on astronomical maps, and the role of the Moon in human activities.

In his *Sacred Moon* (2007, aluminium and copper-wire) the arrangement of colourful and shimmering segments of repurposed metal recreates the effects of celestial bodies rising and setting. The segments themselves are the branded detritus of a consumerist trade born out of colonialism, as majestically expressed in *Behind the Red Moon* (2023), Anatsui's monumental commission in three acts for Tate's Turbine Hall in London. At first glance abstract, the metal hangings that fill the vessel-like hall in fact carry a narrative: the motifs of the Moon, a sail, the Earth and a wall emerge from the intricate stitching of a multitude of repurposed fragments of metal. Seen up close, these fragments reveal logos, the brands expressive of a global trade made possible by colonial exploitation and oppression. In the 'first act', the Moon is conflated with a monumental sail evocative of the ships that transported goods and enslaved African people, the Moon guiding the sailors' journey across the oceans. Crimson in colour, the 'blood' Moon is that seen during a total lunar eclipse of the full Moon, its spiritual and astrological significance pregnant with prophetic undertones of chaos, darkness and destruction, but also of rebirth and renewal. The 'second act' thus represents the world, made from shapes resembling human figures, that unite into a circular form, echoing the Moon. In the final act, the wall is a symbol of both containment and oppression, as well as a veil that conceals better to reveal through the stimulation of curiosity and imagination. Together, these three acts are expressive of oppression and violence, but also of movement and coming together, endurance and connection, cycle and renewal. They also remind us that everything that has lived, and will ever live, has done so under the light of the Moon.

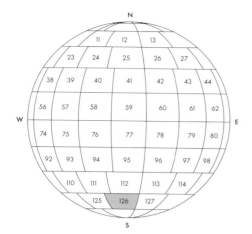

This portion of the southern lunar highlands is densely cratered and without mare. The hilly terra of the highlands is altered by a history of impact cratering. Some of the oldest craters and depressions have been filled and covered by volcanic materials and materials ejected from more recent craters. There are isolated patches of recent volcanic materials that have formed small mare-like plains. The Clavius crater formation in this quadrangle is one of the largest on the Moon. It likely dates from a very early period in the Moon's geologic history, and is heavily worn and altered by subsequent small impacts. A remnant of the crater's original central peak can still be seen amid the many smaller craters that dot Clavius' floor. The Rutherfurd crater that intrudes on Clavius' southern inner wall is the youngest prominent feature in this quadrangle. Its well-defined rim, wall, central peak and rampart, along with a visible ray system, mark it as belonging to the Moon's latest geologic period, the Copernican.

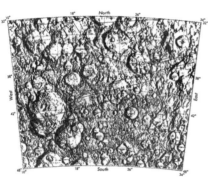

LUNAR ORBITER PHOTOGRAPHIC COVERAGE
OF CLAVIUS QUADRANGLE

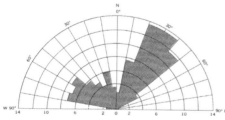

STRIKE ORIENTATIONS OF 643 LINEAMENTS
(FRACTURES AND JOINTS OF TECTONIC ORIGIN)
PLOTTED AGAINST THEIR FREQUENCY

CLAVIUS BASE
Two stills from the set of Stanley Kubrik's sci-fi film *2001: A Space Odyssey* (1968), showing the Clavius base – a fictional research facility based within the large Clavius crater.

CLAVIUS QUADRANGLE: KEY CHARACTERISTICS

1. CLAVIUS CRATER
Latitude: 58.62°S / Longitude: 14.73°W
Diameter: 230.77 km / 143 ½ miles
Depth: 3.5 km / 2 ¼ miles
Photograph: LRO

2. RUTHERFORD CRATER
Latitude: 10.56°S / Longitude: 137.1°W
Diameter: 15.98 km / 10 miles
Photograph: Apollo 16 (USA, 1972)

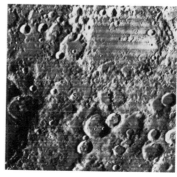

3. MAGINUS CRATER
Latitude: 50.03°S / Longitude: 5.98°W
Diameter: 155.58 km / 96 ¾ miles
Depth: 4.3 km / 2 ¾ miles
Photograph: Lunar Orbiter IV (USA, 1967)

4. HERACLITUS CRATER
Latitude: 49.31°S / Longitude: 6.42°E
Diameter: 85.74 km / 53 ¼ miles
Depth: 3.8 km / 2 ¼ miles
Photograph: Lunar Orbiter IV (USA, 1967)

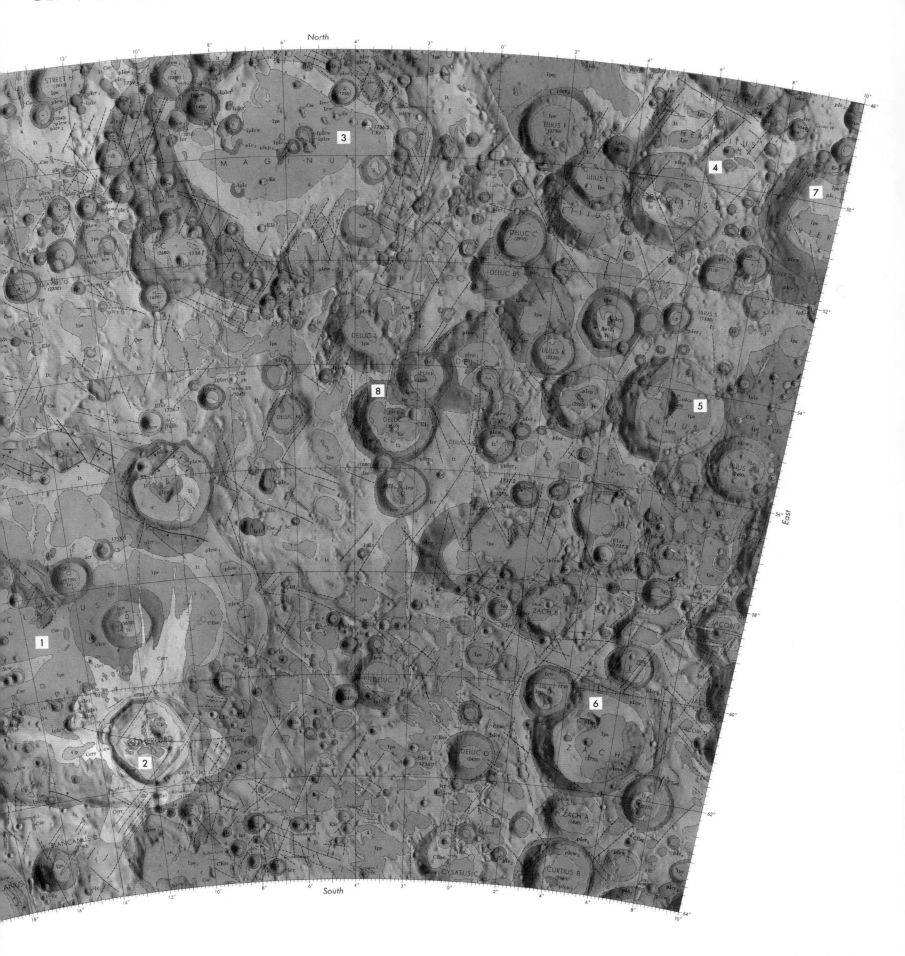

5. LILIUS CRATER
Latitude: 54.6°S / Longitude: 6.1°E
Diameter: 61.18 km / 38 miles
Depth: 3 km / 1 ¾ miles
Photograph: Lunar Orbiter IV (USA, 1967)

6. ZACH CRATER
Latitude: 60.9°S / Longitude: 5.25°E
Diameter: 68.54 km / 42 ½ miles
Depth: 3.7 km / 2 ¼ miles
Photograph: Lunar Orbiter IV (USA, 1967)

7. CUVIER CRATER
Latitude: 50.3°S / Longitude: 9.7°E
Diameter: 77.3 km / 48 miles
Depth: 3.8 km / 2 ¼ miles
Photograph: Lunar Orbiter IV (USA, 1967)

8. DELUC CRATER
Latitude: 55.02°S / Longitude: 2.98°W
Diameter: 45.7 km / 28 ½ miles
Depth: 3.3 km / 2 miles
Photograph: Lunar Orbiter IV (USA, 1967)

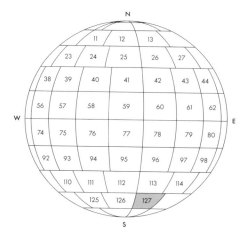

Hommel crater, after which this quadrangle is named, is one of the oldest features shown on this map of a region in the Moon's southeastern highlands. Most of the larger craters in this quadrangle, while not as ancient as Hommel, still date to the Moon's early geologic history. The region is characterized by smooth and ridged terra, isolated plains and densely to moderately spaced impact craters. Many of the smaller, younger craters in this quadrangle were excavated by secondary impacts related to the formation of the crater Tycho (nearby but not in this quadrangle). Bright ray material from the Tycho impact, though not shown in this map, covers most of this area. The terrain in this quadrangle is a result of multiple periods of bombardment and early vulcanism. On 31 July 1999, the US Lunar Prospector spacecraft was deliberately crashed to the south of this quadrangle, inside the crater Shoemaker (not yet named when this map was made) at the coordinates 87.7°S, 42.1°E. This permanently shadowed crater, like others near the Moon's southern pole, was suspected to contain water ice. Astronomers observing the crash from Earth saw no plume that would indicate the release of water vapour from the impact.

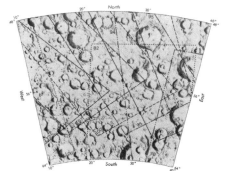

LUNAR ORBITER IV HIGH-RESOLUTION COVERAGE
OF HOMMEL QUADRANGLE

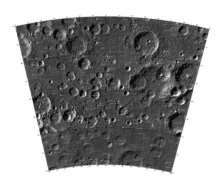

HOMMEL NOMENCLATURE
LAMBERT CONFORMAL PROJECTION

HOMMEL QUADRANGLE: KEY CHARACTERISTICS

1. HOMMEL CRATER
Latitude: 54.74°S / Longitude: 32.93°E
Diameter: 113.6 km / 70 ½ miles
Depth: 2.8 km / 1 ¾ miles
Photograph: Lunar Orbiter IV (USA, 1967)

2. PITTISCUS CRATER
Latitude: 50.6°S / Longitude: 30.57°E
Diameter: 79.85 km / 49 ½ miles
Depth: 3 km / 1 ¾ miles
Photograph: LRO (USA, 2011)

3. ASCLEPI CRATER
Latitude: 55.2°S / Longitude: 25.52°E
Diameter: 40.56 km / 25 ¼ miles
Depth: 2.9 km / 1 ¾ miles
Photograph: Lunar Orbiter IV (USA, 1967)

4. BACO CRATER
Latitude: 51.04°S / Longitude: 19.1°E
Diameter: 65.31 km / 40 ½ miles
Depth: 3.1 km / 2 miles
Photograph: Lunar Orbiter IV (USA, 1967)

5. TANNERUS CRATER
Latitude: 56.44°S / Longitude: 21.92°E
Diameter: 28.07 km / 17 ½ miles
Depth: 1.8 km / 1 mile
Photograph: Lunar Orbiter IV (USA, 1967)

6. JACOBI CRATER
Latitude: 56.82°S / Longitude: 11.3°E
Diameter: 66.28 km / 41 ¼ miles
Depth: 3.3 km / 2 miles
Photograph: Lunar Orbiter IV (USA, 1967)

7. MUTUS CRATER
Latitude: 63.65°S / Longitude: 29.93°E
Diameter: 76.33 km / 47 ½ miles
Depth: 3.7 km / 2 ¼ miles
Photograph: Lunar Orbiter IV (USA, 1967)

8. BREISLAK CRATER
Latitude: 48.31°S / Longitude: 18.31°E
Diameter: 48.64 km / 30 ¼ miles
Depth: 2.6 km / 1 ½ miles
Photograph: Lunar Orbiter IV (USA, 1967)

9. NEARCH CRATER
Latitude: 58.58°S / Longitude: 39°E
Diameter: 72.79 km / 45 ¼ miles
Depth: 2.9 km / 1 ¾ miles
Photograph: Lunar Orbiter IV (USA, 1967)

10. KINAU CRATER
Latitude: 60.75°S / Longitude: 14.94°E
Diameter: 41.87 km / 26 miles
Depth: 2 km / 1 ¼ miles
Photograph: Lunar Orbiter IV (USA, 1967)

11. VLACQ CRATER
Latitude: 53.4°S / Longitude: 38.7°E
Diameter: 89.21 km / 55 ½ miles
Depth: 3 km / 1 ¾ miles
Photograph: Lunar Orbiter IV (USA, 1967)

Emily A. Margolis

THE ARTEMIS MISSIONS

→
The Space Launch System's powerful engines illuminate the night sky as the Artemis I mission lifts off from NASA's Kennedy Space Center in Florida on 16 November 2022.

↳
This illustration imagines the Artemis I mission as it orbits the far side of the Moon with Earth in the distant background. The cone-shaped Orion command module is attached to the cylindrical service module, which generates power from four solar panels.

↳↳
NASA plans for a base camp to support research and exploration of the lunar south pole. As shown in this illustration, the base camp would include living quarters and vehicles for travelling across the Moon's surface. NASA hopes lessons learned from living and working on the Moon will inform future human exploration of Mars.

↪
A set of 360 computer-generated 3D simulations of the phases of the Moon (far side) rendered by Jay Tanner, from 0 to 360 degrees at 1-degree intervals.

ince 1964, the US has sent more than twenty-five uncrewed orbiters, landers and impactors to the Moon. Between 1968 and 1972, twenty-four American astronauts travelled to lunar orbit and twelve descended to the lunar surface. According to NASA, there is still more to explore on the Moon. The US space agency is preparing to send humans back to the surface of the Moon perhaps as early as 2026, more than fifty years after the last Apollo astronauts left their bootprints behind. This time, however, NASA intends to stay.

Project Artemis, as the return-to-Moon programme is known, is named after the Greek goddess of the hunt, Apollo's twin sister. NASA's Apollo and Artemis missions share twin objectives – human exploration of the lunar surface, scientific research and a demonstration of US leadership in spaceflight. However, whereas Apollo represented an end to the Cold War race to the Moon, Artemis represents a beginning – of a permanent human presence offworld, the development of a cislunar economy, and future human exploration of Mars.[1]

NASA has targeted the lunar south pole as the landing site for the Artemis missions. Throughout the programme, astronauts will probe the polar region (which includes some of the only permanently shadowed regions of the lunar surface) to characterize rocks and chemicals, which could be returned to Earth for further study.[2] The space agency is also keen to learn how human beings respond physically to living and working on the surface of another world. Of particular interest is the impact of deep-space radiation on the human body, and mitigation of its harmful effects, as well as adaptation to lunar gravity.[3] Life science and medical data, along with lessons learned more broadly about working offworld, will inform NASA's future Martian exploration programme. In spring 2024, NASA announced some of the first instruments slated for deployment by the crew of Artemis III. They include a seismometer suite to measure ground motion as well as a lunar greenhouse of sorts that will inform the development of crops to support human exploration of the Moon and Mars.[4]

The Orion spacecraft (Lockheed Martin) will transport a crew of four astronauts to lunar orbit, boosted by one of the most powerful launch vehicles to date. The Space Launch System (United Launch Alliance) produces 15 per cent more thrust than Apollo's massive Saturn V rocket. The Human Landing System will transport two astronauts to the surface of the Moon and serve as their temporary home and laboratory, while the other two remain in lunar orbit. Two Human Landing Systems are currently in development, Blue Moon (Blue Origin) and Starship (SpaceX). A next-generation spacesuit (Axiom Space) and a new lunar rover will enable astronauts to explore more of the surface of the Moon. In the future, a permanent base camp will support long-term research and exploration on the lunar surface. The planned Gateway space station will serve as a habitat and research platform in lunar orbit, hosting astronauts and scientists from around the world.

Project Artemis will require significant international cooperation. The labour and technology of other nations will enable NASA to execute its lunar programme safely and effectively. Notably, the European Space Agency (ESA)-designed service module will provide power and propulsion to the Orion spacecraft.[5] The Japan Aerospace Exploration Agency (JAXA) is currently training two astronauts for future Artemis missions. ESA, JAXA and the Canadian Space Agency have all agreed to contribute hardware to lunar Gateway.[6] In anticipation of increased human and robotic activity on the Moon from governments and private companies, NASA instigated a non-binding set of ten shared norms known as the Artemis Accords. Since

2020, more than forty nations have agreed to the tenets of peaceful exploration and scientific data sharing, as well as mitigation of orbital debris.[7]

Project Artemis is well under way. In November 2022, NASA successfully launched the uncrewed Artemis I mission on a twenty-five-day journey around the Moon. This flight demonstrated the Space Launch System and Orion capsule. The crew of Artemis II, announced in April 2023, is preparing for their own circumlunar voyage. They include commander Reid Wiseman, pilot Victor Glover, and mission specialists Christina Koch and Canadian astronaut Jeremy Hansen. A successful Artemis II mission will open the door to a lunar landing by Artemis III no earlier than late 2026.[8] Although NASA has not announced a crew for this mission, it will include a team of experts that looks very different from America's first moonwalkers. The space agency has committed to landing the first woman and first person of colour on the Moon, a corrective to the Apollo-era astronaut selection processes that privileged career paths most accessible to white men.

Even the best-laid plans are, of course, subject to change. The projected timeline for Project Artemis is dependent on many factors beyond hardware development. Historically, the outcome of US presidential elections has shifted directions for NASA; additionally, aligning strategy and schedules with foreign governments and private companies, each with their own priorities, is no mean feat. One thing is certain: whenever the crew of Artemis III touch down on the lunar surface, millions – if not billions – of people will tune in to watch this historic landing.

The China National Space Agency is also preparing to send crewed missions to the lunar south pole before establishing an International Lunar Research Station by the 2030s. Numerous countries, organizations and universities have signed on to participate in this endeavour.[9] Artemis astronauts may not be alone for long. As humans return to the Moon, the impact of these exploration programmes will play out some 384,400 kilometres (238,855 miles) away, here on Earth.

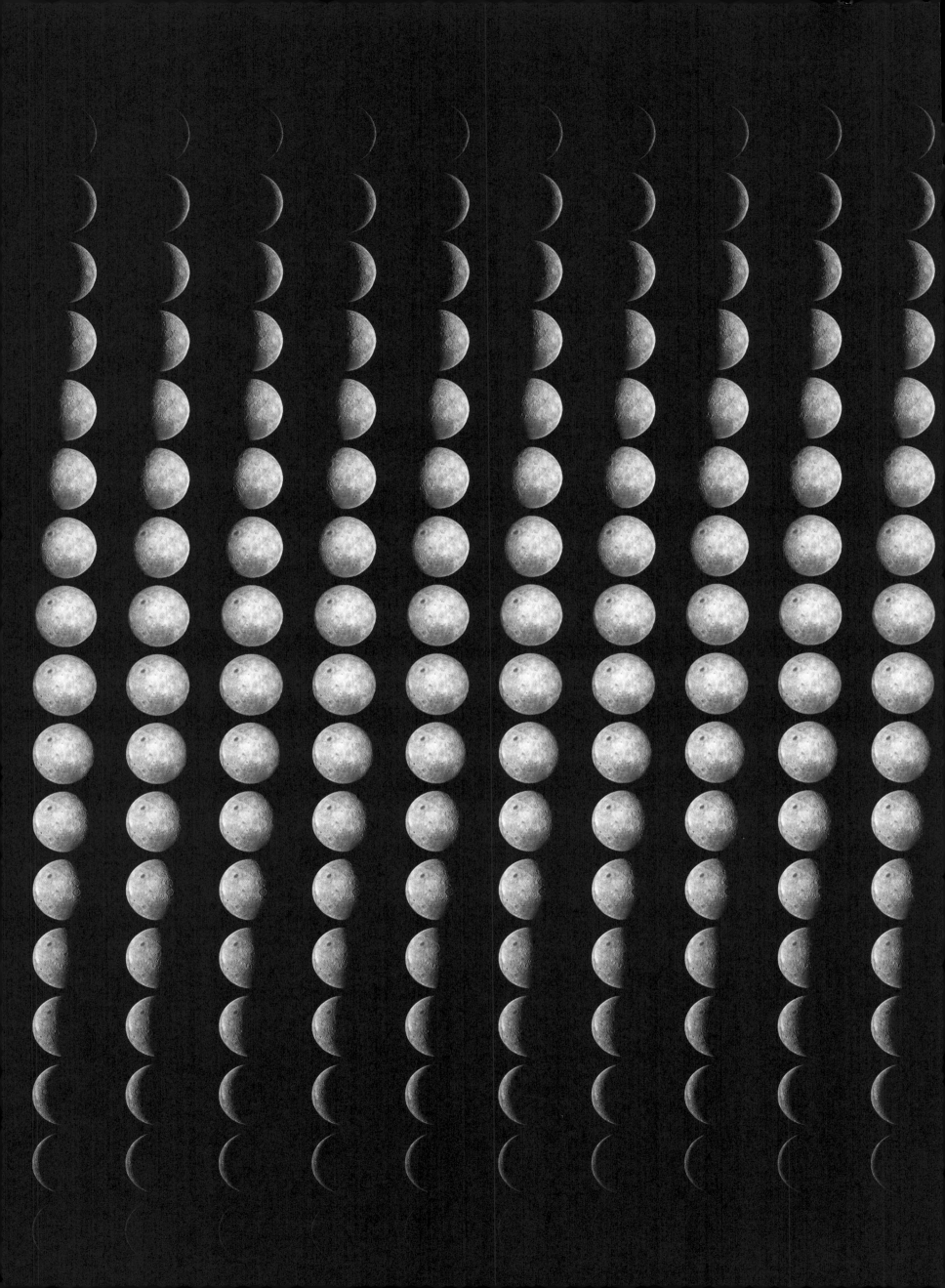

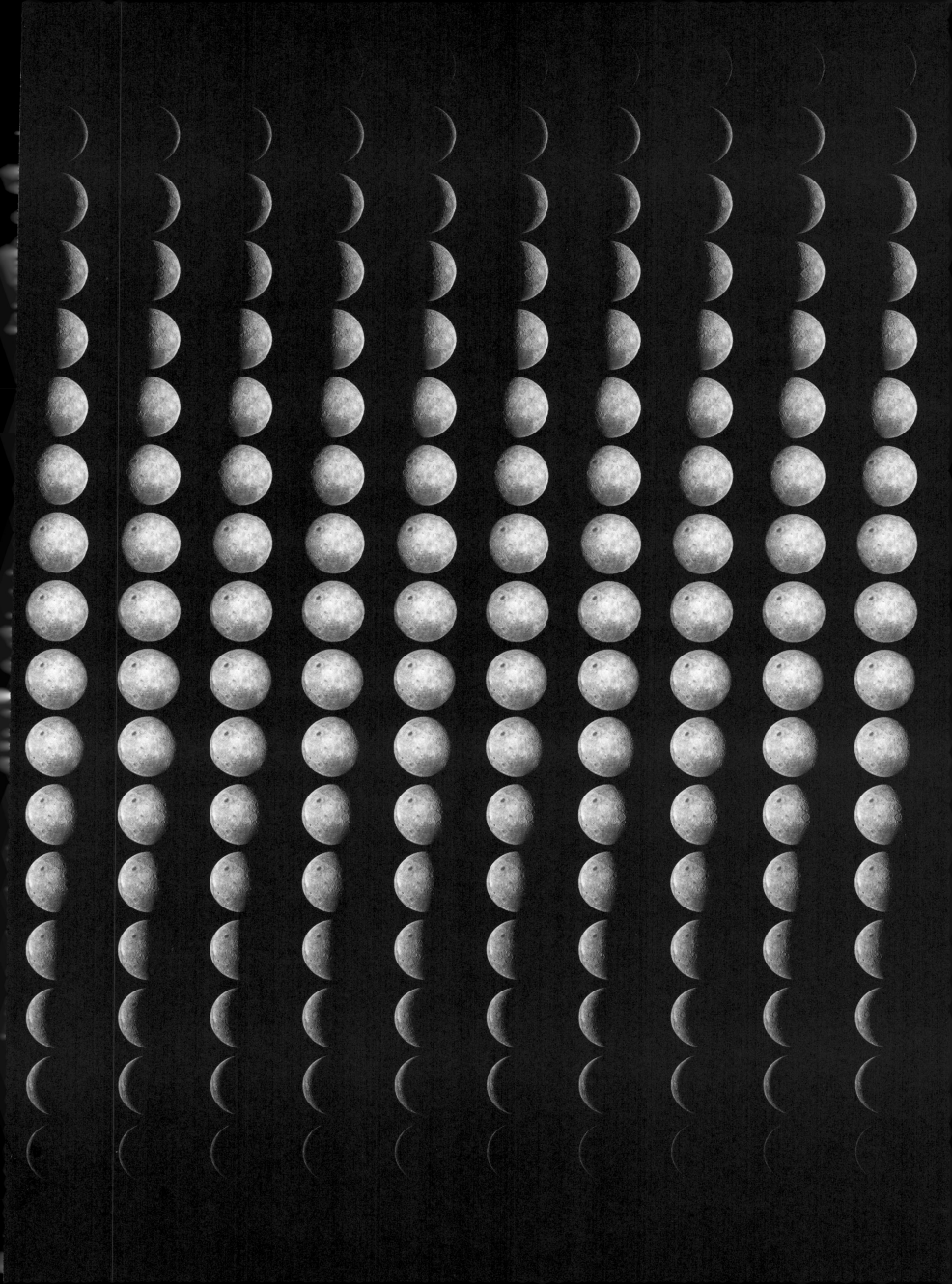

ENDNOTES

INTRODUCTION – THE LUNAR ATLAS

1   Don E. Wilhelms, *To a Rocky Moon: A Geologist's History of Lunar Exploration* (Tucson: University of Arizona Press, 1993), p. 52; Gerald G. Schaber, 'The U.S. Geological Survey, Branch of Astrogeology – A Chronology of Activities from Conception through the End of Project Apollo (1960–1973)', *U.S. Geological Survey Open-File Report* (2005), online at: history.nasa.gov/alsj/Schaber.html

2   Ronald E. Doel, *Solar System Astronomy in America: Communities, Patronage, and Interdisciplinary Science, 1920–1960* (New York: Cambridge University Press, 1996).

3   Matthew Shindell, *The Life and Science of Harold C. Urey* (Chicago: University of Chicago Press, 2019); Matthew Shindell, 'From the End of the World to the Age of the Earth: The Cold War Development of Isotope Geochemistry at the University of Chicago and Caltech', in *Science and Technology in the Global Cold War*, ed. Naomi Oreskes and John Krige (Cambridge, MA: MIT Press, 2014).

4   See, for example, Harold C. Urey, *The Planets: Their Origin and Development* (New Haven: Yale University Press, 1952); Zdeněk Kopal, ed., *Physics and Astronomy of the Moon* (New York: Academic Press, 1962); Zdeněk Kopal and Zdenka Kadla Mikhailov, eds, *The Moon: Proceedings of Symposium No. 14 of the International Astronomical Union, Leningrad, 1960* (New York: Academic Press, 1962); Ralph B. Baldwin, *The Measure of the Moon* (Chicago: University of Chicago Press, 1963).

5   Matthew Shindell, 'Domesticating the Planets: Instruments and Practices in the Development of Planetary Geology', *Spontaneous Generations: A Journal for the History and Philosophy of Science 4*, no. 1 (2010), pp. 191–230.

6   Eugene M. Shoemaker and Robert J. Hackman, 'Stratigraphic Basis for a Lunar Time Scale', in *The Moon: Proceedings of Symposium No. 14 of the International Astronomical Union, Leningrad, 1960*, ed. Zdeněk Kopal and Zdenka Kadla Mikhailov (New York: Academic Press, 1961), pp. 289–300.

7   Schaber, 'The U.S. Geological Survey, Branch of Astrogeology', p. 72.

8   Thomas A. Mutch, *Geology of the Moon: A Stratigraphic View* (Princeton: Princeton University Press, 1970).

9   Shoemaker and Hackman, 'Stratigraphic Basis for a Lunar Time Scale'.

10 Mutch, *Geology of the Moon*, pp. 130–60.

11 Don E. Wilhelms, *The Geologic History of the Moon*, U.S. Geological Survey Professional Paper 1348 (Washington, DC: U.S. Government Printing Office, 1987), online at: pubs.usgs.gov/publication/pp1348

THE PREHISTORIC LUNAR CALENDAR

1   Anthony F. Aveni, *Ancient Astronomers* (Washington, DC: Smithsonian, 1995).

2   Alexander Marshack, *The Roots of Civilization: The Cognitive Beginnings of Man's First Art, Symbol, and Notation* (New York: McGraw-Hill, and London: Weidenfeld & Nicolson, 1972).

3   James Elkins, 'On the Impossibility of Close Reading: The Case of Alexander Marshack', *Current Anthropology* 37, no. 2 (1996), pp. 185–226; Rebecca Boyle, 'Ancient Humans Used the Moon as a Calendar in the Sky', *ScienceNews* (9 Jul. 2019), online at: www.sciencenews.org/article/moon-time-calendar-ancient-human-art (accessed 10 Jan. 2024).

THE MOON IN ANCIENT EGYPTIAN CREATION MYTHS

1   For a translation of the Pyramid Texts, see Raymond O. Faulkner, *The Ancient Egyptian Pyramid Texts* (Oxford: Clarendon Press, 1969).

2   Richard H. Wilkinson, *The Complete Gods and Goddesses of Ancient Egypt* (New York and London: Thames & Hudson, 2003), p. 111.

3   Olaf A. Kaper, 'Myths: Lunar Cycle', in Donald B. Redford, ed., *The Oxford Encyclopedia of Ancient Egypt* (Oxford and New York: Oxford University Press, 2001), pp. 480–82.

4   Diana Delia, 'Isis, or the Moon', in *Egyptian Religion: The Last Thousand Years. Studies Dedicated to the Memory of Jan Quaegebeur*, ed. Willy Clarysse, Antoon Schoors and Harco Willems, Part I (Leuven: Peeters, 2008), pp. 539–50.

5   Jennifer Houser Wegner, 'Khonsu', in Redford, *Oxford Encyclopedia of Ancient Egypt*, p. 233; and Wilkinson, *The Complete Gods and Goddesses*, pp. 113–14.

6   Claas Jouco Bleeker, *Hathor and Thoth: Two Key Figures of the Ancient Egyptian Religion* (Leiden: E. J. Brill, 1973); Denise M. Doxey, 'Thoth', in Redford, *Oxford Encyclopedia of Ancient Egypt*, pp. 398–400; and Wilkinson, *The Complete Gods and Goddesses*, pp. 215–17.

7   J. McKim Malville, Fred Wendorf, Ali A. Mazar and Romauld Schild, 'Megaliths and Neolithic Astronomy in Southern Egypt', *Nature* 392 (1998), pp. 488–91; and J. McKim Malville and Fred Wendorf, 'Megaliths, Neolithic Astronomy, and Emerging Cultural Complexity in Southern Egypt', *American Journal of Archaeology* 106 (2002), pp. 302–3.

8   Ronald A. Wells, 'Astronomy', in Redford, *Oxford Encyclopedia of Ancient Egypt*, pp. 145–51.

9   Richard A. Parker, *The Calendars of Ancient Egypt*, Studies in Ancient Oriental Civilization 26 (Chicago: University of Chicago Press, 1950); and Anthony J. Spalinger, 'Calendars', in Redford, *Oxford Encyclopedia of Ancient Egypt*, pp. 224–27.

THE MOON IN ANCIENT GREEK AND ROMAN MYTH

1   For the Moon's associations with female reproduction, see K. ní Mheallaigh, *The Moon in the Greek and Roman Imagination: Myth, Literature, Science and Philosophy* (Cambridge and New York: Cambridge University Press, 2020), pp. 26–32.

2   The scholiast on Apollonius' *Argonautica* 4.57 tells that Sappho recounted the myth of Selene's love for Endymion. For fuller discussion of the myth, see J. Wang, *Séléné: éclipses, éclat et reflet* (PhD thesis, University of Nanterre-Paris, 2019).

3   See E. Stehle, 'Sappho's Gaze: Fantasies of a Goddess and a Young Man', in *Reading Sappho: Contemporary Approaches*, ed. E. Greene (Berkeley: University of California Press, 1996), pp. 193–225, see pp. 224–25.

4   ní Mheallaigh, *The Moon*, p. 74.

5   Parmenides DK 28 B14 and 15, with discussion in ní Mheallaigh, *The Moon*, pp. 64–65 and (on heliophotism) pp. 68–82.

6   This idea is expounded in detail in Plutarch's dialogue *On the Face of the Moon* (late 1st century AD). Text and English translation are available in the Loeb Classical Library (Vol. 406).

7   For depictions of Endymion and Selene on ancient sarcophagi, see H. Sichtermann and G. Koch, *Griechische Mythen auf römischen Sarkophagen* (Tübingen: E. Wasmuth, 1975), pp. 27–30 and 33–43; further references in ní Mheallaigh, *The Moon*, p. 90; fuller discussion in Wang, *Séléné*, esp. Chapter 8.

8   This myth is recounted in the *Homeric Hymn to Demeter*; text and English translation are available in the Loeb Classical Library (Vol. 496).

9   Plutarch, *On the Face of the Moon*, 942d–f.

10 This myth is recounted in Plutarch, *On the Face of the Moon*, 943a–945d. For discussion, see ní Mheallaigh, *The Moon*, pp. 175–88.

'THE MOON IS MY MOTHER' – LAKOTA PERSPECTIVES

1   William K. Powers, *Oglala Religion*. Reprint edition. Lincoln: University of Nebraska Press, 1977), p. 70.

2   James R. Walker, *The Sun Dance and Other Ceremonies of the Oglala Division of the Teton Dakota*. Vol. 16, Issues 1–7 (Anthropological Papers of the American Museum of Natural History, 1917), p. 167.

3   Ronald Goodman, *Lakota Star Knowledge: Studies in Lakota Stellar Theology* (Mission, SD: Sinte Gleska University Publishing, 2017), p. 1.

4   ibid., p. 54.

5   ibid., pp. 55–56.

6   ibid., p. 56.

7   Dylan Rainforth, 'How Aborigines Invented the Idea of Object-Oriented Ontology', *Un Magazine* 10, no. 1 (2016), online at https://unprojects.org.au/article/how-aborigines-invented-the-idea-of-object-oriented-ontology/

8   Autumn White Eyes, *Instructed by Haŋwí* (Self-published, 2019), p. 32.

9   Eve Tuck and Rubén A. Gaztambide-Fernández, 'Curriculum, Replacement, and Settler Futurity', *Journal of Curriculum Theorizing* 29, no. 1 (2013). https://journal.jctonline.org/index.php/jct/article/view/411.

10   Walker, *The Sun Dance and Other Ceremonies of the Oglala Division of the Teton Dakota*, p.179.

THE MOON AND A GEOCENTRIC UNIVERSE

1   Aristotle, *De Caelo*, trans. J. L. Stocks (Oxford: Clarendon Press, 1922), Book I, Ch. 2, p. 270.

2   Gerard Naddaf, 'On the Origin of Anaximander's Cosmological Model', *Journal of the History of Ideas* 59, no. 1 (Jan. 1998), p 4.

3   Carlo Rovelli, *Anaximander and the Nature of Science*, trans. Marion Lignana Rosenberg (London: Allen Lane, 2023; and New York: Riverhead Books [as *Anaximander and the Birth of Science*], 2023), p. 48.

4   Anaxagoras, Fragment 18, trans. John Burnet, in *Early Greek Philosophy*, 3rd edition (London: A. & C. Black, 1920), p. 261.

5   Daniel Graham, *Science before Socrates: Parmenides, Anaxagoras, and the New Astronomy* (New York and Oxford: Oxford University Press, 2013), pp. 120–21.

6   Aristotle, *De Caelo*, Book II, Ch. 11, p. 291.

UNDERSTANDING THE PHASES OF THE MOON

1   J. L. Berggren and Nathan Sidoli, *Aristarchus's On the Sizes and Distances of the Sun and the Moon: Greek and Arabic Texts*, Archive for History of Exact Sciences 61, no. 3 (2007).

2   T. Hockey, et al., eds, *The Biographical Encyclopedia of Astronomers* (New York: Springer, 2007).

3   Isaac Newton, *The Principia*, trans. B. Cohen and A. Whitman with J. Budenz (Oakland, CA: University of California Press, 2016).

THE FEMALE CYCLE AND THE MOON

1   Lesley Dean-Jones, *Women's Bodies in Classical Greek Science* (Oxford and New York: Clarendon Press, 1996), p. 397.

2   Aristotle, *Generation of Animals*, trans. A. L. Peck, Loeb Classical Library 366 (Cambridge: MA and London: Heinemann, 1942), p. 397.

3   Sara Read, *Menstruation and the Female Body in Early Modern England* (Basingstoke: Palgrave Macmillan, 2013), p. 32.

4   Levinus Lemnius, *The Secret Miracles of Nature in Four Books: Learnedly and moderately treating of generation, and the parts thereof, the soul, and its immortality, of plants and living creatures, of diseases, their symptoms and cures, and many other rarities* (London, 1658); Anon, *Aristotles Master-piece, or, The secrets of generation displayed in all the parts thereof* (London, 1684).

5   Lemnius, *The Secret Miracles of Nature*, pp. 22–24.

6   Sarah Jinner, *An Almanack or prognostication for the year of our Lord 1658 being the second after bissextile or leap year: calculated for the meridian of London, and may indifferently serve for England, Scotland, and Ireland. by Sarah Jinner* (London, 1658); Sarah Jinner, *An Almanack and prognostication: for the year of our Lord 1659.*

Being the third after bissextile or leap year. Calculated for the meridian of London, and may indifferently serve for England, Scotland, and Ireland. By Sarah Jinner student in astrology (London, 1659).

7   Sarah Ginnor, *The Womans Almanack: Or, prognostication for ever: shewing the nature of the planets, with the events that shall befall women and children born under them. With several predictions very useful for the female sex* (London, 1659).

8   Lemnius, *The Secret Miracles of Nature*, pp. 22–24.

9   ibid., pp. 22–23.

10   *The Holy Bible conteyning the Old Testament, and the New: newly translated out of the originall tongues: & with the former translations diligently compared and reuised, by his Maiesties speciall comandement. Appointed to be read in churches* (London, 1611): Leviticus XV.19–27; Pliny, the Elder, *The historie of the vvorld: commonly called, The naturall historie of C. Plinius Secundus. Translated into English by Philemon Holland Doctor of Physicke. The first [-second] tome* (London, 1634), p. 308.

11   Lemnius, *The Secret Miracles of Nature*, pp. 22, 298.

MAYAN LUNAR ASTROLOGY

1   William L. Fash, *Scribes, Warriors and Kings: the city of Copan and the Ancient Maya* (London: Thames & Hudson, 1991), pp. 160–65.

2   David Webster, Barbara Fash, Randolph Widmer and Scott Zeleznik, 'The Skyband Group: Investigation of a Classic Maya Elite Residential Complex at Copán, Honduras', *Journal of Field Archaeology* 25, no. 3 (1998), pp. 319–43, online at https://doi.org/10.2307/530536

3   Robert J. Sharer with Loa P. Traxler, *The Ancient Maya*, Sixth Edition (Stanford: Stanford University Press, 2006), pp. 99–152.

4   Michael D. Coe and Mark Van Stone, *Reading the Maya Glyphs* (London: Thames & Hudson, 2001), pp. 37–56.

5   Gerardo Aldana y Villalobos, *Calculating Brilliance: an intellectual history of Mayan astronomy at Chich'en Itza* (Tucson: University of Arizona Press, 2022), pp. 181–218.

6   Gerardo Aldana, 'Lunar Alliances: Shedding Light on Conflicting Classic Maya Theories of Hegemony', *Proceedings of the Oxford VII Conference on Archaeoastronomy and Astronomy in Culture* (Tuscon: Arizona State Museum, 2006), pp. 237–58.

7   William A. Saturno, David Stuart, Anthony F. Aveni, and Franco Rossi, '2012 Ancient Maya Astronomical Tables from Xultun, Guatemala', *Science* 336, no. 6082 (11 May 2012), pp. 714–17.

8   Anthony F. Aveni, *Skywatchers* (Austin: University of Texas Press, 2001, pp. 67–80.

9   Michael Carrasco and Robert Wald, 'Intertextuality in Classic Maya Ceramic Art and Writing: The Interplay of Myth and History on the Regal Rabbit Vase Princeton Vase', in *Ceramics of Ancient America: Multidisciplinary Approaches*, ed. Yumi Park Huntington, Dean E. Arnold and Johanna Minich (Gainsville: University Press of Florida, 2018), pp. 183–212.

THE MOON AND MEDIEVAL NATURAL PHILOSOPHY

1   These lines from Canto II, translated by Allen Mandelbaum, are taken from Columbia University's website *Digital Dante*, online: digitaldante.columbia.edu (accessed 10 Jan. 2024).

2   Edward Grant, *Planets, Stars, and Orbs: The Medieval Cosmos, 1200–1687* (Cambridge and New York: Cambridge University Press, 1994), p. 452.

3   Claudius Ptolemaeus, *Tetrabiblos*, trans. Frank E. Robbins (Cambridge, MA: Harvard University Press, 1940), p. 7.

4   Ptolemaeus, *Tetrabiblos*, p. 36.

5   Justine Isserles, 'Bloodletting and Medical Astrology in Hebrew Manuscripts from Medieval Western Europe', *Sudhoffs Archiv* 101, no. 1 (2017), pp. 2–41: see p. 22.

ENDNOTES

'MOONSTRUCK' – LUNACY AND THE FULL MOON

1   G. Norevik, et al., 'The Lunar Cycle Drives Migration of a Nocturnal Bird', *PLoS biology* 17, 10 e3000456 (15 Oct. 2019), doi:10.1371/journal.pbio.3000456

2   M. S. Palmer, et al., 'A "Dynamic" Landscape of Fear: Prey Responses to Spatiotemporal Variations in Predation Risk across the Lunar Cycle', *Ecology Letters* 20, issue 11 (12 Sept. 2017), online at: doi:10.1111/ele.12832

3   R. C. Babcock, G. D. Bull, P.L. Harrison, et al., 'Synchronous Spawnings of 105 Scleractinian Coral Species on the Great Barrier Reef', *Marine Biology* 90 (1986), pp. 379–94. Online at: https://doi.org/10.1007/BF00428562

4   C. Plinius Secundus, *The Historie of the World*, trans. Philemon Holland (1601), Book II, pp. 1–49, online at: https://penelope.uchicago.edu/holland/pliny2.html

5   Alina Iosif and Bruce Ballon, 'Bad Moon Rising: The Persistent Belief in Lunar Connections to Madness', *Canadian Medical Association Journal = Journal de l'Association medicale canadienne* 173, no. 12 (2005), pp. 1498–500. Online at: doi:10.1503/cmaj.051119. This paper, in turn, cites C. L. Raison, H. M. Klein and M. Steckler, 'The Moon and Madness Reconsidered', *Journal of Affective Disorders* 53 (1999), pp. 99–106.

6   M. A. Riva, et al., 'The Disease of the Moon: The Linguistic and Pathological Evolution of the English Term Lunatic', *Journal of the History of the Neurosciences* 20, no. 1 (Jan. 2011), pp. 65–73, online at doi: 10.1080/0964704X.2010.481101. PMID: 21253941.

7   William Shakespeare, *Othello*, Act 5, Scene 2, ll.135–8.

8   Iosif and Ballon, 'Bad Moon Rising' (see note 5), pp. 1498–99.

9   J. F. Oliven, 'Moonlight and Nervous Disorders', *American Journal of Psychiatry* 99 (1943), pp. 579–84, cited in Iosif and Ballon, 'Bad Moon Rising' (see note 5).

10 Minjie Su, 'Medieval Werewolves', *Medievalists* (n.d.), online at https://www.medievalists.net/2019/10/5-things-you-might-not-know-about-medieval-werewolves/

11 John William Polidori, *The Vampyre; A Tale* (London, 1819), online at: https://www.gutenberg.org/files/6087/6087-h/6087-h.htm

12 https://www.reddit.com/r/Teachers/comments/qd2gma/the_full_moon_is_making_my_kids_crazy/

13 https://www.reddit.com/r/securityguards/comments/140618u/you_guy_dealing_with_full_moon_bs_too/

14 S. J. Martin, I. W. Kelly and D. H. Saklofske, 'Suicide and lunar cycles: a critical review over 28 years', Psychological Reports 71, no. 3 (Dec. 1992) , pp. 787–95, online at: doi: 10.2466/pr0.1992.71.3.787. PMID: 1454925.

15 J. A. Schafer, S. P. Varano, J. P. Jarvis and J. M. Cancino, 'Bad Moon on the Rise? Lunar Cycles and Incidents of Crime', *Journal of Criminal Justice* 38, no. 4 (Jul.–Aug. 2010), pp. 359–67, online at: https://www.sciencedirect.com/science/article/abs/pii/S0047235210000589?via%3Dihub

16 'Lunacy and the Full Moon', *Scientific American* (1 Feb. 2009), online at: https://www.scientificamerican.com/article/lunacy-and-the-full-moon/

17 Leandro Casiraghi, et al., 'Moonstruck Sleep: Synchronization of Human Sleep with the Moon Cycle under Field Conditions', *Science Advances* 7, no. 5, eabe0465 (27 Jan. 2021), online at: doi:10.1126/sciadv.abe0465

18 T. C. Erren and P. Lewis, 'Hypothesis: Folklore Perpetuated Expression of Moon-Associated Bipolar Disorders in Anecdotally Exaggerated Werewolf Guise', *Medical Hypotheses* 122 (Jan. 2019), pp. 129–33, online at: doi: 10.1016/j.mehy.2018.11.004. Epub 2018 Nov 15. PMID: 30593396.

19 T. A. Wehr, 'Bipolar Mood Cycles and Lunar Tidal Cycles', *Molecular Psychiatry* 23 (24 Jan. 2018), pp. 923–31, online at: https://doi.org/10.1038/mp.2016.263

THE SYMBOLISM OF THE MOON IN ASTROLOGY

1   Hermann Hunger, *Astrological Reports to Assyrian Kings* (Helsinki: Helsinki University Press, 1992), 102, p. 61.

2   From the Hittite 'Removal of the Threat Implied in an Evil Omen' in James B. Pritchard, ed., *Ancient Near Eastern Texts Relating to the Old Testament* (Princeton: Princeton University Press, 1969), p. 355.

3   Ulla Koch-Westenholz, *Mesopotamian Astrology: An Introduction to Babylonian and Assyrian Celestial Divination* (Copenhagen: Carsten Niebuhr Institute of Near Eastern Studies, Museum Tusculunum Press, University of Copenhagen, 1995), p. 153.

4   Vettius Valens, *The Anthology*, Book 1, trans. Robert Schmidt (Berkeley Springs, VA: Golden Hind Press, 1993), p. 2.

5   Claudius Ptolemy, *Tetrabiblos*, trans. F. E. Robbins (Cambridge, MA: Harvard University Press, 1940), Book 1, chapter 2.

6   Hans Dieter Betz, *The Greek Magical Papyri in Translation, Including the Demotic Spells* (Chicago: Chicago University Press, 1992), xiv.1003–14.

7   Margaret Hone, *The Modern Textbook of Astrology* (London: L. N. Fowler, 4th edition reprinted 1973 [1951]), pp. 25–26.

THE THEORY OF LUNAR ECLIPSES

1   Ferdinand Columbus, *The Life of the Admiral Christopher Columbus by his Son Ferdinand*, trans. Benjamin Keen (New Brunswick, NJ: Rutgers University Press, 1959).

2   John North, *The Fontana History of Astronomy and Cosmology* (London: Fontana Press, 1994).

3   Christopher Walker, ed., *Astronomy before the Telescope* (London: British Museum Press, 1996).

4   Herodotus, *The Histories* (1, 73–74), trans. Aubrey de Sélincourt, John M. Marincola (London: Penguin Books, 2003).

5   J. L. Berggren and Nathan Sidoli, *Aristarchus's On the Sizes and Distances of the Sun and the Moon: Greek and Arabic Texts*, Archive for History of Exact Sciences 61, no. 3 (2007), pp. 213–54.

6   Ptolemy of Alexandria, *The Almagest (Syntaxis Mathematica Book VI)*, trans. G. J. Toomer (London: Gerald Duckworth & Co., 1984).

7   Nicolaus Copernicus, *De revolutionibus orbium coelestium, Libri VI*, trans. Edward Rosen (Warsaw: Polish Scientific Publications, 1978).

THE MOON AND THE HELIOCENTRIC WORLDVIEW

1   Noel M. Swerdlow and Otto Neugebauer, *Mathematical Astronomy in Copernicus's De Revolutionibus* (Berlin: Springer-Verlag, 1984).

2   Christopher M. Graney, *Setting Aside All Authority: Giovanni Battista Riccioli and the Science against Copernicus in the Age of Galileo* (Notre Dame: University of Notre Dame Press, 2015).

3   Galileo Galilei and Johannes Kepler, *The Sidereal Messenger of Galileo Galilei and a Part of the Preface to Kepler's Dioptrics Containing the Original Account of Galileo's Astronomical Discoveries*, trans. Edward Stafford Carlos (London: Rivingtons, 1880), p. 15, online at www.loc.gov/item/07005111

4   Galilei and Kepler, *The Sidereal Messenger*, p. 30.

5   Johannes Kepler, *Kepler's Somnium; the Dream, or Posthumous Work on Lunar Astronomy*, trans. Edward Rosen (Madison: University of Wisconsin Press, 1967).

6   Frédérique Aït-Touati, *Fictions of the Cosmos: Science and Literature in the Seventeenth Century*, trans. Susan Emanuel (Chicago: University of Chicago Press, 2011), p. 15.

EARLY TELESCOPIC STUDIES OF THE MOON

1   Galileo Galilei, *Sidereus nuncius* (Venice: Thomas Baglioni, 1610).

2   Michael van Langren, *Plenilunii lumina Austriaca Philippica* (Brussels, 1645).

3 Johannes Hevelius, *Selenographia, sive Lunae descriptio* (Gdansk: Typis Huenefeldianis, 1647).

4 Giovanni Battista Riccioli, *Almagestum novum* (Bologna: Victorius Benatius, 1651).

5 Ed. Lawrence and A. Shimmerman, *NASA Lunar Cartographic Dossier*, Vol. 1 (St Louis: Defence Mapping Agency Aerospace Center, 1973).

6 Tobias Mayer, *Opera inedita* (Göttingen: Johann Christian Dieterich, 1775).

THE MOON IN LITERATURE AND FICTION

1 Plutarch, *Concerning the Face Which Appears in the Orb of the Moon*, in: *Moralia, Volume XII*, trans. Harold Cherniss and W. C. Helmbold (Cambridge, MA: Harvard University Press, 1957), pp. 2–226.

2 First edition: Johannes Kepler (author) and Ludwig Kepler (ed.), *Ioh. Keppleri mathematici olim imperatorii Somnium, seu opus posthumum De astronomia lunari* (Frankfurt, 1634). See also: J. Lear, *Kepler's Dream. With the Full Text and Notes of Somnium, Sive Astronomia Lunaris, Joannis Kepleri*, trans. Patricia Frueh Kirkwood (Berkeley: University of California Press, 1965).

3 See M. H. Nicolson, *A World in the Moon: A Study of the Changing Attitude Toward the Moon in the Seventeenth and Eighteenth Centuries* (Northampton, MA: Smith College, 1936); and *Voyages to the Moon* (New York: Macmillan, 1948).

4 This was one of three satirical novels by Bergerac, published posthumously in 1657.

5 R. E. Raspe, *Baron Münchhausen's Narrative of his Marvellous Travels and Campaigns in Russia* (Oxford, 1785/86).

6 A. Behn, *The Emperor of the Moon: A Farce. As it is Acted by Their Majesties Servants, at the Queens Theatre. Written by Mrs. A. Behn* (London, 1687).

7 *Frankenstein; or, The Modern Prometheus* was written in 1817 and first published anonymously in Londonon 1 Jan. 1818. Mary Shelley's name did not appear until it was reprinted in Paris in 1821. In 1831, an extended version, heavily edited by Shelley, was published in London by Henry Colburn & Richard Bentley. The 1831 text is now the most common version.

8 The full text of the 'Great Moon Hoax', attributed to the journalist Richard Adams Locke, is included in W. N. Griggs, *The Celebrated 'Moon Story,' Its Origins and Incidents; With a Memoir of the Author, and an Appendix* (New York: Bunnell & Price, 1852).

9 E.A. Poe, 'Hans Pfaall, a Tale', first published in *Southern Literary Messenger* 1, no. 10 (June 1835), pp. 565–80. It was included in the posthumous 'Griswold Edition' of Poe's work under the title *The Unparalleled Adventure of One Hans Pfaall*: E. A. Poe, *The Works of the Late Edgar Allan Poe*, Vol. I: Tales (New York: J. S. Redfield, 1850), pp. 1–51.

10 Both *De la Terre à la Lune* and the sequel *Autour de la Lune* were published in Paris by Bibliothèque d'Education et de Récréation, J. Hetzel et Cie. Within a few years translations into English and many editions in England and the USA followed, some in combined formats. The numerous wood engravings by French illustrator Émile-Antoine Bayard and Alphonse de Neuville are among the most recognizable and inventive illustrations of early science fiction.

11 Before it was published as a book by George Newnes Ltd in London in 1901, *The First Men in the Moon* appeared as an eight-part serial in *The Cosmopolitan* and as a ten-part serial in *The Strand Magazine* (from November 1900).

12 T. von Harbou, *Frau im Mond* (Berlin, 1928). While Fritz Lang's film was released in English-speaking countries as *Woman in the Moon*, the novel was translated as *The Rocket to the Moon* (New York, 1930) and *The Girl in the Moon* (London, 1930).

13 It was first published in book form in New York by G. P. Putnam's Sons but had been serialized in the American science fiction magazine *Worlds of If* from Dec. 1965 to Apr. 1966.

MOONSCAPES – THE MOON IN 19TH-CENTURY PAINTING

1 'Exhibition by the Fakirs: Weird Canvases Ready for Eleventh Annual Function', *New York Times* (29 April 1902).

2 Joris Karl Huysmans, *Certains* (Paris: Tresse & Stock, 1889), pp. 67–68.

3 Wolfgang Schivelbusch, *Disenchanted Night. The Industrialization of Light in the Nineteenth Century*, trans. Angela Davies (Berkeley: University of California Press, 1995).

4 Stéphane Le Mentec, 'Les Astronomical Drawings d'Étienne Léopold Trouvelot (1827–1895)', *Nouvelles de l'estampe* 242 (2013), pp. 14–26.

HOW THE MOON GOVERNS THE TIDES

1 Athanasius Kircher, *Mundus subterraneus* (Amsterdam: Johannes Janssen, 1665).

2 René Descartes, *Principia philosophiae* (Amsterdam: Elsevier, 1644).

3 Johannes Kepler, *Somnium* (Frankfurt: Ludwig Kepler, 1634), cited in John Lear, *Kepler's Dream. With the Full Text and Notes of Somnium, Sive Astronomia Lunaris, Joannis Kepleri*, trans. Patricia Frueh Kirkwood (Berkeley: University of California Press, 2022).

4 Isaac Newton, *The Principia*, trans. B. Cohen and A.Whitman with J. Budenz (Oakland, CA, University of California Press, 2016).

5 Pierre-Simon Laplace, *Traité de mécanique céleste* (Paris: Charles Crapelet, 1799–1805).

LUNAR PHOTOGRAPHY – OPPORTUNITY AND CHALLENGE

1 R. A. Proctor, *The Moon: Her Motions, Aspect, Scenery, and Physical Condition* (London: Longmans, Green and Co., 1873), p. 153

2 Phillips to Sabine, 1 Nov. 1856. In Roger Hutchins, 'John Phillips's Astronomy 1852–67, A Pioneering Contribution to Comparative Planetology', *Antiquarian Astronomer* 6 (2012), pp. 44–58, ref. 19.

3 Charles Nevers Holmes, 'Earlier Photography of the Firmament', *Popular Astronomy* 26, no. 2 (Feb. 1918), pp. 80–84, on p. 83.

4 On Whipple, see Bessie Zaban Jones and Lyle Gifford Boyd, *The Harvard College Observatory* (Cambridge: Belknap, Harvard, 1971), pp. 74ff. On the public reception, see Holmes, 'Earlier Photography of the Firmament', p. 83. See also F. E. Harpham, 'The Rutherfurd Photographs', *Popular Astronomy* 8 (1900), pp. 129–36.

5 L. M. Rutherfurd, 'Astronomical Photography,' *American Journal of Science* 117 (May 1865), pp. 304–9, on 309.

6 On Rutherfurd's efforts and success publicizing his work, see Deborah Jean Warner, 'Lewis M. Rutherfurd: Pioneer Astronomical Photographer and Spectroscopist', *Technology and Culture* 12, no. 2 (Apr. 1971), pp. 190–216.

7 Proctor, *The Moon*, pp. 229–30.

8 Frederick E. Wright, F. H. Wright, and Helen Wright, 'The Lunar Surface: Introduction', in Gerard P. Kuiper and Barbara Middlehurst, *The Moon, Meteorites and Comets* (Chicago: The University of Chicago Press, 1963), pp. 1–66, on p. 1.

9 F. G. Pease, 'Astronomical Telescopes', *Publications of the Astronomical Society of the Pacific* 40, no. 233 (Feb. 1928), pp. 11–23, on p. 20.

10 E. A. Whitaker, G. P. Kuiper, and W. K. Hartmann, *Rectified Lunar Atlas. Contributions of the Lunar and Planetary Laboratory, Tucson* (Tucson: University of Arizona Press, 1963). The original method was to coat the sphere with a photographic emulsion. Kuiper reversed the process. Dale Cruikshank kindly provided insights into the rectification process.

11 H. D. Jamieson, et al., 'Crisis in the Lunar Section: A Program for the 1970s', *Strolling Astronomer* 22 (1970), pp. 64–65. Emphasis in the original.

12 Interview of Harold Masursky by Ronald Doel, 19 June 1987. Niels Bohr Library & Archives, American Institute of Physics, College Park,

ENDNOTES

MD (www.aip.org/history, www.aip.org/history-programs/niels-bohr-library/oral-histories/5081-2). On the panoramic camera, see Harold Masursky, G. W. Colton, and Farouk El-Baz, eds, *Apollo over the Moon: A View from Orbit* (NASA SP-362, 1978), online at https://history.nasa.gov/SP-362/contents.htm.

A PAPER MOON – *CARTES DE VISITE*

1 A process that requires photographic material to be coated, sensitized, exposed and developed within a short period of time. Any location work would require the process to be carried out in the field, making it a cumbersome way of producing images.
2 White paper coated with egg white to make it smooth and glossy.

THE FEMININE SYMBOLISM OF THE MOON

1 See D. Hodge, 'Aleksandra Mir: First Woman on the Moon' (Tate, 2015), article and video available at Tate online: www.tate.org.uk/art/artworks/mir-first-woman-on-the-moon-t13704
2 The goddess is included in *The Dinner Party*'s 'Heritage Floor'; see 'Ashtoreth' (Brooklyn Museum, n.d.), online at: www.brooklynmuseum.org/eascfa/dinner_party/heritage_floor/ashtoreth
3 Inana is also referenced within the 'Heritage Floor' of *The Dinner Party* (as Inanna).
4 On the Triple Goddess archetype see D. J. Conway, *Maiden, Mother, Crone: The Myth and Reality of the Triple Goddess* (Minnesota: Llewellyn Publications, 2004).
5 'The Birth Project', *The Judy Chicago Research Portal: Learning, Making, Culture*, available at judychicagoportal.org/projects/birth-project
6 *Yayoi Kusama Discusses 'Soul under the Moon'* (video), Queensland Art Gallery, available at collection.qagoma.qld.gov.au/objects/14351
7 On this body of work please see C. Whiting, 'It's Only a Paper Moon: The Cyborg Eye of Vija Celmins', *American Art* 23, no. 1 (spring 2009), pp. 36–55.

THE MOON IN POPULAR SCIENCE BOOKS

1 For example, Johann Hieronymus Schröter's magnificent *Selenotopographische Fragmente zur genauern Kenntniss der Mondfläche* (Göttingen, 1791).
2 J. Nasmyth and J. Carpenter, *The Moon: Considered as a Planet, a World, and a Satellite* (London: John Murray, 1874).
3 M. Ward, *The Telescope: A Familiar Sketch* (London: Groombridge & Sons, 1870).
4 J. Baikie, *Through the Telescope* (London: Adam & Charles Black, 1906).
5 R. S. Ball, *Atlas of Astronomy* (London: George Philip & Son, 1892).
6 T. Heath, *The Twentieth Century Atlas of Popular Astronomy* (Edinburgh & London: W. & A. K. Johnston, 1903).
7 R. S. Ball, *The Story of the Heavens* (London, Paris and Melbourne: Cassell & Co., 1885).
8 R. S. Ball, *Star-land* (London, Paris and Melbourne: Cassell & Co., 1893).
9 English translation: *Popular Astronomy: A General Description of The Heavens* (London: Chatto & Windus, 1894).
10 English translation: *A Day in the Moon* (London: Hutchinson & Co., 1913).
11 A. Giberne, *Sun, Moon and Stars: Astronomy for Beginners* (London: Seeley & Co., 1879).
12 R. A. Proctor, *The Moon: Her Motions, Aspect, Scenery* (London: Longmans, Green & Co., 1873).
13 M. Proctor, *Romance of the Moon* (New York and London: Harper & Brothers, 1928).
14 M. Proctor, *The Book of the Heavens* (London, Calcutta, Sydney: George G. Harrap, 1924).

THE MOON IN SILENT CINEMA

1 The brothers had organized an earlier screening, the previous March, for members of the nascent film industry. But this December screening was the first public one and therefore regarded as the inception of cinema as we know it.
2 Méliès was the subject of Brian Selznick's 2007 graphic novel *The Invention of Hugo Cabret*, which was adapted by Martin Scorsese in 2011 as *Hugo*. The film featured recreations of Méliès' productions.

SURREALISTS AND THE MOON

1 Letter to Sir Kenneth Clark, 11 March 1941, Imperial Museum War Artists Advisory Committee files.
2 See *René Magritte, The Pleasure Principle*, exhibition catalogue, Tate Liverpool and Albertina, Vienna, 2011.
3 Lynda Roscoe Hartigan, ed., *Joseph Cornell: Navigating the Imagination*, exhibition catalogue, Smithsonian American Art Museum; Peabody Essex Museum and San Francisco Museum of Modern Art, 2007–8
4 Joanna Moorhead and Stefan van Raay, eds, *Surreal Friends: Leonora Carrington, Remedios Varo and Kati Horna* (London, 2010).
5 Werner Spies, ed., *Max Ernst: A Retrospective*, exhibition catalogue, Tate Gallery, London, 1991 and Werner Spies, Iris Müller-Westermann and Kirsten Degel, eds, *Max Ernst: Dream and Revolution*, exhibition catalogue, Moderna Museet, Stockholm, and Louisiana Museum of Modern Art, Humlebæk, Denmark, 2009.

'SELLING' THE MOON IN THE 1950S

1 Michael J. Neufeld, *The Rocket and the Reich* (New York: The Free Press, 1995); Howard E. McCurdy, *Space and the American Imagination* (Washington, DC: Smithsonian Institution Press, 1997), pp. 194–95.
2 Catherine L. Newell, *Destined for the Stars* (Pittsburgh: University of Pittsburgh Press, 2019); Jared S. Buss, *Willy Ley* (Gainesville: University Press of Florida, 2017); Michael J. Neufeld, *Von Braun* (New York: Alfred A. Knopf, 2007).
3 Newell, *Destined*, pp. 113–31; Chesley Bonestell and Willy Ley, *The Conquest of Space* (New York: Viking, 1949).
4 Cornelius Ryan, ed., *Across the Space Frontier* (New York: Viking, 1952) and *Conquest of the Moon* (New York: Viking, 1953); McCurdy, *Space*, pp. 37–41; Newell, *Destined*, pp. 205–20.
5 Neufeld, *Von Braun*, pp. 185–90; Newell, *Destined*, pp. 221–32; McCurdy, *Space*, pp. 41–47.

THE FIRST MOON RACE – LUNA VERSUS PIONEER

1 Wesley T. Huntress, Jr, and Mikhail Ya. Marov, *Soviet Robots in the Solar System* (Chichester, UK: Springer-Praxis, 2011), pp. 36–37, 69–70.
2 Clayton R. Koppes, *JPL and the American Space Program* (New Haven and London: Yale University Press, 1982), pp. 84–85; T. A. Heppenheimer, *Countdown* (New York: John Wiley & Sons, 1997), pp. 148–50; R. Cargill Hall, *Lunar Impact* (Washington, DC: NASA, 1977), pp. 4–5.
3 Huntress and Marov, *Soviet Robots*, pp. 69–71; Hall, *Lunar Impact*, pp. 5–6.
4 Heppenheimer, *Countdown*, pp. 150–53; Huntress and Marov, *Soviet Robots*, pp. 50–51, 70–75; Boris Chertok, *Rockets and People* (Washington, DC: NASA, 2006), Vol. 2, pp. 435–46.
5 Hall, *Lunar Impact*, pp. 8–10; Heppenheimer, *Countdown*, pp. 153–55; Huntress and Marov, *Soviet Robots*, pp. 70, 75; Chertok, *Rockets and People*, Vol. 2, pp. 449–50.
6 Hall, *Lunar Impact*, p. 10; Huntress and Marov, *Soviet Robots*, pp. 77–78; Chertok, *Rockets and People*, Vol. 2, p. 450.

7   Chertok, *Rockets and People*, vol. 2, pp. 446–49, 519–38; Huntress and Marov, *Soviet Robots*, pp. 78–85.

8   Huntress and Marov, *Soviet Robots*, pp. 70, 84.

THE CINEMATIC MOON

1   That mission was the only one of seven, between 1969 and 1972, that failed to reach the Moon. Twelve astronauts have walked on the Moon, the last of whom, Eugene Cernan, was the subject of the 2014 documentary *The Last Man on the Moon*.

2   Emmerich had previously featured the Moon in *Independence Day* and its 2016 sequel *Independence Day: Resurgence*.

FLYBYS, LANDERS AND ORBITERS

1   The history of US and Soviet spacecraft can be found in two sources: Asif A. Siddiqi, *Beyond Earth: A Chronicle of Deep Space Exploration, 1958–2016* (Washington, DC: National Aeronautics and Space Administration, 2018); and Wesley T. Huntress Jr and Mikhail Ya. Marov, *Soviet Robots in the Solar System: Mission Technologies and Discoveries* (New York: Springer, 2011).

2   Hannah Ellis-Petersen, 'Hopes Fade for India's Moon Lander after It Fails to "Wake up" Following Lunar Night', *Guardian* (26 Sept. 2023), sec. World news, www.theguardian.com/world/2023/sep/26/india-chandrayaan-3-vikram-lander-pragyan-sleep-mode-failure

3   Kenneth Chang, 'Moon Landing by Israel's Beresheet Spacecraft Ends in Crash', *New York Times* (11 Apr. 2019), sec. Science, www.nytimes.com/2019/04/11/science/israel-moon-landing-beresheet.html; 'Status Update on ispace HAKUTO-R Mission 1 Lunar Lander', *ispace* (26 Apr. 2023), ispace-inc.com/news-en/?p=4655 (accessed 2 Jan. 2024).

4   Christian Davenport, 'Would-Be Historic Lunar Mission Won't Reach Moon's Surface, Company Says', *Washington Post* (9 Jan. 2024), www.washingtonpost.com/technology/2024/01/08/astrobotic-lunar-lander-failure

5   Jeff Foust, 'Intuitive Machines and NASA Call IM-1 Lunar Lander a Success as Mission Winds Down', *SpaceNews* (blog) (29 February, 2024), https://spacenews.com/intuitive-machines-and-nasa-call-im-1-lunar-lander-a-success-as-mission-winds-down/

6   Matthew Shindell, 'Commercial Lunar Landers and the Promise of Sustainable Exploration', *Quest: The History of Spaceflight Quarterly* 31, no. 1 (2024), pp. 11–15.

THE FAR SIDE OF THE MOON

1   Michael Collins, *Carrying the Fire: An Astronaut's Journeys* (New York: Farrar, Straus & Giroux, 1974), p. 408.

THE APOLLO PROGRAMME

1   Teasel Muir-Harmony, *Apollo to the Moon: A History in 50 Objects* (Washington, DC: National Geographic, 2018), pp. 192–93.

2   John F. Kennedy, 'Special Message to Congress on Urgent National Needs', 25 May 1961, Papers of John F. Kennedy. Presidential papers. President's Office Files. John F. Kennedy Presidential Library.

3   Matthew H. Hersch, *Inventing the American Astronaut* (New York: Palgrave Macmillan, 2012), p. 77.

4   Teasel Muir-Harmony, *Operation Moonglow: A Political History of Project Apollo* (New York: Basic Books, 2020).

5   John Logsdon, *John F. Kennedy and the Race to the Moon* (New York: Palgrave Macmillan, 2010), pp. 2–3.

6   Roger Launius, *Apollo's Legacy: Perspectives on the Moon Landings* (Washington, DC: Smithsonian Books, 2019), p. 24.

7   Hersch, *Inventing*, pp. 14–27.

8   ibid., pp. 75–102.

9   Tyler Peterson, *A Fire to be Lighted* (Burlington, ON: Apogee Books, 2017).

10   Muir-Harmony, *Apollo to the Moon*, pp. 192–219.

CONTEMPORARY ARTISTS AND THE MOON

1   Melanie Vandenbrouck, 'Fly Me to the Moon: From Artistic Moonscape to Destination', in *The Moon*, eds Melanie Vandenbrouck, Megan Barford, Louise Devoy and Richard Dunn (London: Harper Collins, 2019).

2   Hans Werner Holzwarth, *Darren Almond. Fullmoon* (Cologne: Taschen, 2015).

3   Franchesca Cubillo and Wally Caruana, eds, *Aboriginal and Torres Strait Islander Art: collection highlights* (Canberra: National Gallery of Australia, 2010).

4   Wendy Weitman, *Kiki Smith: Prints, Books & Things* (New York: The Museum of Modern Art, 2003).

5   Armin Zweite, ed., *Hiroshi Sugimoto. Revolution*, (Berlin: Hatje Cantz, 2012).

6   Mark Sealy and Gaëtane Verna, eds, *Sasha Huber. You Name It* (Milan: Mousse and Power Plant, 2022).

THE ARTEMIS MISSIONS

1   National Aeronautics and Space Administration, 'The Artemis Plan: NASA's Lunar Exploration Program Overview' (Sept. 2020), online: www.nasa.gov/wp-content/uploads/2020/12/artemis_plan-20200921.pdf

2   National Aeronautics and Space Administration, 'Artemis III Science Definition Team Report' (7 Dec. 2020), online at: www.nasa.gov/wp-content/uploads/2015/01/artemis-iii-science-definition-report-12042020c.pdf

3   Nathan Cranford and Jennifer Turner, 'Step 3, Artemis: Moon Missions as an Astronaut Testbed for Mars' (17 Mar. 2021), online at: www.nasa.gov/humans-in-space/step-3-artemis-moon-missions-as-an-astronaut-testbed-for-mars/

4   Johnny von Einem, 'Plants to be grown on the Moon when humans return,' (3 Apr. 2024), online at: https://www.adelaide.edu.au/newsroom/news/list/2024/04/02/plants-to-be-grown-on-the-moon-when-humans-return; Adriana Fraser, 'NASA selects UMBC-led lunar instrument for implementation on Artemis III Moon landing mission,'(10 Apr. 2024), online at: https://umbc.edu/stories/lems-nasa-moon-instrument/

5   European Space Agency, 'Artemis I' [no date], online at: www.esa.int/Science_Exploration/Human_and_Robotic_Exploration/Orion/Artemis_I

6   Park Si-soo, 'Japan Selects First New Astronauts in 14 years to Support Artemis Program', *SpaceNews* (28 Feb. 2023), online at: spacenews.com/jaxa-astronauts/; National Aeronautics and Space Administration, 'NASA, European Space Agency Formalize Artemis Gateway Partnership' (27 Oct. 2020), online: www.nasa.gov/news-release/nasa-european-space-agency-formalize-artemis-gateway-partnership/; Canadian Space Agency, 'The Artemis Program: Humanity's Return to the Moon' (3 Apr. 2023), online: www.asc-csa.gc.ca/eng/astronomy/moon-exploration/artemis-missions.asp

7   US Department of State, 'Artemis Accords' [no date], online at: www.state.gov/artemis-accords/

8   Abbey A. Donaldson, 'NASA Shares Progress Toward Early Artemis Moon Missions with Crew' (9 Jan. 2024), online at: www.nasa.gov/news-release/nasa-shares-progress-toward-early-artemis-moon-missions-with-crew/

9   Andrew Jones, 'Serbia becomes latest country to join China's ILRS moon base project' (10 May. 2024), online at: https://spacenews.com/serbia-becomes-latest-country-to-join-chinas-ilrs-moon-base-project/

GEOLOGICAL CHARACTERISTICS — LUNAR ATLAS MAPS

This glossary identifies the geological characteristics depicted in the maps of the USGS Lunar Atlas. Each map features the geological elements observed on the near side of the Moon, including mare, craters, mountains and rima. Below is a complete visual glossary of all the key geological characteristics shown on the forty-four maps that comprise the Lunar Atlas.

MARE/MARIA
The Moon's near side is dominated by large roughly circular basins, created by large impacts. The centres of these basins, as well as the depressed troughs that surround them, were later filled with lava that has hardened into smooth basaltic plains that appear dark. These dark areas were named 'maria' (seas) by early astronomers.

TERRA/TERRAE (HIGHLAND)
In contrast to the dark 'seas', early astronomers identified the brightly reflective zones of the Moon as terra/terrae, or 'land'. Primarily composed of light-coloured anorthosite, the terrae formed very early in the Moon's history, when a global magma ocean cooled to form a solid surface. Today these regions are typically referred to as highlands.

PLAINS
Similar to the mare, these are crater floors and terrae depressions that have been filled with lava, giving them smooth texture.

IMPACT CRATERS
Impact cratering is the dominant geologic process on the Moon. Craters comprise more than 90 per cent of the named features on the Moon, and were formed throughout the Moon's history. Based on the degradation of their features, geologists break the Moon's round craters into eight age categories, allowing the relative dating of regions and landforms.

WEATHERING
The lunar surface is subject to constant bombardment by charged particles in the solar wind that break the chemical bonds of surface minerals, altering the composition of rocks and soil. The surface is also struck by micrometeoroids. These processes tend to darken lunar materials over time, meaning that brighter material is generally younger than dark material.

CRATER RIMS, PEAKS AND RAMPARTS
The age of a large complex crater can be subjectively evaluated by looking at the degree to which the rim and other features have been eroded or degraded. A pristine crater will typically have a raised rim with steep interior slopes, a sloped outer rampart, and a central peak rising from the crater floor. A very young crater will also have visible rays.

CRATER CHAIN
A line of secondary impact craters radiating from a larger crater. Crater chains may also be formed through volcanic venting.

CRATER FILLING MATERIALS
Materials that have filled a crater since its formation.

ELONGATED CRATER
A crater that has been stretched by tectonic forces that have deformed the surface since the crater's formation.

IRREGULAR/ELLIPTICAL CRATER
A crater with an irregular shape formed by a low-angle impact or by multiple contemporaneous impacts at the same site.

COMPLEX CRATER
Typically a large crater that has a central peak and a rim that has collapsed inwards to form terraced walls. These attributes are caused by the collapse of the crater cavity after impact. Interior morphology can vary – multiple peaks, peak rings or pits may be present.

BOWL-/CUP-SHAPED CRATERS
Small simple craters with rounded rather than flat floors.

MOUNTAINS
Lunar mountain ranges, montes or massifs, formed during the Moon's early history. In many cases they are the remnants of the rims of giant impact craters or the multi-ringed basins that now hold the lava plains of the lunar maria.

CRATER RAYS
Impacts excavate material from the surface that is typically brighter than the surrounding mare or terra material. This ejected material (ejecta) may extend in radial streaks from the impact site. These rays will become less visible over time, due to space weathering. Large impacts, such as Tycho crater, may throw ejecta over multiple quadrangles.

SUBDUED CRATER
A relatively older crater. As opposed to a fresh crater, a subdued crater has been altered over time. Its rim has been rounded and its walls have degraded either through mantling, infilling, relaxation, and/or space weathering.

CRATER TERRACES
Sections of the walls of a large crater may succumb to gravity and fall inwards, making the crater wall appear terraced.

EJECTA BLANKET
A layer of material ejected from a crater. This blanket is thickest near the rim of the crater and may overlay older nearby surface features.

GLOSSARY

SATELLITE CRATER
Satellite craters are often formed when large materials are ejected from a 'parent' crater and fall back to the surface with enough force to form a visible crater. These are noted with the name of the 'parent' crater followed by a letter.

CRATER CLUSTER
A group of overlapping craters formed by returning ejecta material from a primary impact crater.

DARK MANTLE DEPOSITS
Large lava flows formed the maria, but darker deposits on the maria are believed to be the result of explosive (pyroclastic) volcanic eruptions.

DOMES
Small shield volcanoes or other irregularly shaped volcanic structures that appear mainly in the maria and sometimes the terrae. Because of the Moon's low gravity, cone volcanoes do not typically form – the lava spreads further, forming broad domes. These domes may form hilly terrain in some regions.

RILLES
Grooves or depressions in the Moon's surface that resemble channels. Curved (sinuous or arcuate) rilles may result from the collapse of lava tubes or the contraction of cooled lava flows. Straight (linear) rilles may form where segments of the crust have sunk between parallel faults.

INTRUSIVE ROCK
A rock formed beneath the surface.

EXTRUSIVE ROCK
Rock formed when lava solidifies above the surface.

RIDGES, SCARPS, AND FAULTS
The Moon is cooling, and in the process, it is shrinking and becoming denser. This causes curvilinear ridges, faults and scarps to form like wrinkles on the Moon's surface.

GHOST CRATERS
When the lava eruption from a lunar volcano overfills a crater, it forms a ghost crater where only the rim of the crater is faintly discernable.

ASLEP
Apollo Lunar Surface Experiments Package were a set of scientific instruments that were set up by the astronauts on the Moon at the landing sites of each Apollo mission.

CISLUNAR
Cislunar refers to the space between the Earth and Moon, including the lunar orbit.

COLLODION OR 'WET PROCESS' EMULSIONS
A photographic process which requires photographic material to be coated, sensitized, exposed and developed within a short span of time – often fifteen minutes.

COUDÉ REFRACTOR
French for 'elbow' it's the use of a system of mirrors set at 45-degree angles from of the lens in a mounting that then allows the focus and the observer to be stationary.

ECLIPTIC
The apparent path the Sun travels through the sky over the course of a year. Eclipses only occur when the Moon crosses this path.

EJECTA
Material violently thrown out as the result of a crater-forming impact or a volcanic eruption.

EPICYCLES
Prior to the acceptance of a Sun-centred cosmos with elliptical planetary orbits, astronomers used circular geometric models to reconcile the apparent motion of the planets with their ideal circular paths.

EXOSPHERE
The thin and highly dispersed envelope of gas molecules caught in the Moon's gravity.

EXTRA VEHICULAR ACTIVITY (EVA)
Also known as a space walk, this is when an astronaut in a spacesuit leaves the protective environment of a spacecraft.

IMPACTOR PROBE
A robotic spacecraft that performs a controlled and intentional crash landing on a planetary body.

'INVERSE' SQUARE RELATIONSHIP
In the Newtonian gravity, the attractive force exerted by a massive body is proportional to its mass, but diminishes with distance in accordance with a so-called 'inverse square' relationship (in which doubling the distance to an object diminishes its gravitational influence by a factor of four).

LIBRATION
Although we only ever see 50 per cent of the Moon's surface at any given time, and never see the entire far side, almost 60 per cent of the Moon is visible due to the phenomenon of libration, in which the Moon's orientation relative to Earth changes and makes features on the edges of its disc visible.

LRO
NASA's Lunar Reconnaissance Orbiter spacecraft has been in orbit around the Moon since 2009 and has provided imagery and altimeter data of the lunar surface that has been used to produce high-resolution topographic maps.

LRV
Apollos 15, 16, and 17 each carried a four-wheeled Lunar Roving Vehicle the astronauts utilized to explore and collect samples distant from their landing site.

MANTLE
Between the crust and the core of the Moon is a middle layer known as the mantle. The volcanic materials that filled in the lunar maria came from the mantle.

METONIC CYCLE
A period of 19 years in which 235 lunar months occur. Discovered by the 5th-century BC Greek astronomer Meton of Athens.

MICROMETRIC INSTRUMENTS
Typically a filar micrometer – a telescope eyepiece with fine movable wires in the field of view that the observer uses to measure angular distances between two points in the field of view.

PHOTOGRAM
An image created by placing an object on a piece of photographic paper and exposing the paper to light.

PHOTOGRAVURE
An image reproduction technique in which a photographic image is etched onto a gelatine-coated metal plate and then printed.

RECESSION OF ITS ORBIT
The Moon is moving away from the Earth at a rate of approximately 3.8 cm/year. Scientists believe that this rate has increased over time and will continue to increase in the future.

REFLECTING TELESCOPES / REFLECTORS
Telescopes that use concave mirrors, as opposed to lenses, to bring light to a focus.

REFRACTING TELESCOPES / REFRACTORS
Telescopes that use lenses, as opposed to mirrors, to bring light to a focus.

REGOLITH
On the Moon, the rocks and dust that cover the lunar bedrock. This material was created and distributed primarily by the impacts that formed the Moon's craters.

RETROGRADE MOTION
Because the planets orbit the Sun, their paths across the Earth's sky sometimes appear to reverse course. This phenomenon, most pronounced in the case of Mars, occurs when the Earth passes a planet in its orbit.

SAROS
A period of 223 lunar cycles; the time it takes for the Earth, Sun and Moon to return to the same relative points in space.

SEP TRANSMITTER
A component of the Surface Electrical Properties experiment left on the Moon by Apollo 17. The experiment used electromagnetic radiation to detect properties of subsurface materials and test for the possible presence of water.

SIM BAY
The Apollo 15, 16, and 17 spacecraft carried a Scientific Instrument Module (SIM) Bay in which various cameras and instruments were located. The SIM Bay was located in the Service Module, attached to the Command Module, and was used to study the Moon from orbit.

STRATIGRAPHY
The construction of geologic timescales based on the relative position of strata, or layers of geologic material. Stratigraphy is based on the principle that newer materials will be deposited on top of older materials.

SYNODIC MONTH
The time it takes the Moon to complete a full cycle of its phases, approximately 29.5 Earth days. Also referred to as a lunar month.

SYNODIC PERIOD
The time it takes for any planetary body to return to the same position relative to the Sun as seen from Earth. For the Moon, this is the same as the synodic month.

TERMINATOR
The line that divides day and night/light and dark on the Moon. The terminator moves across the Moon's near side as its phases move from new to full over the course of the synodic month.

SOURCES OF ILLUSTRATIONS

Every effort has been made to locate and credit copyright holders of the material reproduced in this book. The author and publisher apologize for any omissions or errors, which can be corrected in future editions.

a = above / c = centre / b = below / l = left / r = right

ASU: Arizona State University; AIC: Art Institute of Chicago; BnF: Bibliothèque nationale de France, Paris; BL: British Library, London; GSFC: Goddard Space Flight Center; JPGM: J. Paul Getty Museum, Los Angeles; JSC: Johnson Space Center; KSC: Kennedy Space Center; LoC: Library of Congress, Washington, D.C.; LPI: Lunar and Planetary Institute; MET: The Metropolitan Museum of Art, New York; MIA: Minneapolis Institute of Art; MFA: Museum of Fine Arts, Boston; MSFC: Marshall Space Flight Center; NGA: National Gallery of Art, Washington, D.C.; NYPL: New York Public Library; PMA: Philadelphia Museum of Art; SPACE-X: Space Exploration Institute; WAM: Walters Art Museum, Baltimore; WLC: Wellcome Collection, London

All Geological Atlas of the Moon images: Pub. U.S. Geological Survey. Courtesy David Rumsey Map Collection, David Rumsey Map Center, Stanford Libraries

6–7 NASA/Lunar Orbiter Image Recovery Project; **8** Beinecke Rare Book and Manuscript Library, Yale University, New Haven. ENG64; **12** BnF, département cartes et plans, GE DD-2987 (52); **13b** NASA/GSFC/ASU; **16–17** David Rumsey Map Collection, David Rumsey Map Center, Stanford Libraries; **23a** Lionel Bonaventure/AFP/Getty Images; **23bl** GrandPalaisRmn (musée d'Archéologie nationale)/Loïc Hamon; **23br** Illustrations based on a diagram by Alexander Marshack; **25a** kairoinfo4u. www.flickr.com/photos/manna4u; **25b** AIC. Photography Gallery Fund. 1959.608.27; **26al** MET. Gift of Lily S. Place, 1923. 23.6.10; **26ar** MET. Purchase, Edward S. Harkness Gift. 1926. 26.7.992; **26bl** MET. Gift of Ethel McCullough Scott, John G. McCullough, and Edith McCullough Irons, 1972. 1971.272.15; **26br** MET. Bequest of Nanette B. Kelekian, 2020. 2021.41.128; **27al** MET. Purchase, Edward S. Harkness Gift. 1926. 26.7.996; **27ar** MET. Gift of J. Lionberger Davis, 1965. 65.194.1; **27bl** MET. Rogers Fund, 1945. 45.2.11; **27br** Musée du Louvre, Dist. GrandPalaisRmn/Christian Décamps; **31al** GrandPalaisRmn (musée du Louvre)/Stéphane Maréchalle; **31ac** JPGM. 83.AQ.377.493; **31ar** Musée du Louvre, Dist. GrandPalaisRmn/Tony Querrec; **31b** JPGM. 76.AA.8.a; **32–33** Carole Raddato; **34–35** Dir. Irving Pichel. Prod. George Pal Productions/Walter Lantz Productions. Photo Granger/Shutterstock; LPI/NASA/James Stuby; NASA; LPI/NASA/James Stuby; NASA/GSFC/ASU; LPI/NASA/James Stuby; LPI/NASA/James Stuby; NASA/GSFC/ASU; LPI/NASA; **37al** National Museum of the American Indian, Smithsonian Institution, Washington, D.C. 1/617. Photo NMAI Photo Services; **37ar** Courtesy of Kite; **38–39** NASA/GSFC/ASU; LPI/NASA; LPI/NASA/James Stuby; John Johnston; NASA/GSFC/ASU; LPI/NASA/James Stuby; LPI/NASA; NASA/GSFC/ASU; **41l** BnF. Français 9219; **41r** Universität Utrecht; **42–43** NASA/GSFC/ASU; NASA/GSFC/ASU; LPI/NASA/James Stuby; LPI/NASA; ESA/SPACE-X. Photo © ESA; LPI/NASA/James Stuby; LPI/NASA/James Stuby; LPI/NASA/James Stuby; LPI/NASA/James Stuby; LPI/NASA/James Stuby; **44–45** Chine Nouvelle/SIPA/Shutterstock; Xinhua/Shutterstock; Xinhua/Shutterstock; LPI/NASA; NASA/GSFC/ASU; LPI/NASA/James Stuby; NASA/GSFC/ASU; Howard Fink, image based on Lunar Reconnaissance Orbiter data. finkh.wordpress.com; NASA/GSFC/ASU; LPI/NASA; NASA/GSFC/ASU; **46** Private collection; **47l** Harvard Art Museums/Arthur M. Sackler Museum, Gift of Philip Hofer in memory of Eric Schroeder. 1972.3/President and Fellows of Harvard College; **47a** MIA. The Minnich Collection, The Ethel Morrison Van Derlip Fund, 1966. P.14,544; **48–49** LoC; **52** BL Archive/Bridgeman Images; **53a** Private collection; **53b** WLC; **54–55** NASA; DK Images/Science Photo Library; Xinhua/Jiang Hongjing/Alamy Stock Photo; Xinhua/Alamy Stock Photo; CNSA/Chinanews/Ken Kremer/Marco Di Lorenzo; Alec Pipala; NASA/GSFC/ASU; Mike Wirths; Mike Wirths; LPI/NASA/James Stuby; LPI/NASA/James Stuby; NASA/GSFC/ASU; LPI/NASA/James Stuby; **56–57** NASA/GSFC/ASU; Jacques Descloitres, MODIS Land Rapid Response Team; NASA/GSFC/ASU; Damian Peach; NASA/GSFC/ASU; LPI/NASA/James Stuby; Damian Peach; NASA/GSFC/ASU; LPI/NASA; NASA/GSFC/ASU; **58–59** Peter Hess; **59** Lamrc1/Dreamstime.com; **60a** Princeton University Art Museum. Bequest of Gillett G. Griffin. 2016-1123; **60b** Justin Kerr/Dumbarton Oaks Research Library and Collection, PHPC002_K1398; **61a** Museo Nacional de Arqueología y Etnología de Guatemala; **61b** Saxon State and University Library Dresden. Mscr.Dresd.R.310; **62–63** SpaceIL; NASA/GSFC/ASU; NASA/GSFC/ASU; LPI/NASA/James Stuby; Damian Peach; George Tarsoudis; NASA; George Tarsoudis; LPI/NASA; Mario Weigand; LPI/NASA/James Stuby; **64–65** ispace; NASA/GSFC/ASU; Damian Peach; NASA/GSFC/ASU; NASA/GSFC/ASU; NASA/GSFC/ASU; NASA/GSFC/ASU; NASA/GSFC/ASU; NASA/GSFC/ASU; NASA; LPI/NASA/James Stuby; NASA/GSFC/ASU; **66** BL. Royal MS 14 D I, f.337v; **67a** Bridgeman Images; **67b** BL. Yates Thompson 36, f.137and f.133; **70a** MIA. 2012.16; **70b** Musée Carnavalet – Histoire de Paris. G.22102; **71a** Rijksmuseum, Amsterdam. RP-P-OB-4430; **71b** Yale Center for British Art, Paul Mellon Collection, New Haven. B1981.25.853; **72–73** BnF, département Réserve des livres rares. RES-Y2-369; **74–75** Private collection; NASA; NASA; NASA; NASA/GSFC/ASU; LPI/NASA/James Stuby; LPI/NASA/James Stuby; NASA/GSFC/ASU; NASA; LPI/NASA/James Stuby; NASA; NASA; **76–77** NASA; NASA; LPI/NASA/James Stuby; NASA; NASA; NASA; NASA; ASU/NASA; NASA/JSC; NASA/GSFC/ASU; LPI/NASA; **78** BnF, département Estampes et photographie; **79al** MET. Rogers Fund, 1919. 19.68.1; **79ar** JPGM. Ms. Ludwig XII 8, f.53v and f.47v; **79bl** Steve Harris; **79br** Staatsbibliothek zu Berlin. Ms. germ. f.1191; **80** Biblioteca Estense, Modena. Ms. lat. 209; **81** Bibliothèque Médiathèque, Bordeaux; WLC. MS Persian 474; LoC; MET. The Elisha Whittelsey Collection, The Elisha Whittelsey Fund, 1960. 60.576.13; The Museum of Turkish and Islamic Arts, Istanbul; Universitätsbibliothek Heidelberg, Cod. Pal. germ. 833; **82–83** NASA/GSFC/ASU; MODIS Land Rapid Response Team/NASA/GSFC; NASA; LPI/NASA/James Stuby; NASA/JSC/ASU; NASA/JSC/ASU; NASA/GSFC/ASU; LPI/NASA; LPI/NASA; LPI/NASA/James Stuby; NASA/GSFC/ASU; LPI/NASA/James Stuby; NASA; **84–85** NASA/NSSDCA; NASA; NASA; NASA/JSC; NASA/JSC; NASA; NASA; LPI/NASA; Cyril Noger; ASU/NASA/James Stuby; NASA/GSFC/ASU; **86** Biblioteca Apostolica Vaticana. Vat.gr.204, f.116r; **87a** WAM. Acquired by Henry Walters. W.659.10A, W.659.13B; **87b** Yale University Library, Beinecke Rare Book and Manuscript Library. Taylor 17; **88–89** Bayerische Staatsbibliothek, Munich. BSB Cod.icon. 181; **93a** LoC; **93b** Universität Utrecht; **94–95** E.G. Records, 1983. Cover art by Russell Mills; © Luc Viatour/Lucnix.be; illustration by Peter Donahue, based on a photo by Luc Viatour; NASA; LPI/NASA; NASA; ASU/NASA; LPI/NASA; Richard Bosman; LPI/NASA; ASU/NASA; NASA/GSFC/ASU; **96–97** NASA; Sovfoto/Universal Images Group/Shutterstock; Sovfoto/Universal Images Group/Shutterstock; NASA; NASA; NASA; NASA; NASA; LPI/NASA/James Stuby; NASA/GSFC/ASU; NASA; **98** BL; **99a** MET. The Elisha Whittelsey Collection, The Elisha Whittelsey Fund, 1960. 60.634.36, 60.634.38, 60.634.37; **99b** ETH-Bibliothek, Zürich. Poggendorf R, Sp. 628; **100a** MET. The Elisha Whittelsey Collection, The Elisha Whittelsey Fund, 1951, by exchange. 68.746; **100b** Houghton Library, Harvard University, Cambridge, MA. Typ 620.73.451; **101** Petworth House; Petworth House; LoC; LoC; MET. The Elisha Whittelsey Collection, The Elisha Whittelsey Fund, 1960. 60.634.36; Private collection; LoC; LoC; **102–103** IMAGO/SNA; Private collection; NASA/GSFC/ASU; NASA/GSFC/ASU; NASA/GSFC/ASU; NASA/GSFC/ASU; NASA; LPI/NASA; NASA/GSFC/ASU; NASA/GSFC/ASU; **104–105** Private collection; IMAGO/SNA; TASS Archive/Diomedia; TASS Archive/Diomedia; TASS Archive/Diomedia; TASS Archive/Diomedia; LPI/NASA/James Stuby; LPI/NASA/James Stuby; LPI/NASA/James Stuby; NASA/GSFC/ASU; NASA; George Tarsoudis; LPI/NASA; NASA/GSFC/ASU; **106** duncan1890/iStock; **107al** BL; **107ar** Beinecke Rare Book and Manuscript Library, Yale University, New Haven. Hfc39 C999 H5G 1659; **107b** BL. 634.e.4; **108a** LoC; **108bl** Private collection; **108bc** University of Illinois Urbana-Champaign; **108br** The Centre for 19th Century French Studies, University of Toronto; **109l** BnF; **109r** LoC; **110–111** University of Illinois Urbana-Champaign; **114**, **116–117** MET. Wrightsman Fund, 2000, 2000.51/Art Resource/Scala, Florence; **115** NGA. Patrons' Permanent Fund. 1990.6.1; Museo Nacional Thyssen-Bornemisza. Carmen Thyssen Collection. CTB.1994.6; Detroit Institute of Arts/Bridgeman Images; NYPL, Rare Book Division; **116l** MFA. The Hayden Collection—Charles Henry Hayden Fund. 45.201/All rights reserved/Bridgeman Images; **118–119** NASA; NASA; LPI/NASA; NASA; LPI/NASA; LPI/NASA; Howard Fink, image based on Lunar Reconnaissance Orbiter data. finkh.wordpress.com; LPI/NASA/James Stuby; NASA/GSFC/ASU; NASA/GSFC/ASU; **120–121** NASA; LPI/NASA; NASA; NASA/GSFC/ASU; LPI/NASA; NASA/JSC; Luca Zanini; LPI/NASA/James Stuby; NASA/GSFC/ASU; LPI/NASA; LPI/NASA; **122** LoC; **123** Science History Institute, Roy G. Neville Historical Chemical Library, Philadelphia; **124–125** NASA; NASA; NASA; NASA; LPI/NASA; LPI/NASA; NASA/GSFC/ASU; LPI/NASA/James Stuby; LPI/NASA/James Stuby; LPI/NASA/James Stuby; NASA/GSFC/ASU; LPI/NASA/James Stuby; **126 left row, a–b** NASA; NASA; NASA; Smithsonian National Air and Space Museum, Washington, D.C. 2012-0026; NASA; NASA; **126 right row, a–b** Keystone Press/Alamy Stock Photo; NASA/JPL-Caltech; NASA; **126b** NASA; NASA; NASA; National Museum of the US Navy, Washington, D.C. NASA; **127b** LPI/NASA/James Stuby; NASA; NASA; NASA; **128** MET. Gilman Collection, Gift of The Howard Gilman Foundation, 2005, 2005.100.614.0–g; **129al** John G. Wolbach Library, Harvard College Observatory, Cambridge, Massachusetts; **129ar** Harvard University Archives, Cambridge, Massachusetts; **129b** Rijksmuseum, Amsterdam. Purchased with the support of the Mondriaan Stichting, the Prins Bernhard Cultuurfonds, the VSBfonds, the Paul Huf Fonds/Rijksmuseum Fonds and the Egbert Kunstfonds. RP-F-2001-7-315; **130al** Science & Society Picture Library/Getty Images; **130ar** Private collection; **130bl** Science Museum Group. Royal Astronomical Society. 1931-862; **130br** Anna Maria College, Paxton. The Charles Blumsack Collection of Astronomy; **131** Ed. Gerard P. Kuiper. Pub. University of Chicago Press, 1960. B1-d: Pic du Midi, B1-b: Lick, B1-c: Mount Wilson, McDonald, B1-a: Yerkes. Courtesy David Rumsey Map Collection, David Rumsey Map Center, Stanford Libraries. 14057.021. © The University of Chicago 1960; **132–133** Rijksmuseum, Amsterdam. Purchased with the support of the Mondriaan Stichting, the Prins Bernhard Cultuurfonds, the VSBfonds, the Paul Huf Fonds/Rijksmuseum Fonds and the Egbert Kunstfonds; **137–139** Private collection; **140–141** NASA; LPI/NASA; NASA/GSFC/ASU; NASA/GSFC/ASU; NASA; © Song, Shenzhen, China; NASA/GSFC/ASU; LPI/NASA; NASA; NASA; NASA/GSFC/ASU; NASA; **142–143** NASA; NASA; NASA; NASA; NASA; NASA/GSFC/ASU; NASA/GSFC/ASU; LPI/NASA; LPI/NASA/James Stuby; LPI/NASA/James Stuby; NASA; LPI/NASA/James Stuby; **144** © Aleksandra Mir. All rights reserved, DACS 2024/Artimage; **145a** Queensland Art Gallery | Gallery of Modern Art. The Kenneth and Yasuko Myer Collection of Contemporary Asian Art. Purchased 2002 with funds from Michael Sidney Myer and The Myer Foundation, a project of the Sidney Myer Centenary Celebration 1899–1999, through the Queensland Art Gallery Foundation and The Yayoi Kusama Queensland Art Gallery Foundation Appeal. Photo QAGOMA. © YAYOI KUSAMA; **145bl** Brooklyn Museum, New York. Gift of the Elizabeth A. Sackler Foundation, 2002.10. Photo © Donald Woodman/ARS, NY. © Judy Chicago. ARS, NY and DACS, London 2024; **145br** National Portrait Gallery, Smithsonian Institution. NPG.2017.106. © 2014 Aura Satz; **146–147** LPI/NASA; LPI/NASA/James Stuby; LPI/NASA/James Stuby; LPI/NASA/James Stuby; LPI/NASA/James Stuby; LPI/NASA/James Stuby; NASA/GSFC/ASU; LPI/NASA/James Stuby; ESA/SPACE-X. Photo © ESA; LPI/NASA/James Stuby; NASA/GSFC/ASU; NASA/GSFC/ASU; **148–149** NASA; LPI/NASA; NASA, reconstructed by Philip Stooke, University of Western Ontario; NASA/JSC/ASU; LPI/NASA/James Stuby; LPI/NASA/James Stuby; LPI/NASA/James Stuby; **151a** Gerstein Science Information Centre, University of Toronto; **151b** Rijksmuseum, Amsterdam. Purchased with the support of the Mondriaan Stichting, the Prins Bernhard Cultuurfonds, the VSBfonds, the Paul Huf Fonds/Rijksmuseum Fonds and the Egbert Kunstfonds; **152a** Private collection; LoC; Private collection; Private collection; **152b** Wellesley College Library; **153a** Private collection; Wellesley College Library; Gerstein Science Information Centre, University of Toronto; Private collection; **153 left and right rows** Gerstein Science Information Centre, University of Toronto; **153b** Wellesley College Library; **156–157** NASA; NASA; NASA; NASA; NASA/JSC; NASA/MSFC; NASA/JSC; NASA/JSC; NASA; NASA/JSC; NASA/JSC; NASA/JSC; **158–159** NASA; LPI/NASA (all Ranger 9 images); NASA/JSC/ASU; NASA/GSFC/ASU; ASU/NASA/James Stuby; LPI/NASA/James Stuby; U.S. National Archives and Records Administration/NASA/James Stuby; ASU/NASA/James Stuby; NASA; NASA/GSFC/ASU; **160** Prod. Fritz Lang; **161** Prod. Star Film Company. Photo Méliès/Kobal/Shutterstock; Prod. Star Film Company. Photo The Picture Art Collection/Alamy Stock Photo; Prod. Star Film Company; Distr. Gaumont British; **162** Prod. Star Film Company. Photos SuperStock/Méliès/Album/Album Archivo; **163** Prod. Ferdinand Zecca/Pathé Frères; Prod. Latium Films; Prod. UFA and Colonna-film G.m.b.H.; Prod. Fritz Lang. Photo Science Photo Library/Alamy Stock Photo; Prod. Fritz Lang. Courtesy Heritage Auctions; **164–165** NASA; NASA; Associated Press/Alamy Stock Photo; NASA/JSC; NASA/JSC; NASA/JSC; NASA/JSC; NASA/JSC; LPI/NASA; LPI/NASA/James Stuby; NASA/GSFC/ASU; LPI/NASA; **166–167** NASA; NASA; NASA/James Stuby; NASA/GSFC/ASU; LPI/NASA; NASA/GSFC/ASU; ASU/NASA/James Stuby; ASU/NASA/James Stuby; LPI/NASA/James Stuby; NASA; LPI/NASA; LPI/NASA; ASU/NASA/James Stuby; **168–169 b** © Salvador Dalí, Fundació Gala-Salvador Dalí, DACS 2024; **169al** PMA. Purchased with the Alice Newton Osborn Fund, 1987. 1987-86-2. © Man Ray 2015 Trust/DACS, London 2024; **169ar** Sheldon Museum of Art, Nebraska Art Association, Gift of the Peter Kiewit Foundation. N-583.198. Photo Sheldon Museum of Art. © The Feitelson/Lundeberg Art Foundation; **169c** Tate Britain, London. Presented by the War Artists Advisory Committee 1946; **170** Wadsworth Atheneum Museum of Art, Hartford, CT. Purchased through the gift of Henry and Walter Keney, 1938.270. © The Joseph and Robert Cornell Memorial Foundation/VAGA at ARS, NY and DACS, London 2024; **171al** Colección FEMSA, Monterrey. Photo courtesy of Gallery Wendi Norris, San Francisco. © Remedios Varo, DACS/VEGAP 2024; **171ar** Collection of Rowland Weinstein. Photo courtesy of Weinstein Gallery, San Francisco. © Remedios Varo, DACS/ VEGAP 2024; **171b** PMA. Gift of Sylvia and Joseph Slifka, 2004. 2004-45-1. Ernst © ADAGP, Paris and DACS, London 2024; **172–173** Christie's Images/Bridgeman Images. Magritte © ADAGP, Paris and DACS, London 2024; **176l** Private collection; Bembmv; Sipa/Shutterstock; NASA; **176r** IMAGO/SNA; **177b** NASA; NASA/GSFC/ASU; LPI/NASA; LPI/NASA/James Stuby; **178–179** LPI/NASA/James Stuby; LPI/NASA/James Stuby; LPI/NASA/James Stuby; LPI/NASA/James Stuby; LPI/NASA/James Stuby; NASA/GSFC/ASU; LPI/NASA; LPI/NASA/James Stuby; LPI/NASA/James Stuby; LPI/NASA/James Stuby; LPI/NASA; LPI/NASA/James Stuby; LPI/

NASA/James Stuby; **180** Reproduced Courtesy of Bonestell LLC; **181** Marshall Space Flight Center/NASA; ullstein bild/Getty Images; TopFoto; Courtesy of RR Auction. Photo Nikki Brickett. Reproduced Courtesy of Bonestell LLC; Dir. Irving Pichel. Prod. George Pal. Photo Universal History Archive/Shutterstock; Private collection; **182–183** National Air and Space Museum, Smithsonian Institution, Washington. D.C. Gift of the Charles Hayden Planetarium of the Museum of Science. A19772550000; **182a** Walter Leporati/Scott McPartland/Getty Images; **183** Private collection; **184–185** NASA/GSFC/ASU; NASA/GSFC/ASU; LPI/NASA/James Stuby; LPI/NASA/James Stuby; LPI/James Stuby; LPI/NASA/James Stuby; LPI/NASA/James Stuby; LPI/NASA/James Stuby; LPI/NASA; Howard Fink, image based on Lunar Reconnaissance Orbiter data. finkh.wordpress.com; LPI/NASA; **186–187** LPI/NASA/James Stuby; ASU/NASA/James Stuby; LPI/NASA/James Stuby; LPI/NASA/James Stuby; Damian Peach; Rüdiger Proske, Ansbach, Germany; NASA/GSFC/ASU; NASA/GSFC/ASU; LPI/NASA; LPI/NASA/James Stuby; LPI/NASA/James Stuby; **188** Private collection; **189** IMAGO/SNA; NASA; NASA; NASA; **190–191a** Marshall Space Flight Center/NASA; **190–191b** NASA; **194–195** ASU/NASA/James Stuby; LPI/NASA/James Stuby; LPI/NASA/James Stuby; LPI/NASA; LPI/NASA/James Stuby; LPI/NASA/James Stuby; LPI/NASA/James Stuby; NASA/GSFC/ASU; LPI/NASA/James Stuby; NASA/GSFC/ASU; LPI/NASA/James Stuby; LPI/NASA/James Stuby; **196–197** NASA; NASA; NASA; NASA/GSFC/ASU; LPI/NASA/James Stuby; LPI/NASA/James Stuby; LPI/NASA/James Stuby; © 2004 Wes Higgins; LPI/NASA/James Stuby; NASA/GSFC/ASU; LPI/NASA; LPI/NASA; LPI/NASA/James Stuby; LPI/NASA; **198** Prod. Stanley Kubrick Productions; **199** Prod. Lennauchfilm; Prod. Universal Pictures, Imagine Entertainment. Photo Universal/Kobal/Shutterstock; Prod. Summit Entertainment, UK Moonfall LLP, Centropolis Entertainment, Huayi Brothers International, AGC Studios, Huayi Tencent, Street Entertainment. Photo Collection Christophel/Alamy Stock Photo; Prod. Regency Enterprises, Bona Film Group, New Regency, Plan B Entertainment, Keep Your Head Productions, RT Features, MadRiver Pictures, TSG Entertainment. Photo 20th Century Fox/Kobal/Shutterstock; **200** Prod. Stanley Kubrick Productions. Artwork by Robert McCall; **201** Dir. Pavel Klushantsev. Prod. Lennauchfilm; Dir. Robert Altman. Prod. A. William Conrad Production; Dir. John Sturges. Prod. M. J. Frankovich; Dir. Al Reinert. Prod. Betsy Broyles Breier, Al Reinert, Ben Young Mason, Fred Miller; Dir. Ron Howard. Prod. Universal Pictures, Imagine Entertainment; Dir. Roland Emmerich. Prod. Centropolis Entertainment. Photo 20th Century Fox/Kobal/Shutterstock; Dir. Rob Sitch. Prod. Summit Entertainment, Dish Film Productions; Dir. Aleksei Fedorchenko. Prod. Kinokompaniya Strana, Sverdlovskaya Kinostudiya; Dir. David Sington. Prod. Film4 Productions, Passion Pictures, Discovery Films; Dir. Duncan Jones. Prod. Stage 6 Films, Liberty Films, Xingu Films, Limelight; Dir. Matthew Avant. Prod. Media Savant; Dir. Michael Bay. Prod. Hasbro Studios, Di Bonaventura Pictures. Photo Paramount/Kobal/Shutterstock; Dir. Gonzalo López-Gallego. Prod. Bazelevs. Photo BFA/Alamy Stock Photo; Dir. Timo Vuorensola. Prod. Blind Spot Pictures, 27 Films Production, New Holland Pictures; Dir. Roland Emmerich. Prod. 20th Century Fox, TSG Entertainment, Centropolis Entertainment, Electric Entertainment. Photo 20th Century Fox/Kobal/Shutterstock; Dir. Damien Chazelle. Prod. DreamWorks Pictures, Perfect World Pictures, Temple Hill Entertainment. Photo Universal/Kobal/Shutterstock; Dir. Todd Douglas Miller. Prod. CNN Films, Statement Pictures; Dir. James Gray. Prod. Regency Enterprises, Bona Film Group, New Regency, Plan B Entertainment, Keep Your Head Productions, RT Features, MadRiver Pictures, TSG Entertainment. Photo BFA/Alamy Stock Photo; Dir. Zhang Chiyu. Prod. Mahua FunAge, China Film Co., Ltd., Alibaba Pictures, Shanghai Ruyi Film and Television Production Co., Ltd., Tianjin Maoyan Weiying Culture Media, Xihongshi Film and Television Culture (Tianjin) Co., Ltd.; Dir. Roland Emmerich. Prod. Summit Entertainment, UK Moonfall LLP, Centropolis Entertainment, Huayi Brothers International, AGC Studios, Huayi Tencent, Street Entertainment. Photo Collection Christophel/Alamy Stock Photo; **202–203** NASA; NASA; LPI/NASA/James Stuby; NASA/GSFC/ASU; LPI/NASA/James Stuby; LPI/NASA/James Stuby; LPI/NASA/James Stuby; NASA/GSFC/ASU; LPI/NASA/James Stuby; NASA/GSFC/ASU; LPI/NASA; LPI/NASA/James Stuby; NASA/GSFC/ASU; **204–205** NASA/GSFC/ASU; NASA/GSFC/ASU; LPI/NASA/James Stuby; NASA/GSFC/ASU; LPI/NASA/James Stuby; LPI/NASA/James Stuby; NASA/GSFC/ASU; LPI/NASA/James Stuby; NASA/GSFC/ASU; NASA; **206–207** NASA; **208–209** LPI/NASA; **212–213** ESA/SMART-1/SPACE-X/AMIE camera team. Photo © ESA; ESA/CNES/Arianespace – Service optique CSG. Photo © ESA; ESA/SPACE-X. Photo © ESA; LPI/NASA/James Stuby; LPI/NASA/James Stuby; NASA/GSFC/ASU; LPI/NASA/James Stuby; LPI/NASA/James Stuby; LPI/NASA/James Stuby; USGS/Mark Rosiek; Canada-France-Hawaii Telescope, 2006; **214–215** LPI/NASA; LPI/NASA; LPI/NASA/James Stuby; Gerard ter Horst; NASA/GSFC/ASU; LPI/NASA/James Stuby; LPI/NASA/James Stuby; LPI/NASA/James Stuby; LPI/NASA/James Stuby; LPI/NASA; LPI/NASA/James Stuby; LPI/NASA/James Stuby; **216** Society for Co-operation in Russian & Soviet Studies/TopFoto; **217l** LPI/NASA; **217r** NASA/GSFC/ASU; **218–219** NASA; NASA/Philip Stooke, University of Western Ontario; NASA; NASA/GSFC/ASU; Steve Mandel, Deep Space Remote Observatories; NASA/GSFC/ASU; NASA/GSFC/ASU; LPI/NASA/James Stuby; LPI/NASA/James Stuby; LPI/NASA/James Stuby; LPI/NASA; **220–221** NASA/GSFC/SwRI/JHU-APL/Tod R. Lauer (NOIRLab); LPI/NASA/James Stuby; LPI/NASA/James Stuby; NASA/GSFC/ASU; LPI/NASA/James Stuby; NASA/GSFC/ASU; LPI/NASA/James Stuby; LPI/NASA/James Stuby; LPI/NASA/James Stuby; LPI/NASA/James Stuby; LPI/NASA; LPI/NASA; **223b** NASA; **223a** NASA/JSC; **224–225** All images NASA/JSC except first image/second row NASA/Langley Research Center/Bob Rye; fourth image and fifth image/fourth row, sixth image/fifth row, fourth image/sixth row NASA/KSC; **228–229** ISAS/JAXA; ISAS/JAXA/Ted Stryk; ISAS/JAXA; NASA/GSFC/ASU; LPI/NASA; LPI/NASA; NASA/GSFC/ASU; LPI/NASA/James Stuby; LPI/NASA/James Stuby; NASA/GSFC/ASU; LPI/NASA/James Stuby; **230–231** LPI/NASA; LPI/NASA; LPI/NASA/James Stuby; LPI/NASA; LPI/NASA/James Stuby; LPI/NASA/James Stuby; LPI/NASA/James Stuby; LPI/NASA/James Stuby; NASA/GSFC/ASU; LPI/NASA/James Stuby; LPI/NASA/James Stuby; LPI/NASA/James Stuby; LPI/NASA/James Stuby; **232** NGA. Gift of Gemini G.E.L. and the Artist. 1981.5.74. © Robert Rauschenberg Foundation/VAGA at ARS, NY and DACS, London 2024; **233a** Christie's Images/Bridgeman Images. © 2024 The Andy Warhol Foundation for the Visual Arts, Inc./Licensed by DACS, London; **233b** Courtesy of the artist and Mitchell-Innes & Nash, New York. © Martha Rosler; **234b** National Gallery of Australia, Canberra. Purchased 1993. 93.181. © Mick Namarari Tjapaltjarri/Copyright Agency. Licensed by DACS 2024; **234a** Installation view, Light Show, Hayward Gallery, London, 2013. Photo © Marcus J Leith. © Katie Paterson; **235b** © Kikuji Kawada, courtesy of PGI; **235a** Photo courtesy Lisson Gallery. © Richard Long. All Rights Reserved, DACS 2024; **236–237** Prod. Stanley Kubrick Productions. Photo Moviestore/Shutterstock; Prod. Stanley Kubrick Productions; NASA/GSFC/ASU; LPI/NASA/James Stuby; LPI/NASA; LPI/NASA; LPI/NASA/James Stuby; LPI/NASA/James Stuby; LPI/NASA/James Stuby; LPI/NASA/James Stuby; **238–239** LPI/NASA/James Stuby; NASA/GSFC/ASU; LPI/NASA/James Stuby; LPI/NASA/James Stuby; LPI/NASA/James Stuby; LPI/NASA/James Stuby; LPI/NASA/James Stuby; LPI/NASA/James Stuby; LPI/NASA/James Stuby; LPI/NASA/James Stuby; LPI/NASA/James Stuby; **240** NASA/Joel Kowsky; **241a** NASA/Liam Yanulis; **241b** NASA; **242–243** Jay Tanner

FURTHER READING

1 Aveni, Anthony F., *Stairways to the Stars: Skywatching in Three Great Ancient Cultures* (New York: Wiley, 1997).

2 Donald A., Beattie, *Taking Science to the Moon: Lunar Experiments and the Apollo Program* (Baltimore: Johns Hopkins University Press, 2001).

3 Boyle, Rebecca, *Our Moon: How Earth's Celestial Companion Transformed the Planet, Guided Evolution, and Made Us Who We Are* (Random House, 2024).

4 Brush, Stephen G., *Fruitful Encounters: The Origin of the Solar System and of the Moon from Chamberlin to Apollo* (New York: Cambridge University Press, 1996).

5 Campion, Nicholas, *Astrology and Cosmology in the World's Religions* (NYU Press, 2012).

6 Chaikin, Andrew, *A Man on the Moon: The Voyages of the Apollo Astronauts* (New York: Penguin Books, 2007).

7 Cokinos, Christopher, *Still As Bright: An Illuminating History of the Moon, from Antiquity to Tomorrow* (New York: Pegasus Books, 2024).

8 Collins, Michael, *Carrying the Fire*, Anniversary edition (Farrar Straus & Giroux, 2019).

9 Doel, Ronald E., *Solar System Astronomy in America: Communities, Patronage, and Interdisciplinary Science, 1920–1960* (New York: Cambridge University Press, 1996).

10 Evans, James, *The History and Practice of Ancient Astronomy* (New York: Oxford University Press, 1998).

11 French, Bevan M., *The Moon Book* (New York: Penguin Books, 1977).

12 Gingerich, Owen, *The Eye of Heaven: Ptolemy, Copernicus, Kepler* (New York: American Institute of Physics, 1993).

13 Goodman, Ronald, *Lakota Star Knowledge: Studies in Lakota Stellar Theology* (Mission, SD: Sinte Gleska University Publishing, 2017).

14 Graney, Christopher M., *Setting Aside All Authority: Giovanni Battista Riccioli and the Science against Copernicus in the Age of Galileo* (Notre Dame: University of Notre Dame Press, 2015).

15 Hersch, Matthew H., *Inventing the American Astronaut* (New York: Palgrave Macmillan, 2012)

16 Heilbron, John L., *Galileo* (Oxford: Oxford University Press, 2012).

17 Kopal, Zdenek, and R. W. Carder, *Mapping of the Moon: Past and Present* (Springer Science & Business Media, 2013).

18 Logsdon, John M., *John F. Kennedy and the Race to the Moon* (New York: Palgrave Macmillan, 2010).

19 Morton, Oliver, *The Moon: A History for the Future* (London: *The Economist*, 2019).

20 Muir-Harmony, Teasel, *Operation Moonglow: A Political History of Project Apollo* (New York: Basic Books, 2020).

21 Muir-Harmony, Teasel, *Apollo to the Moon: A History in 50 Objects* (Washington, D.C: National Geographic, 2018).

22 Nasmyth, J., and J. Carpenter, *The Moon: Considered as a Planet, a World, and a Satellite* (London: John Murray, 1874).

23 Neufeld, Michael J., *Von Braun: Dreamer of Space, Engineer of War* (New York: Vintage Books, 2008).

24 Neufeld, Michael J., *Spaceflight: A Concise History* (Cambridge: The MIT Press, 2018).

25 Nicolson, Marjorie Hope, *Voyages to the Moon* (New York: Macmillan Company, 1948).

26 Proctor, R. A., *The Moon: Her Motions, Aspect, Scenery, and Physical Condition* (London: Longmans, Green and Co., 1873)

27 Scott, David Meerman, and R. Jurek, *Marketing the Moon: The Selling of the Apollo Lunar Program* (Cambridge: MIT Press, 2014).

28 Shindell, Matthew, *For the Love of Mars and The Life and Science of Harold C. Urey* (Chicago: University of Chicago Press, 2019)

29 Siddiqi, Asif A., *Beyond Earth: A Chronicle of Deep Space Exploration, 1958–2016* (Washington, DC: National Aeronautics and Space Administration, 2018).

30 Siddiqi, Asif A., *Sputnik and the Soviet Space Challenge* (Gainesville: University Press of Florida, 2003).

31 Siddiqi, Asif A., *The Soviet Space Race with Apollo* (Gainesville: University Press of Florida, 2003).

32 Sobel, Dava, *The Glass Universe* (New York: Viking, 2016).

33 Swift, Earl, *Across the Airless Wilds: The Lunar Rover and the Triumph of the Final Moon Landings* (New York: Mariner Books, 2021).

34 Westman, Robert, *The Copernican Question: Prognostication, Skepticism, and Celestial Order* (Berkeley: University of California Press, 2011).

35 Voelkel, James R., *The Composition of Kepler's Astronomia Nova* (Princeton: Princeton University Press, 2001).

36 Wilhelms, Don E., *To a Rocky Moon: A Geologist's History of Lunar Exploration* (Tucson: University of Arizona Press, 1993).

37 Worden, Al, Francis French and Dick Gordon, *Falling to Earth: An Apollo 15 Astronaut's Journey to the Moon* (Washington, DC: Smithsonian Books, 2012).

INDEX

ABOUT THE CONSULTANT EDITOR

Matthew Shindell is a historian of science whose work focuses on the history of the Earth and planetary sciences, in particular on the development of these fields during the Cold War. He is the curator of Planetary Science at the Smithsonian National Air and Space Museum and co-hosts the Museum's podcast, AirSpace. Shindell received a PhD in History of Science from the University of California, San Diego, and has taught at the University of Southern California and Harvard University. In addition to writing poetry, he is the author of *For the Love of Mars* and *The Life and Science of Harold C. Urey*, co-author of *Spaceships and Discerning Experts: The Practices of Scientific Assessment for Environmental Policy* and co-editor of *Smithsonian American Women*.

ABOUT THE FOREWORD WRITER

Dava Sobel is the author of *Longitude, Galileo's Daughter, The Planets, A More Perfect Heaven, The Glass Universe*, and *The Elements of Marie Curie*. She edits the 'Meter' poetry column in *Scientific American* and has an asteroid named after her (30935davasobel).

CONTRIBUTORS

JENNIFER HOUSER WEGNER is a curator in the Egyptian section at the University of Pennsylvania Museum of Archaeology and Anthropology in Philadelphia, PA. A specialist in ancient Egyptian languages, Wegner has also participated in archaeological fieldwork in Egypt as an epigrapher at the sites of Giza, Bersheh, Saqqara and Abydos.

KAREN NÍ MHEALLAIGH is Andrew W. Mellon Chair in the Humanities and Professor of Classics at Johns Hopkins University, Baltimore, MD. Her research focuses on the intersections between ancient science and the literary imagination, and she is fascinated by ancient cosmology and the Moon.

KITE (DR SUZANNE KITE) is an Oglála Lakȟóta performance artist, visual artist and composer, concerned with contemporary Lakota ontologies through research-creation, computational media and performance practice.

SAMANTHA THOMPSON is the Phoebe Waterman Haas Astronomy Curator at the Smithsonian Institution National Air and Space Museum. Thompson received a BA in Astrophysics and Physics from the University of California, Berkeley, an MSc in the History, Philosophy and Sociology of Science, Medicine and Technology from Imperial College London, and a PhD in History and Philosophy of Science from Arizona State University.

GILES SPARROW, FRAS, is a science writer, journalist and consultant specializing in astronomy, spaceflight and the history of how we understand the universe. His recent books include: *Phaenomena: Doppelmayr's Celestial Atlas* and *The History of Our Universe in 21 Stars*.

RHIANNA ELLIOT is a PhD candidate in the Department of History and Philosophy of Science at the University of Cambridge, UK. Her current research explores connections between menstruation and the Moon in early modern England. She has held research fellowships at the Science Museum, London, and the Huntington Library in San Marino, CA.

GERARDO ALDANA is a professor of Chicana/o Studies at UC Santa Barbara, CA. His most recent book, *Calculating Brilliance: an intellectual history of Mayan astronomy at Chich'en Itza*, explores the historical development of the Dresden Codex Venus Table, culminating in its final version deployed at Mayapan.

KATE GOLEMBIEWSKI is a science writer based in Chicago. Her work has been published in the *New York Times, The Atlantic, Scientific American, CNN, Discover Magazine, National Geographic Voices* and *Atlas Obscura*. She earned her Master's degree in science writing from Johns Hopkins University, Baltimore, MD.

NICHOLAS CAMPION is Associate Professor in Cosmology and Culture at the University of Wales Trinity Saint David where he is Programme Director of the MA in Cultural Astronomy and Astrology. His books include the two-volume *History of Western Astrology* and *Astrology and Cosmology in the World's Religions*.

ALEXANDRA LOSKE is a British-German art historian and museum curator. She has written, co-authored and edited many books and articles on colour, art and other subjects, including *The Artist's Palette, Moon: Art, Science, Culture* and *Pale Fire: New Writing on the Moon*.

HÉLÈNE VALANCE is a scientific advisor at Institut National d'Histoire de l'Art and InVisu (Centre National de la recherche Scientifique). She is the author of *Nocturne: Night in American Art, 1890–1917*.

DR DAVID H. DEVORKIN is senior curator, emeritus, of the history of astronomy at the National Air and Space Museum, Smithsonian Institution. DeVorkin's research, collections, exhibits and publications centre on the origins and development of modern astrophysics and the space sciences during the 20th century. In his teenage years, he relished photographing the Moon.

IAN HAYDN SMITH is a London-based writer and editor, and author of *Well Documented, The Short Story of Photography, Cult Writers, Selling the Movie* and *A Chronology of Film*.

DR ELIZABETTA FABRIZI is a curator and writer. Her work primarily focuses on artists' film and video practices.

DR MELANIE VANDENBROUK is a curator, writer and art historian. In 2019, she curated the exhibitions The Moon at Royal Museums Greenwich, London, and Moonlight at the Hasselblad Foundation, Gothenburg, Sweden. She is now Chief Curator at Pallant House Gallery in Chichester, UK.

MICHAEL J. NEUFELD is a retired senior curator in the Space History department of the Smithsonian's National Air and Space Museum. He was the lead curator of the Destination Moon exhibit and has written or edited nine books, notably *The Rocket and the Reich* and *Von Braun: Dreamer of Space, Engineer of War*. Asteroid 329018 Neufeld is named for him.

DR EMILY S. MARTIN is a Research Geologist in the Center for Earth and Planetary Studies at the Smithsonian Institution's National Air and Space Museum.

TEASEL MUIR-HARMONY is a historian of spaceflight and the curator of the Apollo Collection at the Smithsonian National Air and Space Museum. She is the author of *Operation Moonglow: A Political History of Project Apollo* and *Apollo: To the Moon in 50 Objects*.

DR EMILY A. MARGOLIS is a social and cultural historian of spaceflight. She is Curator of Contemporary Spaceflight at the Smithsonian National Air and Space Museum and is also responsible for the Museum's Mercury and Gemini collections. Margolis is writing a book on the history of space tourism.

FROM THE CONSULTANT EDITOR

Editing this book has been a wonderful opportunity to tap into the expertise of colleagues, some of whom I already knew, others whom I came to know during this project; I thank them all for their thoughtful contributions to this expansive volume. The tireless work of the T&H team, Tristan de Lancey, Jane Laing and Georgina Kyriacou, kept this project moving forward; Tristan, Jane and Georgina are responsible for the initial vision we worked toward collectively, and I credit them for seeing the lunar equivalent of the forest through the trees while the rest of us dove deep into our explorations of the Moon. Finally, thanks to Karen Darling, my editor at the University of Chicago Press, for recommending me for this project.

COVER, SPINE & ENDPAPERS: Map details taken from the NASA Lunar Atlas, Pub. U.S. Geological Survey. Courtesy David Rumsey Map Collection, David Rumsey Map Center, Stanford Libraries; Moon image courtesy NASA

First published in the United Kingdom in 2024 by
Thames & Hudson Ltd, 181A High Holborn, London WC1V 7QX

LUNAR: A History of the Moon in Myths, Maps + Matter
© 2024 Thames & Hudson Ltd, London

Introductions, map texts and four essays by Matthew Shindell
© 2024 Smithsonian's National Air and Space Museum

Foreword by Dava Sobel © 2024 John Harrisson & Daughter, Ltd

For image copyright information see p. 252

Consultancy by Matthew Shindell

Designed by Daniel Streat, Visual Fields

British Library Cataloguing-in-Publication Data
A catalogue record for this book is available from the British Library

ISBN 978-0-500-02714-1

Printed in China by Shenzhen Reliance Printing Co. Ltd

FSC
www.fsc.org

MIX
Paper | Supporting
responsible forestry
FSC® C102842

Be the first to know about our new releases, exclusive content and author events by visiting
thamesandhudson.com
thamesandhudsonusa.com
thamesandhudson.com.au

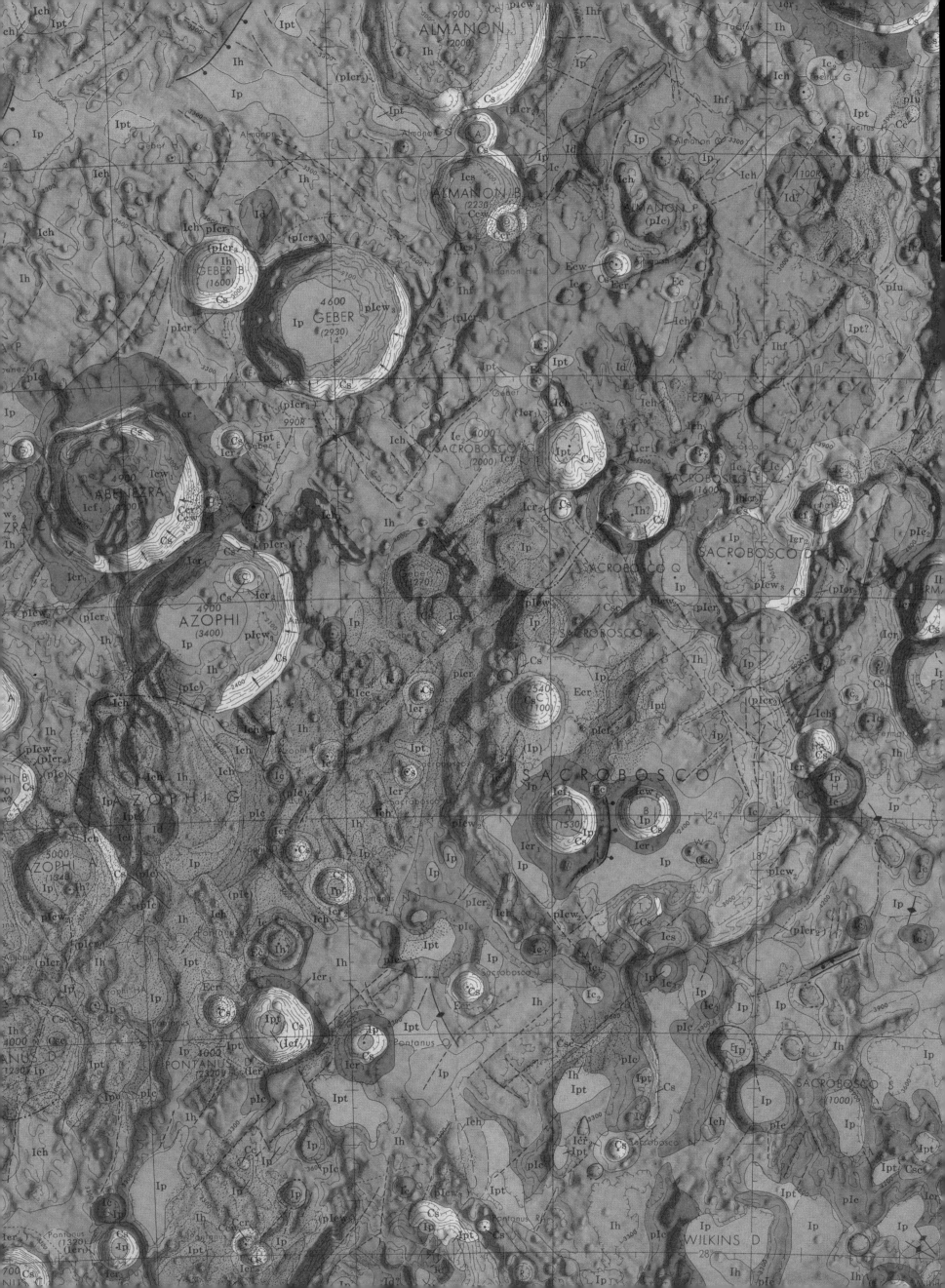